NATIONAL ACADEMY PRESS
The National Academy Press was created by
the National Academy of Sciences to publish
the reports issued by the Academy and the
National Academy of Engineering, the Institute of Medicine, and the National Research
Council, all operating under the charter
granted to the National Academy of Sciences
by the Congress of the United States.

STUDIES IN GEOPHYSICS

Sea-Level Change

Geophysics Study Committee
Commission on Physical Sciences, Mathematics,
   and Resources
National Research Council

NATIONAL ACADEMY PRESS
Washington, D.C.   1990

NATIONAL ACADEMY PRESS 2101 Constitution Avenue, N.W. Washington, DC 20418

Support for the Geophysics Study Committee was provided by the Department of Energy, the National Science Foundation, and the National Oceanic and Atmospheric Administration.

Sea-level change / Geophysics Study Committee, Commission on Physical
    Sciences and Resources, National Research Council.
        p. cm. — (Studies in geophysics)
    Includes bibliographical references.
    ISBN 0-309-04039-6
    1. Sea level.   I. Geophysics Research Forum (U.S.). Geophysics
Study Committee.   II. Series.
    551.4'58—dc20   90-5839
                                                                   CIP

Printed in the United States of America

Panel on
Sea-Level Change

ROGER R. REVELLE, University of California, San Diego, *Chairman*
TIM P. BARNETT, Scripps Institution of Oceanography
ERIC J. BARRON, Pennsylvania State University
ARTHUR L. BLOOM, Cornell University
NICHOLAS CHRISTIE-BLICK, Lamont-Doherty Geological Observatory
C. G. A. HARRISON, University of Miami
WILLIAM W. HAY, University of Colorado
ROBLEY K. MATTHEWS, Brown University
MARK F. MEIER, University of Colorado
WALTER H. MUNK, University of California, San Diego
W. RICHARD PELTIER, University of Toronto
DEAN ROEMMICH, Scripps Institution of Oceanography
W. STURGES, Florida State University
ERIC T. SUNDQUIST, U.S. Geological Survey
KEITH R. THOMPSON, Dalhousie University
STARLEY L. THOMPSON, National Center for Atmospheric Research

Geophysics
Study Committee

Commission on Physical Sciences, Mathematics, and Resources

Studies in Geophysics[*]

*Published to date.

Preface

This study is part of a series, *Studies in Geophysics*, that has been undertaken to provide assessments from the scientific community to aid policymakers in decisions on societal problems that involve geophysics. An important part of such assessments is an evaluation of the adequacy of current geophysical knowledge and the appropriateness of current research programs as a source of information required for those decisions.

This study on sea-level change was initiated by the Geophysics Study Committee in consultation with the liaison representatives of the agencies that support the committee, relevant boards and committees within the National Research Council, and members of the scientific community.

The study addresses our current scientific understanding of sea-level change—particularly the processes of sea-level change, their rates, and the record of past change. For example, how much of apparent sea-level change is related to global changes in the volume and mass of the ocean basins (eustatic signal) and how much is related to tectonic factors that might contaminate the eustatic signal? The object of the study is to present an integrated picture of sea-level change—its causes, feedbacks, and record.

The preliminary scientific findings of the authored background chapters were presented at an American Geophysical Union (AGU) symposium. In completing their chapters, the authors had the benefit of discussions at this symposium and comments from several scientific referees. Ultimate responsibility for the individual chapters, however, rests with their respective authors. Although a fair amount of time has elapsed since the symposium, the authors made efforts to incorporate up-to-date information within their respective chapters.

The Overview and Recommendations of the study summarizes the highlights of the chapters and formulates conclusions and recommendations. In preparing the Overview and Recommendations, the panel chairman and the Geophysics Study Committee made use of comments from meetings at the AGU symposium, the members of the panel, several meetings of the committee, and the reviews of scientists, who were approved by the National Research Council's Report Review Committee. Responsibility of the Overview and Recommendations rests with the Geophysics Study Committee and the chairman of the panel.

Contents

PROCESSES AND FEEDBACKS

FUTURE MEASUREMENTS

SEA-LEVEL CHANGE

Overview and Recommendations

EXECUTIVE SUMMARY

This study addresses current scientific understanding of sea-level change—particularly the processes of sea-level change, their rates, and the record of past change. An important part of such an assessment is an evaluation of the adequacy of the geophysical knowledge base and the opportunities to improve upon it. Discussion of engineering and societal responses to sea-level change is not included in this study as these issues are fully discussed in a report, *Responding to Changes in Sea Level: Engineering Implications* (NRC, 1987).

Average sea level over the oceans has never been constant throughout earth history, and it is changing slightly today. The entirety of civilization has occurred within a single high stand of the sea, and yet global sea level was more than 100 m lower than it is at present only 18,000 yr ago. And during the geologic past there have been repeated variations of more than 100 m from present sea level, both during times of intense glaciation and during times of an ice-free Earth.

Relative sea level (RSL; i.e., sea level relative to a fixed point on land) at any particular place in the ocean varies over a wide range of time and space scales. The direct causes of these variations are vertical motions of the land to which the tide gauge or other measuring instrument is attached, and changes in the volume of sea water in which the tide gauge is immersed. But changes in climate, plate tectonics, ice and snow, and ocean circulation are all indirect causes of changing sea level. The relative importance of the forcing functions varies with the time scales of interest.

On the basis of estimates of global warming of the atmosphere and ocean resulting from increasing concentrations of carbon dioxide and other greenhouse gases, it is possible to make an approximate forecast of global rise in sea level during the next 100 yr. Two processes will be principally involved: thermal expansion of ocean waters as they become warmer and changes in the mass of land ice in both continental ice sheets and mountain

3

glaciers. One hundred years from now it is likely that sea level will be 0.5 to 1 m higher than it is at present.

There are two principal uncertainties at present about global sea level. (1) What if any is the value of the long-term trend over the next few centuries? (2) If a secular trend exists, what proportion of this trend results from changes in the specific volume of sea water (called steric changes) and what from changes in the total mass of water in the oceans?

The apparent trend of sea level at any particular place as measured by a tide gauge is the sum of the trend in motion of the gauge itself as the land on which it is mounted moves vertically, the trend of change in steric sea level, and the trend of change in water mass under the tide gauge. To understand what is happening, one needs to be able to make measurements that will separate these three components of the observed sea level. In principle, a combination of inverted echo sounders (which in effect measure thermal expansion or contraction of the water column) or systematic observations of ocean temperature as a function of depth with conductivity-temperature-depth recorders to determine the steric component, plus one or more of three methods (very-long-baseline interferometry, global positioning system, and absolute gravimetry) for measuring the vertical motions of the tide gauge, plus tide-gauge measurements at the sea surface should allow us to separate the three major components of changes in RSL. Within the next few years, it should be possible to measure accurately the combined effects on global sea level of steric changes and changes in the mass of sea water from laser or radar altimeters mounted on Earth-orbiting satellites.

Recommendation 1. Long-term sea-level measurements of sufficient accuracy over the world's oceans could provide one of the most significant data sets for understanding global change, particularly climatic change resulting from greenhouse warming. It is for this reason that the planning committees for the World Climate Research Program and the Intergovernmental Oceanographic Commission of UNESCO have given a very high priority to extending the global sea-level network in the Indian, South Atlantic, and South Pacific oceans. **We strongly recommend that national oceanographic and meteorological communities lend moral and intellectual support to this sea-level program.**

Recommendation 2. Possible changes in the mass balance of the Antarctic and Greenland ice sheets are fundamental gaps in our understanding and are crucial to the quantification and refinement of sea-level forecasts (the probable contribution from ice wastage makes up more than half of various forecasts). **A polar-orbiting satellite altimeter would be invaluable in monitoring the mass balance of these ice sheets.**

Recommendation 3. To refine estimates of sea-level change related to greenhouse warming, it is necessary to **develop and improve coupled atmosphere–ocean–cyrosphere global circulation models in which greenhouse gas concentrations in the atmosphere are gradually increased.**

Recommendation 4. The Cretaceous period offers special opportunities to understand global processes and their variations, in particular, large, long-term changes in sea level. One of the major projects of the Global Sedimentary Geology Program, under the auspices of the International Union of Geological Sciences, is entitled "Cretaceous Resources, Events, and Rhythms." **We urge national and international support of this and similar programs that will improve our understanding of past sea-level changes and the processes that produced them.**

Recommendation 5. To separate epeirogeny from eustasy and steric components, it is important to measure repeatedly the absolute heights of tide gauges. **We recommend that global measurements of absolute heights of these gauges be undertaken using absolute gravimetry and space-based techniques.**

INTRODUCTION

It is easy to think of sea level as a stable baseline against which changes on land can be measured, and it is so used by politicians, engineers, land planners, and households. But as we have become more aware of the Earth and its metabolism, we have recognized that sea level is highly unstable both in time and in space. We can regain our former confidence in its usefulness only by learning how and why it varies and compensating for these variations.

Average sea level over the globe has never been constant throughout earth history; and it is changing slightly today. The entirety of civilization has occurred within a single high stand of the sea, and yet global sea level was more than 100 m lower than it is at present during the maximum of the last glacial period only 18,000 yr ago (yrBP).

Relative sea level (RSL; i.e., sea level relative to a fixed point on land) at any particular place in the ocean varies over a wide range of time and space scales. Among the causes of these variations are vertical motions of the land to which the tide gauge or other measuring instrument is attached and changes in the volume of sea water in which the tide gauge is immersed. Wind-driven waves produce the shortest-period variations in the height of the sea surface. Variations with periods of 12 hours or more are caused by lunar and solar tides. Variations in atmospheric pressure cause inverse variations in sea level—an atmospheric pressure differential of 1 mbar is equivalent to a sea-level differential of 10 mm. A series of depressions in atmospheric pressure can cause a rise in sea level in a shallow ocean basin of 0.3 m or more (Hekstra, 1988). Variations in the runoff of large rivers can result in local sea-level changes of as much as 1 m. In relatively shallow water, large variations in sea level are also caused by offshore and onshore winds that pile up water against the shore or drive it away from the shore. (This process is called wind setup.) In exceptional circumstances, in the North Sea, along the Chinese coast, and in the Bay of Bengal, sea level may rise by 5 m or more in a "storm surge" under the action of strong winds (Hekstra, 1988). Both irregular and seasonal variations in temperature or salinity of the upper ocean layers cause expansion or contraction of the water volume in different regions. These relatively short-term steric changes in sea level may persist for a few days, several months, or even several years, and the magnitude may be as much as 50 to 150 mm.

Changes in sea level have many practical consequences, often disruptive but sometimes beneficial. The disruptive consequences can sometimes be avoided and the benefits enhanced if the ways and means by which sea level varies are understood. Although disruptive consequences, especially from sea-level rise, are far more common, occasional examples of benefits are being obtained as a result of changes in policy. For example, in the Wadden Sea on the northwest coast of the Netherlands, where the land has subsided as much as 260 mm during the past 20 yr, the Dutch government has decided not to follow age-old tradition by attempting to reclaim more land for agriculture. The area has been designated as a "Declared UNESCO Biosphere Reserve" in which only those economic activities are allowed that do not conflict with natural conditions or processes. In view of agricultural surpluses in the Netherlands, and indeed throughout the European community, the Wadden Sea can be used much more effectively as a nursery for young fish and shrimp and other valuable invertebrates (Hekstra, 1988) than as reclaimed land for agriculture.

Changes of sea level on the order of 300 mm can have significant implications for coastal communities and coastal engineering practices. Engineering responses to sea-level change are largely a function of the rate of change. Many of these issues and engineering responses are described in the report *Responding to Changes in Sea Level: Engineering Implications* prepared by the Marine Board of the National Research Council (NRC, 1987). A 1-m change in average sea level can translate into major shifts in shoreline positions, positions that have both economic and legal significance.

Sea-level change, seemingly so simple and straightforward, is in fact the product of many interrelated processes. Insight into these processes can be gained by intensive study of sea-level change in the context of related environmental phenomena, remembering that changes in sea level are an integrated measure of environmental change, in terms of both causes and consequences. Changes occur on all space and time scales, from local to global, and from a few seconds to geologic ages.

This volume is primarily concerned with future sea-level changes over the next few centuries and past changes over a wide range of times from which greater understanding can be gained and used as an aid in the prediction of future changes and in the search for fossil fuels and other natural resources. Also covered are the mechanisms and processes involved in past changes, in order to gain greater knowledge of the Earth as a dynamic system, i.e., how the Earth works.

Climate, plate tectonics, the cryosphere, and ocean circulation all contribute to changing sea level. The relative importance of the forcing functions varies with the time scales of interest. The effects of changing sea level are also broad—on the one hand with direct feedbacks to the causative forcing functions, e.g., albedo change, and on the other hand with effects on other processes such as sedimentation or coastal ecology.

PROCESSES AND FEEDBACKS

Many processes can cause a change in RSL at any particular location. They include the following:

1. local or regional uplift or subsidence of the land;
2. changes in atmospheric pressure, winds, or ocean currents;
3. changes in the mass of ocean water brought about by wastage or accumulation of ice sheets and mountain glaciers (glacio-eustatic) or by increased or decreased retention of liquid water within or upon the continents, and possibly also by a slow release through geologic time of juvenile water from the Earth's interior;
4. steric changes in the volume of ocean water without changes in water mass (Patullo *et al.*, 1955) in response to temperature or salinity changes (also called thermohaline changes in Table 1); and
5. changes in the volume of the ocean basins owing to changes in the rate of plate divergence (seafloor spreading), plate convergence (subduction, overthrusting), epeirogenic changes in the elevation of the seafloor (largely from mid-ocean volcanism), marine sedimentation, or isostatic adjustment of the Earth's crust under the sea resulting from glaciation or deglaciation on land.

The latter three processes can affect global mean sea level or eustatic sea level. But all processes need to be considered, even though, depending on the time scale of interest or the magnitude of the sea-level change, some of them may be insignificant. In a following section on forecasting sea-level change due to greenhouse-induced climate warming, a projected global sea-level rise of 0.5 ± 1 m by the year A.D. 2100 is ascribed to a combination of thermal expansion of ocean water and melting of glaciers and ice sheets.

Time Scales of Sea-Level Change

Sea-level change encompasses a broad range of time scales, with different mechanisms associated with change over different times. The oceanographer concerned with storm tides will not have much interest in the factors explaining Cretaceous sea levels; likewise, the geologist's glacio-eustatic theories have little application to seasonal events. The problem of sorting out time scales and processes afflicts studies of climate change in general. Complexities multiply in attempts to link different processes together.

Table 1 summarizes mechanisms of sea-level change by time scale and magnitude.

TABLE 1 Some Mechanisms of Sea-Level Change

	Time Scale (years)	Order of Magnitude of Change (mm)
Ocean Steric (thermohaline) Volume Changes		
Shallow (0 to 500 m)	10^{-1} to 10^2	10^0 to 10^3
Deep (500 to 4000 m)	10^1 to 10^4	10^0 to 10^4
Glacial Accretion and Wastage		
Mountain Glaciers	10^1 to 10^2	10^1 to 10^3
Greenland Ice Sheet	10^2 to 10^5	10^1 to 10^4
East Antarctic Ice Sheet	10^3 to 10^5	10^4 to 10^5
West Antarctic Ice Sheet	10^2 to 10^4	10^3 to 10^4
Liquid Water on Land		
Groundwater Aquifers	10^2 to 10^5	10^2 to 10^4
Lakes and Reservoirs	10^2 to 10^5	10^0 to 10^2
Crustal Deformation		
Lithosphere Formation and Subduction	10^5 to 10^8	10^3 to 10^5
Glacial Isostatic Rebound	10^2 to 10^4	10^2 to 10^4
Continental Collision	10^5 to 10^8	10^4 to 10^5
Sea Floor and Continental Epeirogeny	10^5 to 10^8	10^4 to 10^5
Sedimentation	10^4 to 10^8	10^3 to 10^5

Heat exchange is relatively rapid within the uppermost few hundred meters of the oceans. A one-dimensional treatment implies that thermal expansion of these waters can occur on time scales of months to decades. Sea level will rise about 100 mm for every degree of temperature increase throughout the uppermost 500 m. Heat exchange with ocean deep waters is slower (Chapter 13). If the deep ocean were to warm everywhere by 10°C, as was perhaps the case during the early Tertiary and Cretaceous, sea level could rise by about 10 m.

The time scales and magnitudes of melting ice can be estimated from both historical data and mass balance considerations. The present Greenland and Antarctic ice caps are remnants of the late Pleistocene ice sheets that increased sea levels about 100 m by disintegrating over a period of several thousand years encompassing the end of the Pleistocene (Chapters 4 and 5). Mass balance estimates suggest modern ice residence times on the order of 10^2 to 10^5 yr. The Antarctic Ice Sheet contains enough water to raise sea level by about 60 m, and the Greenland Ice Sheet contains water equivalent to a 6-m sea-level rise.

Sea Level and the Geoid

The sea surface departs significantly from the geoid, owing to waves, the tidal attraction of the Sun and Moon, ocean circulation that tilts the ocean surface, atmospheric disturbances, and steric regional variations in water temperature and salinity. If these effects can be taken into account, the resulting mean sea level accurately follows the geoid, which however is far from being an ideal spheroidal shape. Deep mantle phenomena and gravity anomalies associated with subduction can cause deviations from the spheroid of many tens of meters. Local crustal geological features, such as seamounts, fracture zones, and other abrupt topographic forms that are not in isostatic equilibrium, can cause deviations of several meters. In fact, it is possible to derive a partial picture of ocean floor topography from a knowledge of the mean shape of the sea surface.

If the effects of vertical tectonic movements, in addition to the local or regional variations of sea-surface topography, are removed from tide-gauge records, presumably any remaining trend is global (eustatic sea-level change). Eustatic changes can result from processes that change the mass or volume of water in the ocean basin or those that change the volume of the ocean basin itself. These processes are discussed in later sections.

Land Elevation Changes

Tide gauges are anchored to the land, which itself can be moving vertically at rates comparable to sea-level change. These vertical tectonic movements of the land will result in an RSL change as measured by a tide gauge. If the land where the tide-gauge station is located is subsiding, RSL will show a rise; likewise, uplift will result in an RSL fall. The Viking city now called Old Uppsala was a major port for ships sailing Lake Malaren; however, the port and the city of Uppsala were forced to move downstream around the eleventh century because of a 10-mm/yr uplift in the Fennoscandian region while the lake remained connected with the worldwide mean sea level.

Vertical tectonic motions can be very localized, such as subsidence along a coastline owing to sediment load or from the withdrawal of groundwater or hydrocarbons. Subsidence is occurring along the Louisiana gulf coast because of the deposition and dewatering of sediment from the Mississippi River system. On the other hand, vertical motions can be systematically related to one another through isostasy.

Land areas that were ice covered during the last glacial episode (~18,000 yrBP) were depressed; this depression created a forebulge in adjacent areas (see Peltier, Chapter 4). The North American or Laurentide ice sheets began disintegrating about 15,000 yrBP and by about 7000 yrBP had all but vanished. Along the east coast of North America, those areas that were ice covered are still being uplifted due to isostatic rebound at rates of up to about 10 mm/yr. In the peripheral area where the forebulge is collapsing, the land is subsiding at more than 1 mm/yr. The east coast of the United States is perhaps the best location illustrating this "drowning" effect, which is responsible for many of the unique features of its nearshore environment, including the extensive occurrence of salt marshes. In these areas and with current models of isostasy and the Earth, the relative vertical motions are predictable; thus, their relative contribution to sea-level changes as measured by tide gauges can be taken into account in arriving at eustatic sea-level changes.

Effects of Atmospheric Pressure, Winds, and Ocean Currents

Local RSL variability can result from several forcing functions including air pressure, wind stress, ocean circulation, and thermohaline changes. These processes contribute to the high-frequency noise in tide-gauge data and need to be compensated for in extracting eustatic changes.

The difference between the average air pressure over the world ocean and the local barometric pressure can result in a change of sea level at annual and shorter periods. One millibar of pressure differential is equivalent to 10 mm of sea-surface change.

Wind stress can have an important effect on the sea level. The coastal sea-level response to a steady longshore wind stress can be similar in magnitude to the air-pressure effect. The time to achieve a steady state is of the order of a few days. Sea-surface changes caused by wind stress are localized and are not a major contributor to low-frequency sea-level change.

Ocean circulation can also result in sea-level variability from long-period waves, as Sturges (Chapter 3) shows. Sea-level signals of about 50 to 150 mm with periods of 5 to

10 yr and longer are coherent between the U.S. Pacific coast and Hawaii, and on both sides of the Atlantic. These signals are out of phase, with delays of several years, and are consistent in part with baroclinic wave propagation across the ocean.

Also coherent are sea-surface variabilities resulting from El Niño and related effects (Figure 1). From these, sea level can vary on the time scale of a few years by 50 to 150 mm resulting from warming of the ocean subsurface waters by a few degrees Centigrade (°C).

Changes in the Mass of Ocean Water

For all intents and purposes, the mass of water at or near the Earth's surface is constant over time scales of less than 10^4 yr. It is how the water is partitioned between the major hydrologic reservoirs that is of importance to sea-level change. The four major reservoirs, in order of abundance, are the oceans (1370×10^6 km³), ice (30×10^6 km³), ground and surface waters (8×10^6 to 19×10^6 km³), and atmospheric moisture (0.01×10^6 km³). The principal exchange of water over the past several million years involved ice. Sea level was over 100 m lower during the peak of the most recent glacial 18,000 yrBP. The melting of the northern continental ice sheets between 15,000 and 7000 yrBP probably accounted for most of the rise of the sea to present levels. Indeed, sea-level change within the next few thousand years probably will also be dominated by water within the global ice budget and by ice sheet dynamics. If the polar ice sheets were to disappear, sea level would be some 60 to 70 m higher than at present.

Mountain glaciers make up about 1 percent of the volume of land ice. Their potential contribution to sea level is about 1 m if they were to melt totally and all the meltwater reached the sea. Data cited by Meier (Chapter 10) suggests that wastage of the world's

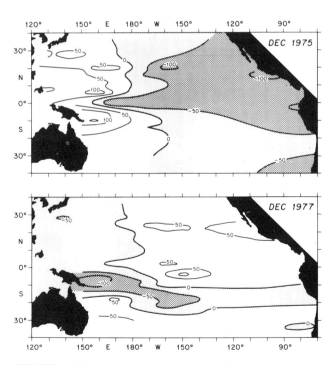

FIGURE 1 Maps of sea-level anomaly for December 1975 and December 1977. Contours show sea-level anomalies in millimeters after removal of seasonal cycle. The two cases were selected for their large contrast. From Wyrtki and Nakahoro (1984).

mountain glaciers and small ice caps contributed about 0.46 ± 0.26 mm/yr to higher sea level between 1900 and 1961 corresponding to a total sea-level rise of 28 ± 16 mm, which is about a third of the estimated sea-level rise during that period.

The Antarctic and Greenland ice sheets gain material mainly through the accumulation of snow and lose material through several processes. These processes include surface melt and runoff of meltwater, calving (discharge) of icebergs, and melting of the underside of floating ice shelves. Surface melt/runoff is a minor process for the Antarctic Ice Sheet, but is important to the balance of the Greenland Ice Sheet. Iceberg calving is an important loss process for both ice sheets, and is predominant in Antarctica. Melting of the underside of ice shelves has no effect on sea level, but does control the speed of discharge of land-based ice from Antarctica and is important for predicting the behavior of that ice sheet in the next century.

The increase in air temperature caused by a rise in concentration of CO_2 and other greenhouse gases may cause increased snow and ice melting, and some of this meltwater may run off to the oceans, causing a rise in sea level. Increased air temperature and/or meltwater production may also cause some outlet glaciers and ice streams to flow faster, transferring land-based ice to the ocean and causing a further rise in sea level. On the other hand, a rise in CO_2 concentration may, in some regions, lead to increased snow precipitation on glaciers and ice sheets, which will have the opposite effect on sea level. Predicting the effects of climate change on ice growth and wastage is a complex problem because several different, interacting processes must be considered.

Five important factors in future changes of glaciers and ice sheets are (1) the variation of energy and mass balance components with altitude, (2) the warming of cold firn to allow meltwater runoff, (3) the dynamic response of ice masses to changes in thickness, (4) increased flow and iceberg calving of tidal glaciers due to increased meltwater, and (5) the stability of ice-sheet/ice-stream/ice-shelf systems. The time frame is restricted to the next 100 yr, approximately the time of doubling of the present level of CO_2. The first and fifth points are briefly discussed below; see Chapter 10 for additional details on all five processes.

For glaciers, many of the mass and energy fluxes related to melting are known or observable functions of altitude. The two most sensitive of these are absorbed solar radiation and precipitation of snow. The first is critically dependent on the albedo (reflectivity) of the surface, which in turn depends on how long the surface is covered with high-albedo snow during the course of the melt season. The persistence of the snow cover depends, of course, on the amount of snowfall and the intensity of melt processes, both of which also depend on altitude. The altitudinal dependence of the mass and energy fluxes leads to a potential instability, which is one of the reasons for the sensitivity of glaciers to slight climate changes. An increase in melting causes a lowering of the ice surface, which in turn may cause a further increase in melting or decrease in snow accumulation, leading to further changes accentuating the melting.

Some have suggested that a climatic change due to increased CO_2 in the atmosphere could lead to disintegration of the West Antarctic Ice Sheet, most of which is grounded below sea level, causing a 6-m rise in global sea level. The discharge of ice from this ice sheet is mainly through rapidly moving ice streams, which flow into floating ice shelves. An ice sheet that rests on a flat bed situated below sea level can be inherently unstable. Floating ice shelves at its seaward edge, which are "pinned" in position by shallow bottom areas, act as buttresses that prevent the ice sheet from quickly flowing out into the ocean. The rise of 100 m in sea level during the past 15,000 yr caused a substantial collapse of a large part of the West Antarctic Ice Sheet until it approximately stabilized at its present size. A relatively small further sea-level rise could act to "unpin" the buttressing ice shelves that allow the remnant ice sheet to exist (Thomas and Bentley, 1978).

If the climate in the future becomes warmer with the result that warmer ocean water intrudes under the ice shelves causing increased melting under the shelves, then the back pressure exerted on the ice streams by the shelves will be reduced and the ice streams will accelerate, draining the ice sheet itself.

The critical questions then are: (1) How rapidly will the temperature of Antarctic subsurface waters rise in response to increased atmospheric concentrations of greenhouse gases? (2) How much sub-ice melting will be caused by the circulation of this warmer water under the ice shelves? (3) How will the changed conditions affect calving rates and thus the dimensions of ice shelves? (4) How rapidly will the ice streams react to changes in the back pressure? For the next one or two centuries we need to consider only the first question because warming of the ocean south of the Antarctic convergence is likely to be markedly delayed. Bryan *et al.* (1988) used a combined oceanic and atmospheric general circulation model to show that convective mixing from the surface down to 4000 m will slow the rate of ocean warming because the entire water column must be warmed by the same amount. If we consider the added heat energy of around 4 watts/m^2 transferred from the atmosphere to the surface ocean layers, a time of several hundred years would be required to accomplish this warming. Bentley (1985) has summarized a number of other reasons why disintegration of the West Antarctic Ice Sheet should not occur within the next one or two centuries. Accepting Bentley's arguments, sea level will not rise catastrophically in the near future resulting from the demise of this ice sheet.

Scientific concern about possible disappearance of the West Antarctic Ice Sheet has largely been based on Mercer's (1978) hypothesis that this body of ice disappeared during the last interglacial, 125,000 yrBP, with the result that a terrace about 5 m above present sea level was created around many shorelines around the world. An alternative hypothesis is that the surface of the East Antarctic Ice Sheet was lower by some 300 to 350 m than today (Robin, 1987). This idea is supported by the investigation by Lorius *et al.* (1985) of $\delta^{18}O$ from the ice core collected by Soviet engineers at Vostok in the East Antarctic Ice Sheet. At 125,000 yrBP, the oxygen isotope values were about 2 ‰ higher than present. Robin points out that this apparently higher temperature could be caused by a reduction in surface elevation of about 300 m. Surface lowering of this amount for the East Antarctic Ice Sheet would correspond to a volume of ice of about 2 million km^3 and a corresponding rise of sea level of 5 to 7 m during the last interglacial.

Eustatic Effects of Changes in Liquid Water on Land

In the absence of large-scale glaciation and deglaciation, a possible mechanism for relatively rapid eustatic sea-level change could be changes in the mass of liquid water sequestered on the continents, both above and below the ground surface. Such a mechanism is needed to explain the apparently eustatic sea-level changes in a virtually ice-free Earth during Mesozoic and early Cenozoic time described in papers by Haq *et al.* (1987) and by Christie-Blick *et al.* (Chapter 7). The topic presented first is the possible variations in groundwater.

A global climate change toward less precipitation will lower the water table in groundwater aquifers, transfer water from the land to the sea, and raise sea level. Less precipitation on land should result from atmospheric cooling, lower wind velocity, or changes in atmospheric circulation which would alter the balance between precipitation over the ocean and over the land. Increased precipitation resulting from atmospheric warming or other causes will raise the water table and lower sea level.

Removal of fresh-water-bearing porous sediments—chiefly sands and calcareous deposits—by erosion should have the same effect as a decline of precipitation, while accretion of such sediments will create a greater potential aquifer volume and hence be roughly equivalent to a rise in the water table.

Land subsidence, which may now be occurring in many coastal areas, will reduce the volume of sedimentary aquifers above sea level, and thus have much the same effect as erosion of sediments or a decline in the level of the water table due to decreased precipitation. Emergence of previously submerged aquifers should have an opposite effect, especially in carbonate terrains where karst formation can occur.

It is also possible that infiltration or discharge rates could change with time, leading to larger or smaller accumulations of groundwater as the balance between infiltration and discharge approaches a new equilibrium determined by changes in hydrostatic pressure in the aquifer. For example, canyon cutting during times of low sea level should increase discharge rates while deposition of unconsolidated coarse sediments should increase infiltration rates.

According to Hay and Leslie (Chapter 9) the late Cenozoic fluctuations in sea level caused by glaciation and deglaciation resulted in an offloading of sediments from coastal plain regions and continental shelves into the continental slopes and abyssal plains of the oceans. Thus the potential groundwater reservoir that exists today may well represent a minimum for much of geologic history. In past times, the potential change in sea level resulting from fluctuations in groundwater storage could have been double that which exists today.

These statements may be roughly quantified by assuming an equivalent rise or fall of the water table by 100 m, 13.3 percent of the average height of continental surfaces above sea level. Hay and Leslie (Chapter 9) estimate that porosities are close to 40 percent in the upper layers of the sands and calcareous sediments resting on the continents in geosynclines (intracratonic basins), coastal plains, and cratonic platforms. They calculate that these two types of sediments make up 38 to 47 percent of the total deposits in these three sedimentary environments, and that they are the only sediments that take part in significant water exchange with the environment outside the aquifers.

Table 2 shows the volumes and pore space of the top 100 m of sands and calcareous sediments in cratonic platforms, geosynclines, and coastal plains, computed from these estimates by Hay and Leslie and supplemented by estimates made by Southam and Hay (1981) and by Ronov (1982).

The total volume of pore space in the top 100 m of the continental sediments—2.5×10^6 km³—is equivalent to a rise or fall of sea level by 7 m. Hay and Leslie assume that the rates of filling or discharge in these coarse sediments would be less than 13.5×10^3 km³/yr. Thus more than 185 yr would be required to fill or empty an aquifer 100 m thick, corresponding to a rate of sea-level change of less than 4 mm/yr. Where only a slight imbalance exists between infiltration and discharge the times required for filling or emptying an aquifer 100 m thick could be tens to hundreds of thousands of years. For example, Meier (1984)

TABLE 2 Aquifers and Pore Space in Top 100 Meters of Sediments on Land

Location	Area covered by sediments (10^6 km²)	Area covered by sandy or calcareous sediments (percent)	Volume of sandy and calcareous sediments in top 100 m of aquifers (10^6 km³)	Volume of pore space in top 100 m of aquifers[a] (10^6 km³)	Sea-level equivalent (meters)
Cratonic platforms	55	47	2.6	1.0	2.9
Geosynclines	59	38	2.3	0.9	2.6
Coastal plains and shelves	31	47	1.4	0.6	1.7
Total	145		6.3	2.5	7.2

[a]Assuming 40 percent porosity

estimates that global depletion of groundwater during this century has been between 1600 and 2400 km^3/yr or 20 to 30 km^3/yr. At this rate, filling or emptying a 100-m-thick global aquifer would take 85,000 to 130,000 yr and the corresponding rate of rise or fall of sea level would be less than 0.1 mm/yr.

Baumgartner and Reichel (1975) and Woods (1984) estimate the global volume of groundwater at 8×10^6 km^3, about 22 percent of the Earth's fresh water, equivalent to 22 m of sea level. This may be compared to 70 m of sea-level equivalent for the Greenland and Antarctic ice sheets.

From data given by Hay and Leslie, assuming an average porosity of 20 percent for sandy and calcareous sediments, a pore volume in sediments has been computed at 64.8×10^6 km^3. But 55.5×10^6 km^3 of this pore space is below sea level and hence presumably saturated with water. The pore space above sea level, which could be drained or filled by the various mechanisms discussed herein, is 9.4×10^6 km^3, equivalent to 27 m of sea level, very close to the estimate of groundwater volume given by Baumgartner and Reichel. However, as Hay and Leslie suggest, the porosity of the 750 m of sediments above sea level may be 30 to 40 percent, corresponding to a pore volume of 15×10^6 to 19×10^6 km^3.

Geological evidence shows that large quantities of shallow water carbonates and non-marine sands were laid down in mid-Paleozoic, late Paleozoic to early Mesozoic, and mid-Cretaceous times. During these periods, their abundance was probably more than twice that of the present time and the potential for groundwater storage was higher. Insofar as these porous sediments occupied a greater percentage of the land area than their present counterparts, infiltration rates must also have been higher than today, compared with rates of discharge, which can occur only at the edges of sedimentary columns.

The mass of liquid water above ground in lakes and rivers is only a small fraction of the mass of groundwater. Residence times in rivers vary from less than a week to about a year depending on size and length and on the slope of the river bed.

The volume of water stored in lakes is about 0.22×10^6 km^3 (Robin, 1987), probably about 50 times the volume in rivers but only about 1 percent of the volume of groundwater above sea level. Changes in lake volume result from climate variation and change and from human activities, primarily diversion of inflows for irrigation or other purposes. The Aral Sea in the Uzbek Republic of the Soviet Union is a striking example (Micklin, 1971). This lake without an outlet, fed by the Amur Darya and Syr Darya rivers, was formerly the world's fourth largest lake, behind the Caspian Sea, Lake Superior, and Lake Victoria. In 1960 its area was 68,000 km^2, its average depth was 16 m, and its volume was 1090 km^3. Beginning in 1960, there was a large increase in diversions of the river flows for irrigation caused by expansion and intensification of the irrigated areas. These increased diversions were not compensated for by conservation measures as previously, and the lake began to shrink rapidly. By the beginning of 1970, the area had decreased by 40 percent, the volume had decreased by 66 percent, and the water depth had dropped to 9 m. This change in lake volume must have been accompanied by a eustatic rise in sea level of slightly less than 2 mm. Destruction of the lake is still occurring; without drastic changes in irrigation practices, it will have largely disappeared by the early part of the twenty-first century, and there will be a further eustatic rise in sea level of about 1 mm.

On a worldwide basis, Robin (1987) estimates lake volumes are diminishing by 72 km^3/yr. He bases his estimates on the observed annual decline of the Caspian Sea by 10.9 km^3, assuming that this decline in the Caspian, the world's largest lake, is 15 percent of the global diminution of lake volumes. The corresponding eustatic rise in sea level should be 0.2 mm/yr or 20 mm/century. To this should probably be added the sea-level rise of about 1 mm due to the future decline of the Aral Sea.

Robin (1987) also points out that a long-term warming of the atmosphere by 3°C at low latitudes and 6°C at higher latitudes should result in an increase of the atmospheric content of water vapor, and a corresponding fall in sea level of about 7 mm.

Changes in the Volume of Water Without Changes in Mass

Steric height change (i.e., temperature and/or salinity variation) contributes to global sea-level fluctuations from year to year and decade to decade. It has been shown in regional studies that steric height changes account very well for sea-level variations (on the order of 100 mm) observed on seasonal and interannual time scales. On the other hand, sea-level fluctuations over periods of tens of thousands of years are so large that steric expansion can be ruled out as a major contributor. A warming of the entire ocean from 0°C to the currently observed temperatures would involve a thermal expansion of only about 10 m.

The thermal expansion coefficient increases significantly both with increasing temperature and with increasing pressure. The steep increase with increasing temperature has sometimes been used as an argument that deep cold water may be ignored in steric height calculations relative to warm surface water. As Roemmich (Chapter 13) shows, this is not so, because a parcel of water at 4°C at about 2000 m depth has a thermal expansion coefficient 60 percent as large as that of a parcel at the ocean surface at 20°C.

We need to discover how much variability in steric height is contained in time scales of decades or longer and to determine what vertical and horizontal scales of variability correspond to these long-term changes. These essential questions must be answered first before we can ask how long the ocean must be sampled at a single location, how many locations must be sampled, and to what depths the sampling should extend to determine the relationship of steric height to the long-term trend in sea level. Trends in steric height over three decades in the central north Pacific appear to be about 1 mm/yr, but these trends are so small in comparison with the large interannual variations of 10 to 100 mm that a much longer time series is needed to confirm the trend (Thomson and Tabata, 1987).

On time scales of tens to hundreds of years, the effects of salinity changes on global sea level are slight. The total salt content of the oceans remains approximately constant, and sea-level changes due to vertical redistribution of salt in the water column are likely to be small. Heat, unlike salt, is freely exchanged with the atmosphere, and even the combined heat content of the ocean–atmosphere system may change significantly over a few decades or centuries. Thus global expansion or contraction of ocean waters must be dominated by temperature change rather than by salinity change except under some circumstances in high latitudes and semi-isolated basins. In general, there is a fairly tight temperature/salinity correlation; a temperature increase of 1°C or above in the thermocline can be accompanied by a salinity increase of approximately 0.1 parts per thousand. This temperature change produces an increase in specific volume whose magnitude is more than three times the corresponding decrease in specific volume caused by the salinity change. Thus the effects are of opposite sign, and although salinity is not negligible, the temperature effects dominate.

Steric changes in the water column on time scales of a decade and longer are not confined to or concentrated in the upper ocean. In the subtropical North Atlantic such changes extend to depths of at least 3000 m (Chapter 13). For observational programs the large vertical extent of the signal means that the entire depth of the ocean should be sampled. In the upper water layers, both vertical diffusion and advection are important, as are water mass transport along quasi-horizontal isopycnal (constant density) surfaces from subsurface regions in higher latitudes toward the mid-latitude ocean interior. Critical processes are air-sea interactions in the regions where the isopycnals intercept the sea surface, and such cross-pycnal mixing processes as double diffusion (diffusion of heat and salinity at different molecular diffusion rates) and intermittent internal wave breaking. In the deeper layers (below the permanent thermocline in the North Atlantic) we are concerned with the formation of North Atlantic deep water by convective mixing and sinking in the Greenland and Norwegian seas and by horizontal advection and mixing as this water mass moves southward. The critical problems are the mechanisms and rates by which waters of lower density can gradually replace higher density water in the deep ocean basins.

Roemmich (Chapter 13) shows that at mid-latitudes in the North Atlantic the thermocline fluctuations in steric height exhibit greater variance in the time domain than the deeper changes and also more variability around the subtropical gyre. For the 27 yr of Panuliris hydrographic data taken by the Bermuda Biological Station (discussed by Roemmich, Chapter 13), the thermocline fluctuations make a greater contribution to steric height change than the deep changes, but on a longer time scale, it is not clear whether mid-depth or deep variations would dominate. The thermocline comes to the surface along the northern rim of the subtropical gyre, and these waters are therefore subject to atmospheric forcing relatively near the Panuliris area. The deep water is farther removed from contact with the atmosphere, which can account for the more gradual changes and larger spatial scale of the deep signal.

Volume of Ocean Basins

Changes in the volume of the ocean basins can have very significant long-term effects on sea level—effects that can be as great as or greater than that caused by ice. The dominant cause of these large-scale changes is the variable volume (Figure 2) of ocean ridge material from seafloor spreading processes. If spreading increases, the volume of the ridge crest increases, displacing water and flooding the continental areas. The critical quantity is the area of seafloor produced per unit time, which can be changed either by an increase in spreading rate over a ridge crest of constant length, or by increasing the length of the ridge crest, or by a combination of the two. Harrison (Chapter 8) shows that during the past 80 million yr (m.y.) a diminution in the volume of ocean-ridge crests by 9.55×10^{16} m^3 could have produced an increase in freeboard (essentially equal to a fall in sea level) of about 180 m, with a very large uncertainty of +140 and −180 m.

The continents themselves can undergo vertical motion termed epeirogeny. Evidence shows that epeirogenic movements of the continents can result in several hundreds of meters of elevation over very long time periods. Similar elevation changes occurring in ocean basins can significantly affect the volume of the ocean basin and, thus, sea level. For example, there is evidence for a thermally induced swelling or uplift in the equatorial Pacific during the Cretaceous, which was accompanied by large amounts of extrusive

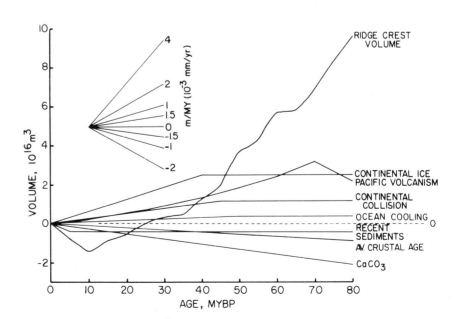

FIGURE 2 Volume estimates for six tectonic processes affecting the volume of the ocean basins. An increase in volume as time progresses produces a sea-level fall (or freeboard increase). See Harrison (Chapter 8) for additional detail.

volcanic activity. A decrease of the volume of the ocean basins of about 2×10^{16} m^3 was produced by this "ocean-floor epeirogeny" between 110 m.y. ago (Ma) and 70 Ma.

Many other possibilities exist for changing sea level by tens to hundreds of meters. For instance, a change of the pattern of subduction can cause the ocean basins to become, on average, younger or older and so produce a sea-level change that is in principle similar to that caused by varying the amount of seafloor produced per unit time interval. If the area of the ocean basin changes due to continental growth or continental destruction (e.g., India colliding with Asia and decreasing the continental surface), a change in ocean volume will also occur.

Harrison (Chapter 8) suggests three processes by which the quantity or distribution of sediments in the deep sea could be changed: (1) the evolution of pelagic foraminifera in the Late Cretaceous, (2) the greater average age of the ocean basins today than 80 Ma, and (3) higher sedimentation rates during the past 5 m.y. than earlier times. The last two processes would tend to counteract about one-eighth of the effect of greater ridge areal volume in earlier times compared with today. The evolution of pelagic foraminifera may not have changed the rates of deposition of calcareous deposits, which depend almost entirely on the rate of weathering of calc-silicates and volcanic emission of CO_2, but it should have resulted in a redistribution of these sediments from mainly shallow-water reef deposits to a more or less uniform blanket over the deep-sea floor. This would have had a significant isostatic effect, replacing marked subsidence of shallow water areas and probably a forebulge under both adjacent land areas and the offshore deep ocean, with a relatively uniform sinking of the deep-sea floor and a rise under the continents.

Harrison (Chapter 8) calculates that the total increase in the average freeboard of the continents from all causes over the past 80 m.y., not counting the possible effects of redistribution of calcareous sedimentation, should have been about 260 m. But estimates based on paleogeographic sediment distribution for the past 80 m.y. indicate a much smaller value—150 m. He suggests that the discrepancies could be resolved if the continental hypsometric curves were steeper in the past than they are today. However, the uncertainties of the estimates of the various volume effects are sufficient to account for this freeboard discrepancy.

RECORD OF CURRENT AND PAST CHANGE

Sea level has changed dramatically over the course of earth history, with repeated variations of more than a 100 m from present sea level. These large sea-level changes occurred both during times of intense glaciation and during times of an ice-free Earth. Even during times of glacio-eustatic change, there were other broad-scale variations resulting from long-term climatic and tectonic changes. The details of sea-level variations are well documented for much of the record of the past few hundred thousand years but gradually deteriorates with older records, mainly because the time resolutions of dating and correlation methods become less precise. In addition, the numbers of assumptions needed to extract sea-level change from proxy records increase with age.

Historic and Tide-Gauge Records—The Past 100 Years

Large sections of the world's coastlines are experiencing changes in RSL that have been analyzed as being spatially and temporally coherent, but there is no unique way to average the existing sea-level data from different regions to obtain a global measure of RSL change. Variations on the order of 50 percent in the estimate of relative change can be induced simply by use of different averaging methods.

Barnett (Chapter 1), using both a key tide-gauge station approach and a grouping of records from different regions, has estimated that RSL has increased by 1 to 2 mm/yr over

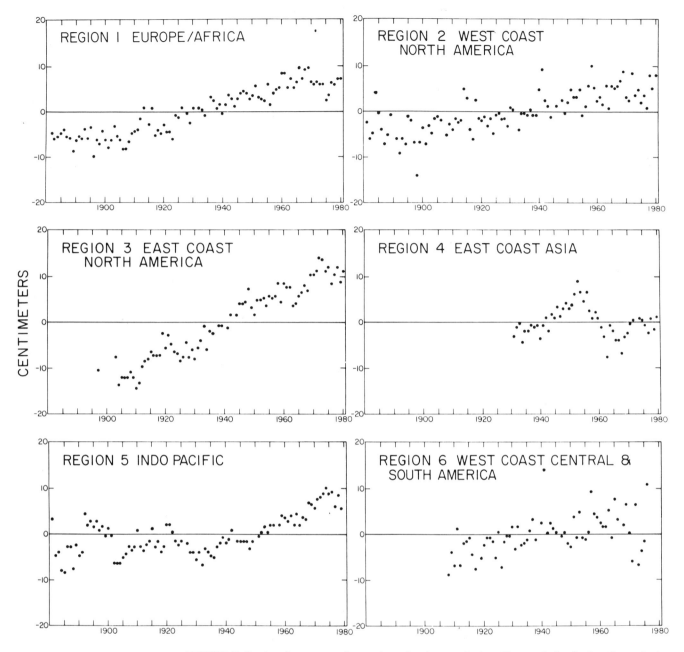

FIGURE 3 Regional averages of annual sea-level anomaly (see Chapter 1, for further discussion). From Barnett (1984) with permission of the American Geophysical Union.

the past 80 yr or so (Figure 3). There is no strongly coherent change on a global scale (coherence is primarily regional), nor is there an obviously accelerating rate of RSL increase in recent years.

Interpretation of sea-level records from tide-gauge measurements faces many problems. The principal problems are summarized below:

1. Station locations have been subject to position changes as, for example, during harbor development, which can result in discontinuity of the time-series record.

2. Tectonic uplift/subsidence of the land mass on which the tide gauge sits will induce

an RSL change. These vertical tectonic movements are further discussed in a previous section and in the discussion by Peltier (Chapter 4).

   3. A variety of physical processes that include changes in water temperature and salinity, river runoff, local and nonlocal currents, winds, waves, and atmospheric pressure all can affect RSL, sometimes in conflicting ways. Thompson (Chapter 2) found that the standard error of an estimate of longer term trend in North Atlantic sea-level records can be halved (from 0.4 to 0.2 mm/yr) on removal of short-term meteorological and oceanographic effects. Munk *et al.* (Chapter 14) discuss subsidiary measurements that can be used to take account of some of the oceanographic effects.

   Sturges (Chapter 3) finds that the dominant sea-level signals at periods of 5 to 10 yr and longer are coherent between the U.S. Pacific coast and Hawaii, and on both sides of the Atlantic. The amplitudes, at these periods, are ~50 to 150 mm. The signals are out of phase, with delays of several years, but are partly consistent with baroclinic wave propagation across the ocean. The longest records appear to be "contaminated" by this ocean-wave propagation. A wave of 10-yr period and 50-mm amplitude has a rate of rise of the order of 20 mm/yr, which is about an order of magnitude larger than Barnett's estimate of the mean rate of sea-level rise during the past 80 yr. Consequently, records of short duration, even if they have high precision, can hide the existence of long-term variability or change.

The Past 18,000 Years

Eighteen thousand years before present, huge continental ice sheets covered most of Canada, parts of the northern United States, Greenland, northwestern Europe and Antarctica, and sea level was more than 100 m lower than it is today; this is known as the Weichsel-Wisconsin glacial maximum (see discussion by Matthews, Chapter 5). The ice sheets stored the water that corresponded to nearly all of this lowering of global sea level. The mass of the ice sheets also deformed the shape of the Earth, and affected the volume of the ocean basins.

   The ice sheets in North America and Europe began to dissipate by 15,000 yrBP. The meltwater was returned to the ocean basins with a concomitant rise in sea level. The causes of ice-sheet disintegration are not clearly understood, but cyclical Earth-orbital variations in the inclination of the Earth's axis and in the seasonal Earth-Sun distance (COHMAP, 1988) together with ice-sheet dynamics and changes in ocean circulation probably formed a complex set of interrelated forcing functions that initiated breakup and continued wastage of the ice sheets between 15,000 and 7000 yrBP.

   The evidence for past sea levels is partly geological—the existence of coral reef terraces containing the littoral zone coral, *A. palmata*; submerged or buried strand-line complexes; peat deposits just above subaerial exposure surfaces; and articulated valves of shallow water or tidal zone pelecypods. Evidence of this kind includes a series of paired pelecypod valves off the Texas gulf coast at –57 m with a carbon-14 date of 12,900 ± 400 yrBP and a brackish water peat from New England at almost the same depth and age. Similar data are found at younger ages, but the only high-quality data between about 12,500 and 18,000 yrBP are the deep-sea oxygen-isotope records in planktonic and benthic foraminifera of pelagic sediments. The oxygen-isotope ratios reflect both the quantity of light oxygen sequestered in the ice sheets and the temperature of the waters in which the foraminifera lived. The relative importance of these two effects is uncertain. Matthews shows that this uncertainty gives a range of 120 to 108 m for the maximum sea-level lowering in the tropical North Atlantic near Barbados. Chappell and Shackleton (1986) estimate a value of –130 m for the Huon Peninsula in New Guinea, and they cite a maximum sea-level lowering estimate of 150 m on the broad shelf off northern Australia. They state that such variations around the globe "are to be expected as a result of adjustment to glacial/interglacial change of ice and ocean volumes."

During most of the time between 15,000 and 7000 yrBP, the rate of dissipation of the ice caps was very rapid, and the average rate of rise in sea level was 12.5 mm/yr. But there were two brief periods of renewed cold in the European, though not the American, record at about 11,000 to 12,000 yrBP. These are called the Older and Younger Dryas. Pollen records from European bogs show that the forests that had begun to repopulate northern Europe after the ice departed were destroyed and replaced by Arctic shrubs. These colder intervals lasted for several hundred years. They may have resulted from a reduction in the temperature of the near-surface water lying to the west of Europe (Broecker, 1987).

By 7000 to 8000 yrBP the North American and European ice sheets had disappeared. In high northern latitudes, RSL was 10 to 30 m higher than it is today because of the isostatic depression of the land under the previous weight of ice. In England and southern New England and in lower latitudes, the isostatic "forebulge" had produced an RSL 10 to 20 m lower than today's sea level.

Bloom and Yonekura (Chapter 6) examined 13 well-documented Holocene sea-level records from published literature that covers the past 8000 radiocarbon years and a range of latitudes from the Arctic to Panama. In the north, RSL declined exponentially at initial rates of ≥10 mm/yr and a final rate during the past 1000 yr of the order of 1 or 2 mm/yr. South of the "hinge line" in central New England, RSL rose at an exponentially declining rate. Interestingly enough, the amount of emergence or submergence for each 1000-yr interval at any particular place is proportional to the total amount of emergence or submergence at that place over the past 8000 yr. This would imply that there has been no eustatic change in sea level during the later Holocene, only isostatic adjustment—upward RSL rise in the south and downward RSL fall in the north, which are strictly proportional to each other. If so, in the absence of climatic change, future sea-level change could be controlled by residual isostatic response to the loading and unloading completed 7000 to 8000 yrBP.

This suggestion is consistent with Peltier's model of isostatic adjustment in the mantle (see Chapter 4) and with a recent discussion by Stewart (1989). It is given further credibility by Barnett's (1984) observation that sea level is falling on the east coast of Asia, where presumably different isostatic adjustments are taking place.

The Past 250,000 Years

The record of sea-level change for this time period can be interpreted from two principal proxy sources: (1) oxygen isotope records in sediment cores from the deep sea and (2) the study of ancient shorelines (coral reefs) that have been tectonically uplifted. Although each type of record presents several interpretive problems and requires assumptions in deriving sea level, a recent correlation of an oxygen isotope record and a flight of raised coral terraces in the southwest Pacific has provided improved estimates of sea level (Table 3 and Figure 4) (Chappell and Shackleton, 1986).

Coral terraces from the Huon Peninsula (Papua New Guinea) are described by Bloom and Yonekura in Chapter 6. These uplifted terraces are like a continuous tape recorder; each coral reef developed when the rising sea level overtook the rising land, and the reef crests represent approximately the peaks of each sea-level rise. RSL can be estimated from the current heights of the terraces. It has been assumed that sea level at ~125,000 yrBP was +6 m above the present sea level—a number derived from several different terrace sequences around the world. On the basis of this assumption, the uplift rates for different portions of the Huon Peninsula can be derived. Further from the assumption that these uplift rates have remained essentially constant for the past 125,000 yr, sea level can be calculated (Figure 4) for the other Huon Peninsula terraces as described in Chapter 6.

Oxygen isotope ratios, $\delta^{18}O$ [$\delta = (^{18}O/^{16}O \div {}^{18}O/^{16}O_{SMOW}) - 1$, where SMOW is standard

TABLE 3 Sea Level and Calculated Rates of Average Change for the Past 135,000 Years

Stage	Age[a] (1000 yrBP)	Sea Level[a] (m)	$\delta^{18}O$[a]	Change	Rate (mm/yr)
Modern	0	0	3.42		
				—	—
I	6	0[b]	3.47		
				+118	+10.7
I/II	17	−130	5.09		
				−84	−7.0
II	29	−46	4.48		
				+20	+3.3
II/IIIb	35	−65	4.67		
				−24	−4.8
IIIb	40	−41	4.47		
				+11	+5.5
IIIb/a	42	−52	4.66		
				−7	−3.5
IIIa	44	−45	4.51		
				(−15)	(−2.5)
IVb	53	−30	4.30		
				+14	+4.7
IVb/a	56	−44	4.58		
				−16	−5.3
IVa	59	−28	4.20		
				+33	+6.6
IVa/Vb	64	−61	4.66		
				−25	−3.1
Vb	72	−36	4.21		
				+8	+4.0
Vb/a	74	−44	4.33		
				−25	−3.6
Va	81	−19	3.99		
				+25	+3.6
Va/VIb	88	−44	4.37		
				−18	−2.2
VIb	96	−26	4.09		
				(+7)	(+0.7)
VIa	106	−19	4.03		
				+43	+7.2
VIa/VIIb	112	−62	4.38		
				−62	−10.3
VIIb	118	0	3.49		
				+8	+2.0
VIIb/a	122	−8	—		
				−14	−7.0
VIIa	124	+6	3.37		
				+136	+12.4
VII/VIII	135	−130	5.20		

[a]Data from Chappell and Shackleton (1986).

[b]Sea level at 6000 yrBP has been within a meter or so with fluctuations arising from isostatic and tectonic adjustments (Bloom and Yonekura, Chapter 6).

mean ocean water], of foraminifera vary with the temperature, salinity, and δ^8O of the water from which their carbonate tests are deposited. Ocean water $\delta^{18}O$ varies with the quantity of isotopically light ice stored on the continents so that the record from foraminifera is a blend of global ice volume and local water temperature and salinity components. For the late Pleistocene, temperature and salinity components of the signal are demonstrably small compared to the ice volume signal. A $\delta^{18}O$ record from the east equatorial Pacific (core V19-30; Shackleton *et al.*, 1983) is shown in Figure 4. Peaks represent the influx of isotopically light water derived from continental ice and are correlated with high stands of sea level indicated by the Huon Peninsula terraces.

If these sea-level change magnitudes and timing are representative of global sea level, then rates of eustatic sea-level change during the period from 250,000 yrBP to about 6000 yrBP varied between –10 mm/yr and +12 mm/yr, with typical rates being on the order of ±3 to ±7 mm/yr. These rates exceed those that have prevailed over the past few thousand years but may be approached by those projected for the next millennium or so (see section on Forecasting Changes in Sea Level Related to Greenhouse Gases).

At the peak of glaciation 18,000 and 135,000 yrBP, sea level near New Guinea was lower than it is today by about 130 m (Table 3). But between these extremes, during glacial episodes, sea level oscillated between 19 and 65 m below the interglacial levels over time intervals of 2000 to 12,000 yr. These oscillations resulted from variations in the area and thickness of both the continental ice sheets in northern Europe and northern North America and the Antarctic Ice Sheet. As shown by Labeyrie *et al.* (1986), at least some intervals of rising sea level may have been initiated by surges of ice streams behind ice shelves in Antarctica, which greatly increased the rates of iceberg calving. The resulting rise of a few meters in global sea level may have then destabilized the marine portions of the Northern Hemisphere ice sheets, leading to the much larger rise in sea level shown in the oxygen isotope and Huon Peninsula terrace records.

The Past 250 Million Years

The sea-level record becomes much more cloudy as we look back in time. Much of the past 250 m.y. was dominated by an ice-free Earth or at least a world lacking extensive continental ice caps (see Frakes and Francis, 1988).

Oxygen isotope ratios of tropical planktonic foraminifera show that continental ice volumes during only the past 40 to 45 m.y. were consistently as large as or larger than that of today (Prentice and Matthews, 1988). Between 50 and 65 Ma the world was intermit-

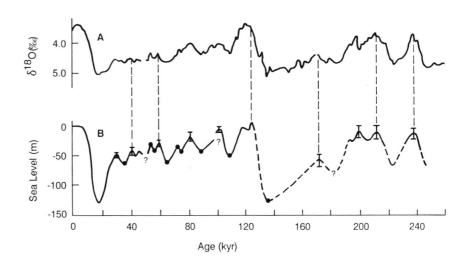

FIGURE 4 (A) $\delta^{18}O$ record for the past 240,000 yr from east equatorial Pacific core V19-30; and (B) sea-level curve for the Huon Peninsula recalculated to correlate with the $\delta^{18}O$ record from core V19-30. Modified after Chappell and Shackleton (1986).

tently ice free; and continental ice was apparently absent throughout the Mesozoic (65 to 250 Ma). Consequently, sea level was considerably higher than in later times, and vast continental areas on the order of 50 million km^2 (one third of present land area) were flooded, e.g., the Cretaceous Interior Seaway in North America.

Sea-level changes were dominated by three major forcing factors: climatic change, tectonic processes, and sedimentation. Climate was much warmer than it is at present, with ocean-bottom temperatures at some 10 to 15°C. Thermal contraction of seawater (steric effect) could account for a sea-level fall of more than 10 m between the climatically warm and equable Cretaceous (144 to 66 Ma) and the onset of an ice-dominated system in the Cenozoic. Tectonic processes such as seafloor spreading, subduction of oceanic crust, broad-scale ocean-floor epeirogeny, continental collisions, and other activity affected the volume of the ocean basins. Epeirogenic uplift or subsidence of the continents also affected RSL. The volume of the ocean basins was also changed by changes in sedimentation. The evolution of the calcareous algae called *Coccolithophoridae* at the end of the Jurassic and of pelagic foraminifera in the Middle Cretaceous resulted in deposition, perhaps for the first time, of a thick blanket of calcareous sediments over the deep seafloor and in a possible rise in sea level.

Direct evidence for sea level at different times during the past 250 m.y. comes from paleogeographic maps, from which the area of the continents covered by marine sediments can be estimated. Sea level was highest, about 180 ± 40 m above the present level, at 100 Ma, having risen from 80 m about 180 Ma. There was also a pronounced regression of sea level (commonly associated with a fall) during the Paleocene at 60 Ma.

For the past 80 m.y., there is evidence concerning changes in the volume of the ocean basins, particularly in the volume of ancient mid-ocean ridges, based on estimates of spreading rates between times of magnetic reversals and estimates of ridge lengths. From the data analyzed by Kominz (1984), ridge-crest volume changes since 80 Ma would have resulted in a fall of sea level of $180 \pm \sim 150$ m. If ice volume changes, the effects of Pacific volcanism, and changes in sedimentation are added, the calculated sea level fall since 80 Ma is about 230 m, some 40 percent greater than that computed from modern paleogeographic maps. Although the errors in both sets of estimates are more than sufficient to account for the discrepancy, it is possible, as Harrison has suggested (Chapter 8), that the continental hypsographic curves were steeper in the past than they are today, so that a greater change of freeboard was necessary to produce a given increase in the areas of flooding.

Little or no oceanic crust older than about 80 m.y. still exists in the present oceans; hence there is little direct evidence concerning the volume of the ocean basins. We must rely on paleogeographic reconstructions of the continents and analysis of their sediment covers to infer ancient sea levels.

Within the past dozen or so years, seismic stratigraphic methods have been developed and applied to the analysis of scores of events of coastal onlap/offlap within stratigraphic successions. P. R. Vail and his colleagues, e.g., Haq *et al.* (1987), have ascribed most of the boundaries between offlapping and onlapping stratal units to sea-level change (Figure 5) and have implied that the vast majority reflect eustatic or global sea-level variation. Although this approach has proven valuable in exploration for hydrocarbon resources, many investigators question whether these boundaries are necessarily of eustatic origin. Some boundaries may be caused by variations in the local rate of subsidence or by changes in the rate of sediment supply. Vail and his co-workers have estimated magnitudes of sea-level change that are on the order of 10 to 100 m and are superimposed on the long-term trends from climatic and tectonic processes discussed above, but questions also remain about whether such large variations are indicated by the data in hand, even if the signal is a eustatic one. It is particularly difficult to explain sea-level changes in excess of 100 m in intervals of 1 m.y. or less during the Cretaceous, a time when the Earth was largely ice free. These and related issues are discussed by Christie-Blick *et al.* in Chapter 7.

Unconformity-bounded sequences, which form largely in response to changes in de-

23

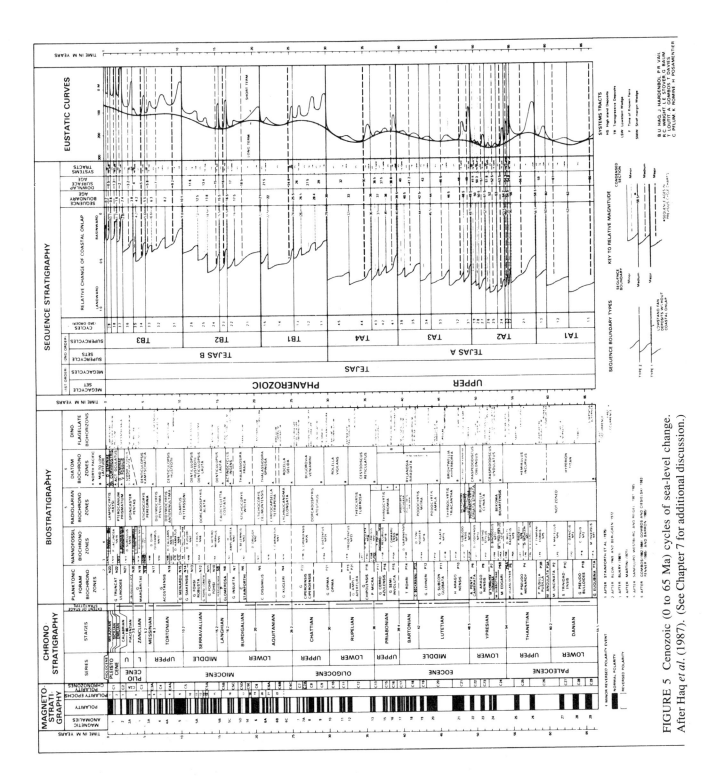

FIGURE 5 Cenozoic (0 to 65 Ma) cycles of sea-level change.
After Haq *et al.* (1987). (See Chapter 7 for additional discussion.)

positional base level, can be identified in outcrops and boreholes as well as in seismic profiles. Unlike transgressions and regressions of the shoreline or changes in paleobathymetry, which are used in classical stratigraphy to gauge sea-level fluctuations, the formation of regional unconformities is relatively insensitive to the rate of sediment supply, and this constitutes the main advantage of the sequence stratigraphic approach. According to Christie-Blick *et al.* (Chapter 7) and Christie-Blick (1989) most sequence boundaries record times at which the rate of sea-level fall increased (or reached a maximum) or the rate of tectonic subsidence decreased. Those of eustatic origin should be formed in all basins connected to the open ocean and should be nearly correlative, whatever the local tectonic history. Unconformities of tectonic origin may also be present, but these are not expected to extend beyond a region more than a few hundred or a few thousand kilometers across. The identification of a global sea-level signal therefore depends on demonstrating that particular unconformities are present in widely separated basins, and are synchronous (ideally to within 0.5 m.y.).

More difficult than establishing the timing of sea-level change is the problem of estimating amplitudes and rates of change. Amplitudes of sea-level oscillations may be estimated through a combination of sequence stratigraphy and geophysical modeling of subsidence history, but practical difficulties and sensitivity of results to model assumptions may limit estimates to no better than a factor of 2 or 3 larger or smaller than true amplitudes (Christie-Blick *et al.*, Chapter 7). The principal uncertainties are in age control, paleobathymetry, compaction history, the effects of sediment loading on the lithosphere, and the tectonic subsidence that must be subtracted from corrected stratigraphic data to obtain the sea-level signal.

Harrison (Chapter 8) has suggested that 10 transgressive-regressive cycles of sedimentation with average periods of 7 to 10 m.y. during the Cretaceous resulted from alternating increases and decreases in the rates of seafloor spreading. These would have caused eustatic rises and falls of sea level of about 15 m. The eustatic sea-level changes were greatly amplified in the Western Interior Seaway by subsidence of the basin resulting from subduction of the Pacific plate under the western part of the North American continent. In Chapter 9, Hay and Leslie discuss possible changes in the volume of liquid water stored as groundwater as a cause of some of these Cretaceous sea-level changes. These and many other problems reviewed in Chapters 7 and 8 need to be solved before these curves can be read as eustatic changes.

If stratigraphic sequence boundaries do turn out to be related to eustatic sea-level change, the record from the Triassic (about 250 Ma) through to the present becomes rich with events that need to be considered in the light of possible causal processes.

From 250 to 2500 Ma

Although there is some direct stratigraphic record of sea-level change that extends back into the Paleozoic and Precambrian (Christie-Blick *et al.*, 1988; Bond *et al.*, 1988), most of the data are derived from continental freeboard consideration. Geologic evidence on the area of the continents covered with marine sediments indicates that continental freeboard, the average elevation of the continents above sea level, has been approximately constant since the Archean, 2500 Ma. Continental freeboard today, taking the mean of the continents, is about 750 m, and changes of more than a few hundred meters are apparently ruled out by the geologic evidence.

Provided that the mass of seawater (plus water on land) has not varied significantly, an approximately constant continental freeboard since the Archean, combined with a 50 percent decrease of heat flow from the mantle, has been shown by Schubert and Reymer (1985) to require a net growth of the continents of about 25 percent (1 km^3/yr) over the past 2500 m.y. They believe that post-Archean changes in the mean thickness of the continental crust are highly unlikely, hence the continents must have grown in area. If freeboard has been within ±200 m of its present value since the end of the Archean, the continents have grown in area by anywhere from 10 to 40 percent. On the other hand, if the mass of seawater has increased significantly since the Archean, isostasy requires that the thickness of the continental crust should have likewise increased. For example, if the flux of water to the ocean from the Earth's interior has been constant during the Earth's lifetime, the entire growth of continental volume estimated by Schubert and Reymer can be accounted for by an increase in continental thickness rather than continental area. Continental growth is required as long as freeboard at the end of the Archean was not more than 400 m smaller than it is today.

Continental ice ages, which must have caused a fall in sea level, occurred at intervals during the Proterozoic and the Paleozoic. According to Crowell (1982), evidence from what is now southern Ontario shows that there were large-scale glaciations between 2500 and 2100 Ma. One ice sheet may have reached southern Wyoming. A similar glaciation probably occurred in South Africa sometime between 2700 and 2200 Ma, and possible tillites are reported from Western Australia. During the span of 600 m.y. between 2700 and 2100 Ma, a period as long as the entire time in which multicelled animals and plants have existed, several widespread glaciations, and presumably concomitant falls in sea level, apparently occurred, separated by ice-free eras.

There is no record of glaciation anywhere on Earth during the ensuing 1100 m.y. until the Late Precambrian, when continental glaciation flourished intermittently on all the continents, except possibly Antarctica, from about 950 to 560 Ma. Apparently there were three glaciation peaks, one about 940 Ma, another around 770 Ma, and a third about 615 Ma, in regions around the present North Atlantic, in central and southwestern Africa, and in Brazil, western North America, and Australia. Each of these extreme glacial times must have been a period of markedly lower sea level in comparison with earlier or later periods.

These Late Precambrian glaciations apparently occurred at low latitudes, distant from the Earth's poles, for which no satisfactory explanation has come forth. Perhaps, there was a reduced greenhouse effect or the patterns of oceanic and atmospheric circulation were markedly different during Precambrian times from those of today, or of the past half billion years.

The next widespread glacial epoch began in the supercontinent Gondwana near the end of the Ordovician lasting into the Silurian, from about 450 to 400 Ma. The record extends in scattered localities from northern Europe to South Africa and from the Sahara region to Bolivia and Peru.

For 90 m.y., from 330 Ma to 240 Ma, a Late Paleozoic ice age existed in large areas of the Gondwana supercontinent, beginning in what is now South America, reaching a climax in South America, Africa, and Antarctica, and ending in Australia. There were apparently several moderately sized ice caps instead of one huge one, because it does not appear that sea level was drastically lowered. The center of the glaciated regions was near the South Pole, as the supercontinent of Gondwana drifted across it.

There was no continental glaciation between Late Permian and Paleogene time, even though Antarctica lay near the South Pole. This may have been the result of a greenhouse effect caused by a high atmospheric CO_2 content, related to high rates of seafloor spreading and undersea volcanism (Berner *et al.*, 1983).

DO CHANGES IN SEA LEVEL CAUSE CHANGES IN CLIMATE?

Climate variability and change cause local or regional variations in sea level over months, years, or decades. Quantitatively the seasonal oscillation is the most important of these, followed by interannual variations of 5 to 10 yr, related to the Southern Oscillation in atmospheric pressure between the two sides of the Pacific Ocean, and manifested in El Niño in the eastern Pacific and in many other phenomena elsewhere.

Longer-term global climatic changes can be reflected in eustatic sea-level change through three processes: the waxing and waning of continental ice caps and alpine glaciers; variations in the quantity of liquid water stored on land in lakes, rivers, and underground aquifers; and steric changes in the volume of seawater resulting from warming or cooling.

It is also often suggested that a change in sea level can produce a change in climate. As Barron and Thompson (Chapter 11) point out, this suggestion is consistent with the apparent correlation of climate and sea level over geologic time. Several lines of evidence show that during the past 70 m.y. globally averaged surface temperatures have declined by 6° to 12°C, global sea level has fallen by perhaps 200 m, and the total land area above sea level has increased by about one-third.

There is a variety of physical mechanisms by which sea-level changes could directly affect global climate: changes in albedo, regional changes in atmosphere-surface coupling, changes in ocean circulation, changes at ice-sheet ocean margins, and changes in ocean/atmosphere chemical composition.

Virtually every aspect of the Earth's climate is affected by the exchange of heat, moisture, and momentum between the atmosphere and the underlying surface. Changes in the surface energy balance are determined by changes in surface albedo and surface wetness. A rise of sea level will increase the area covered by ocean water, which has a much lower albedo and a much higher wetness than the land. A change in global albedo by 0.01 will produce about a 1°C change in surface temperature. Such an albedo change would require a major change in the land/sea ratio, caused by a rise or fall of 100 m or more in sea level.

The change in wetness resulting from this assumed change in relative areas of sea and land would lead to marked changes in evaporation, and therefore in precipitation, although not necessarily in the same region, and also to a significant change in average summer surface temperatures because of the much larger thermal inertia of the ocean compared to the land. However, model calculations indicate that when both surface albedo and moisture availability are altered simultaneously, they produce nearly complete compensating effects. For example, modeled deforestation of the Amazon Basin, which would produce the same effects on albedo and wetness as a fall in sea level, produced only a small net surface temperature change in the deforested area, and no detectable global scale climate effects (Henderson-Sellars and Gornitz, 1985).

In general, surface roughness is an order of magnitude higher over the land than over the ocean. Consequently, a change in ocean/land proportions caused by a rise or fall of sea level may result in a marked redistribution of the surface areas of momentum exchange between the land and ocean and the atmosphere. This could have a marked effect on atmospheric circulation systems.

In some regions, for example, the Blake Plateau in the Western North Atlantic, it can be shown from studies of seafloor erosion that the position of the Gulf Stream has varied repeatedly by hundreds of kilometers with changes of sea level during the past 20 m.y. But the extent of possible climatic change caused by such movements in current position is unknown. The same lack of knowledge about climatic effects afflicts examples of seafloor subsidence such as that of the Greenland-Iceland-Faroe Ridge between the Arctic and the Atlantic oceans, and the Walvis Ridge off South Africa. Sea-level change can also isolate or reconnect small basins along the ocean margins. Enhanced evaporation in partially isolated basins can produce water masses of markedly different density than those of the main ocean, and thereby affect deep water and mid-water formation.

A marked increase in atmospheric CO_2 accompanied the deglaciation of the Northern Hemisphere ice sheets about 10,000 yrBP; it probably had a significant warming effect on the lower atmosphere. One hypothesis to explain at least part of this increase in atmospheric CO_2 involves the submergence of the continental shelves by the rise in sea level. In tropical waters this resulted in a large increase in the growth of coral reefs and precipitation of other calcareous sediments with a corresponding release of CO_2 to the subsurface ocean layers and the atmosphere.

If bottom water temperatures were unchanged, the fall of the sea level by 20 to 130 m during the last glaciation should also have resulted in the release of methane (a potent greenhouse gas) from methane ices (clathrates) in the upper layers of continental slope sediments. These sediments are believed to contain 2.2 mg/cm^3 of methane, or a total of 3400 gigatons of methane (Revelle, 1983a). A fall of 100 m in sea level without a change in ocean-bottom temperature should release 425 gigatons of methane. If the sea-level change took 5000 yr, 0.085 gigatons would have been released each year. Methane in the atmosphere has a half life of about 8 yr, implying that the release of methane from continental slope sediments should have increased the preindustrial methane level of less than 2 gigatons by about 50 percent. In fact, evidence from ice cores indicates that the methane content of the atmosphere decreased by nearly 50 percent during the latter half of the last glacial epoch and was at no time higher than the postglacial, preindustrial level of 650 parts per billion (Stauffer, 1988). This information suggests that during the glacial period the temperature of the ocean waters bathing the continental slope sediments decreased by more than 1°C than it is at present. This would prevent destabilization of the sedimentary clathrates by the release of pressure that resulted from the drop in sea level.

As far as the apparent correlation between the 70-m.y. fall in eustatic sea level and the drop in temperature over the same period are concerned, it now seems most reasonable to ascribe both phenomena to the same set of processes within the Earth and not to attempt to forge a causal relation between them. The high sea levels of late Cretaceous time were most probably caused by intense oceanic lithosphere formation and seafloor spreading from the mid-ocean ridges (see Harrison, Chapter 8). This same process resulted in perhaps a tenfold higher level of atmospheric CO_2 and a correspondingly warmer climate, perhaps 10°C warmer (Berner *et al.*, 1983).

FORECASTING CHANGES IN SEA LEVEL RELATED TO GREENHOUSE GASES

On a time scale of 10^2 to 10^5 yr, global changes in sea level (eustatic changes) can result mainly from the buildup or decay of alpine or continental glaciers, and from long-term ocean volume or steric changes caused by temperature or salinity changes in waters below the thermocline. The concern here is with forecasting eustatic changes related to the rise of CO_2 and other greenhouse gas concentrations in the atmosphere.

The past few thousand years have been a time of a high and relatively stable stand of sea level after 100 millennia of rapidly varying levels during the last ice age. Regional steric and other irregular variations of 10 to 20 cm from year to year, or from decade to decade, have been common, but there has been at most only a small unequivocally detected long-term trend in eustatic sea level. This situation can be expected to change with the advent of greenhouse-gas-induced climate change.

Sundquist (Chapter 12) examined the probable future course of atmospheric CO_2 concentrations over the next 1000 yr. It might be expected that oceanic biogeochemical processes related to the dissolution of calcium carbonate sediments on the deep-sea floor would, within a few centuries, reduce the content of free CO_2 in seawater, and hence the atmospheric content. This turns out not to be so, provided sufficient CO_2 has been generated by fossil fuel combustion.

The total remaining reserves of coal, oil, natural gas, oil shale, and oil sands that are ultimately recoverable for human use are believed to correspond to about 7500 billion tons of carbon. Of this amount approximately 3500 billion tons have already been identified. Sundquist assumes that 2500 billion tons will ultimately be consumed in human activities, with a peak rate of production in the middle of the next century of 16.8 billion tons/yr compared to the annual carbon emissions in 1985 of 6 billion tons. He assumes that the production rate will decline to 6 billion tons by about A.D. 2150 and go nearly to zero by A.D. 2350. With this assumed history of hydrocarbon combustion, Sundquist finds that the

CO_2 content of the air rises from 350 parts per million by volume (ppmv)(700 billion tons) in 1985 to 800 ppmv (1600 billion tons) around A.D. 2160, and slowly declines thereafter to 550 ppmv (1100 billion tons) by A.D. 2700. The peak concentration is approximately 2.9 times the base concentration of 280 ppmv in 1880, from which increases of atmospheric CO_2 content are usually calculated. Also to be taken into account are the increasing concentrations of methane, nitrous oxide, tropospheric ozone, and other minor greenhouse gases that can be expected 150 yr from now.

General circulation models of the atmosphere constructed by the National Center for Atmospheric Research, the NASA Goddard Institute of Space Sciences, the Geophysical Fluid Dynamics Laboratory of NOAA, and others indicate that the estimated increased concentrations of greenhouse gases should result in an average global temperature rise of 3° to 6°C in the atmosphere near the Earth's surface in the next 100 yr. [Recent modeling studies show that more sophisticated parameterization of clouds, taking into account both ice and liquid droplets, greatly reduces the sensitivity of the climate models to increased CO_2 (Cess et al., 1989; Mitchell et al., 1989). These newer models indicate a smaller temperature increase than the 3° to 6°C range given above.]

The next question to ask is: How will this atmospheric temperature change affect the world's oceans. This question has recently been studied by Frei et al. (1988). They use two kinds of models: a "pure diffusion" (PD) model in which heat is carried downward by eddy diffusion, assuming vertical diffusion coefficients of 1.3 and 2 cm²/s and a modified diffusion model in which cold, polar water sinks to the bottom of the ocean and mass is conserved by assuming a slow, global upwelling. Frei et al. (1988) call this model an upwelling-diffusion (UD) model; they assume that the coefficient of vertical eddy diffusion is about 0.65 cm²/s and that the global average upwelling rate to balance deep and bottom water formation is about 4 m/yr. A considerably smaller rate of upwelling and, correspondingly, a higher rate of downward vertical diffusion would correspond to estimates by Whitehead (1989) that cold deep and bottom water is formed at a rate of only 5 million to 10 million m³/s instead of the rate of 40 million m³/s assumed by Frei et al. (1988). Measurements of tritium distribution in the North Atlantic made by the "GEOSECS" Expedition in 1972 (Ostlund et al., 1974) and 10 yr later by the "Transient Tracers in the Ocean" Expedition (PCODF, 1981; Ostlund, 1983) indicate that the tritium "front" in the Atlantic deep water between depths of 2500 and 5000 m moved about 800 km south during this 10-yr period, indicating that the mass of deep water sinking in the Norwegian Sea and cascading downward through the Denmark Strait, is 10 million to 20 million m³/s. Hence, the rate of upwelling is probably smaller and the rate of downward diffusion greater than assumed by Frei et al. (1988).

Basically, a PD model transports heat relatively rapidly into the oceans, which slows the atmospheric temperature response to the rising CO_2 concentration but increases the rate of sea-level rise. The UD model reduces heat penetration into the ocean, allowing the climate to warm relatively rapidly but reducing the sea-level response.

Frei et al. estimate that the rise in sea level during the past 100 yr caused by thermal (steric) expansion was between 3 and 8 cm, with the lower value corresponding to the UD model. They project a rise of 10 to 50 cm during the next century resulting from thermal expansion, the range arising from uncertainties in CO_2 and trace gas concentrations and in estimates of climate sensitivity to greenhouse warming.

To this estimate of steric sea-level rise in the next century must be added the NRC (1985) Committee on Glaciology's estimate of the contribution to sea-level rise by ice wastage in a CO_2-enhanced environment. This would come from three sources: glaciers and small ice caps, the Greenland Ice Sheet, and the Antarctic Ice Sheet. This NRC committee estimates a sea-level rise by the year 2100 (the assumed time for a doubling of atmospheric CO_2) from ice wastage of 0.1 to 1.6 m with "most likely" values of 0.2 to 0.9 m; the "most likely" scenario can be expressed as 0.55 ± 0.21 m if one assumes that this range expresses a standard deviation from the mean and if the "errors" are considered to be independent.

Revelle (1983b), using a two-dimensional vertical diffusion model for ocean thermal expansion, estimated a total sea-level rise of about 0.7 m. Also, in 1983, Hoffman *et al.* (1983) forecast a larger global rise, between 1.44 m and 2.17 m by A.D. 2100. This was estimated to result from thermal expansion of ocean waters and from ablation and partial melting of alpine glaciers and the ice caps of Antarctica and Greenland. Of the total rise, an average 0.72 m was estimated to result from ocean thermal expansion, and 0.72 to 1.45 m were added from ice discharge to maintain the relative contributions thought to have contributed to sea-level rise over the past 100 yr. Because the range of rise related solely to ocean thermal expansion was calculated to be 0.28 to 1.15 m, the ratio approach led to an extreme upper limit of more than 3 m (of this amount, mountain glacier melting could contribute, at most, 0.3 to 0.5 m). Robin (1987) forecast a rise of 0.80 m by A.D. 2100 with a range of 0.20 to 1.65 m. Gornitz *et al.* (1982) calculated the component of sea-level rise resulting from ocean thermal expansion between 1980 and 2050 as 0.20 to 0.30 m. MacCracken *et al.* (1989), incorporating the Frei *et al.* (1988) model results and, with some adjustments, the findings of Revelle (1983b), estimate a total sea-level change by the year 2100 of less than 0.5 to 1 m.

In all these estimates, the possible change, largely a result of human activities, in the quantity of water stored in lakes, man-made reservoirs, and underground aquifers has been neglected. Robin (1987) estimated the net effect of these three sources together at the present time as causing an annual rise in sea level of 0.08 mm, or 8 mm/century.

Most calculations indicate that sea level will continue to rise for at least several hundred years at average rates of centimeters per year or less. Of course, a considerably more rapid rate would ensue if the West Antarctic ice cap should disintegrate during the next 1000 yr. As we have seen, such a dissolution is probably impossible during the next several hundred years because, as Bryan *et al.* (1988) have shown, deep ocean convection on the southern side of the great circumpolar Antarctic Current may markedly delay much Antarctic temperature change in the upper ocean layers.

The assumed rate of carbon combustion is highly uncertain. In order to reach an atmospheric level of 800 ppmv by A.D. 2180, an average rate of CO_2 production corresponding to 9 billion tons of carbon per year during the next 200 yr is assumed. This is at least 50 percent higher than the present rate of 6 billion tons/yr, which is estimated to include tropical deforestation of greater than about 1 billion tons/yr (Machta, 1983). On the one hand, economic and social development of the now developing countries, which make up the vast majority of mankind, may well require a considerable increase above present levels of carbon combustion, and consequently increase the worldwide rate, perhaps by a factor of 3 or more. On the other hand, as Goldemberg *et al.* (1987), Mintzer (1987), and Bach (1988) have persuasively argued, it may be possible, through increases in energy use efficiency and substitution of other energy sources for fossil fuels, to reduce considerably the influx of CO_2 to the atmosphere and ultimately to the oceans. With these alternative energy sources in use, the atmospheric burden of CO_2 might remain at all times much below our estimated figure of 800 ppmv, and the consequent rise in air and sea temperature and steric sea level would be considerably reduced.

In these calculations it has been tacitly assumed that the circulation of the deep water of the world ocean will continue relatively unchanged, except that the deep water will be somewhat warmer because of vertical and lateral mixing. In other words, the future ocean circulation will be "surprise free." However, the warming of the atmosphere in high latitudes, and the corresponding warming of the subsurface ocean waters, may greatly reduce the volume of water sinking to the depths. The projected rise in sea level from steric expansion would then be intermediate between the range estimated from the UD model (10 to 50 cm), and the range computed from the PD model (20 to 110 cm).

HOW CAN WE IMPROVE THE MEASUREMENT OF SEA-LEVEL CHANGE?

The apparent trend of sea level at a particular place as measured by a tide gauge is the sum of trends in motion of the gauge itself as the land on which it is mounted moves vertically, the trend of change in steric sea level, and the trend of change in water mass under the tide gauge. To understand what is happening one needs to be able to make measurements that will separate these three components of the observed sea level. This problem is addressed by Munk *et al.* (Chapter 14). Each component of the observed sea level is considered separately.

The vertical motion of the tide gauge can be measured with fair accuracy in four different ways: by measuring changes in the acceleration of gravity at the sea surface under the gauge, by very-long-baseline interferometry (VLBI), by satellite laser ranging, and by use of the satellite signals of the global positioning system (GPS). Without repeated measurements over several years, none of these methods is sufficiently accurate to determine the vertical position of the tide gauge.

The acceleration of the Earth's gravity, g, can be measured with an uncertainty of about 1 part in 10^8 by the methods described in Chapter 14. This accuracy corresponds to a sensitivity to height changes of the tide gauge of about 30 mm. VLBI observing stations yield estimates of intercontinental baselines with an rms scattering of 20 to 30 mm. Up to the present, the scatter of the vertical components has been 3 or 4 times larger than this. One of the major sources of error is atmospheric refraction caused by water vapor. Improved water vapor radiometers are being developed and placed in operation. The rms scatter of all components of the baseline should then be reduced to the 10- to 20-mm level. However, reducing the scatter below the 10-mm level may be very difficult (Carter *et al.*, 1986).

Global Positioning System surveys of benchmarks separated by 8 to 50 km agree in the height components by ±10 to 30 mm. Carter *et al.* (1986) believe that these results are about as good as can be expected reliably for the foreseeable future. However, there are plans for increasing the accuracy of space-based geodetic techniques. It is possible that satellite laser ranging (SLR) can improve position accuracy from the present 1-cm value to 1 mm in the next decade. If vertical positions can achieve this accuracy, we should be in a position to improve our knowledge of eustatic change considerably. The Geodynamic Laser Ranging System (GLRS) of the Earth Observing System (EOS) will allow a large number of tide gauges to be included in the network.

At present, the single measurement errors in all four methods of measuring the changes in the elevation of the tide gauge are comparable at ±10 to 30 mm, and it seems likely that they will remain for some time to come. In contrast, it is desirable that the motion of the tide gauge be determined to a fraction of 1 mm/yr. Hence observational strategies will have to rely on repetitive measurements spanning intervals of several years and even then the desired accuracy can only be achieved by GLRS.

The change in the steric height of sea level can perhaps best be monitored with bottom-mounted upward-looking fathometers plus tide gauges. (The alternative of measuring bottom pressure with sufficient accuracy presents many difficulties, because of seemingly inevitable unpredictable drift of the pressure gauges at pressures of a few tens of atmospheres.) Because of the marked variation in the velocity of sound in water with changes in temperature (some 23 times larger than the changes of specific volume or density with temperature), it should be quite practical to estimate the integrated changes in temperature over the water column above the fathometer. The main problems in measuring steric changes, as Roemmich (Chapter 13) points out, are that they tend to extend over great depths and are not confined to or concentrated in the upper ocean. In the subtropical North Atlantic, steric changes extend to at least 3000 m with maxima in the thermocline at depths of 300 to 700 m and below the thermocline at about 1800 m. Hence the inverted fathometer is likely to be most useful in the depths of the open ocean far from land. The logistical problems of maintaining operating fathometers in such depths and locations and combining their measurements with surface mounted tide gauges may be difficult to solve.

In principle, however, the combination of inverted echo sounders plus one or more of the four methods for measuring the vertical motions of the tide gauge, plus tide-gauge measurements at the sea surface should allow us to separate the 3 major components of changes in RSL (volume of water, mass of water, and changes in the elevation of the tide gauge).

A further difficulty arises, as Roemmich shows in Chapter 13: steric height variations occur over such a range of space and time scales that possible trends cannot be identified, even in a 30-yr time series at a single station, no matter how dense the sampling. However, the major steric changes seem to be relatively coherent over distances of several thousands of kilometers. Munk *et al.* (Chapter 14) therefore suggest that 10 spatially independent stations, each forming 5 independent samples over a 25-yr period, could be combined to give a ±7-mm standard deviation in a long-term trend of 10 to 25 mm in steric level. It should be possible within the next few years to measure change in sea level with the precision of 1 cm or less from satellite altimeter measurements such as those planned for the TOPEX/POSEIDON experiment (B. Tapley, University of Texas, personal communication, 1989). Born *et al.* (1986) listed the magnitude of different sources of error in such measurements. The accuracy of the measurements will be limited by the calibration of the altimeter and of the estimated heights of fixed points on land. The effects of ionospheric and tropospheric refraction can be eliminated through use of two-frequency altimeters and a water vapor radiometer in the satellite. The drift of the altimeter, and not the absolute calibration, is important for measurement in changes of sea level during the life of the instrument.

RECOMMENDATIONS

1. Long-term sea-level measurements of sufficient accuracy over the world's oceans could provide one of the most significant data sets for understanding global change, particularly climatic change resulting from the greenhouse effect. It is for this reason that the planning committees for the World Climate Research Program and the Intergovernmental Oceanographic Commission of UNESCO have given a very high priority to extending the global sea-level network in the Indian, South Atlantic, and South Pacific oceans. This effort is being supported, insofar as available funds will allow, by the U.S. National Oceanic and Atmospheric Administration (NOAA). **We strongly recommend that the national oceanographic and meteorological communities lend moral and intellectual support to this sea-level program and to develop satellite altimeter methods for changes in sea level.**

2. **A polar orbiting satellite equipped with a radar, preferably a laser, altimeter should be operated on a continuing basis to measure changes in volume of the Antarctic and Greenland ice sheets.** These ice sheets may be the principal sources of variations in sea level during the next century.

As reviewed by the Topographic Science Working Group (1988), detailed and repeated height measurements by near polar-orbiting satellites are required to study the mass balance and dynamics of ice sheets. Repeat surveys of the ice sheets at 1- to 5-yr intervals with a vertical resolution of 10 cm are required for determination of elevation changes indicative of changes in ice volume, thus providing a measurement of net mass balance. Only refined radar altimeters or laser altimeter systems are capable of global coverage with the requisite accuracy.

The output of ice in the mass balance equation occurs through iceberg discharge, surface melting near the margin, melting at the bottom of ice shelves, evaporation, and ablation. A 1-m difference in surface elevation of the ice shelves reflects a nearly 10-m difference in ice shelf thickness (Topographic Science Working Group, 1988). The position of the grounding line can also be observed in elevation data because of a marked change in slope. The output of the ice sheets is not known better than 30 to 100 percent of the total snow accumulation, thus its measurement is critical in an assessment of the mass balance of the ice sheets.

3. A geological record of sea-level change is well preserved in numerous basins for at least the past 200 m.y., a span that includes the nearly ice-free Cretaceous Period and the present ice age. Comparison of the record between glacial and nonglacial times will provide an improved understanding of how depositional systems respond to sea-level change, as well as insights about nonglacial mechanisms of sea-level change.

Support of programs, both national and international, that address the questions of the sea-level record during the past 250 m.y. should be vigorously pursued. One such program, the Global Sedimentary Geology Program (of the International Union of Geological Sciences) on Cretaceous Resources, Events and Rhythms, addresses a variety of questions that we have raised, viz., (a) Is there a global correlation of sequences? (b) Are sequences caused by eustatic fluctuations and/or global tectonic variations, or are sequences developed as a result of regional and local tectonic adjustments? (c) What are the relationships between subsidence, sea level, sediment supply, erosion, and other factors in mid-Cretaceous sedimentary basins?

4. To improve estimates of future steric changes in ocean volume caused by greenhouse warming of the ocean water, **coupled ocean-atmospheric general circulation models should be improved and used to trace probable changes in ocean and atmospheric temperature as the greenhouse gas concentrations in the atmosphere gradually increase.**

5. **To measure absolute elevation of tide gauges, measurements of position using satellite laser ranging, the global positioning system, and very-long-baseline interferometry techniques and absolute gravity should be started.**

REFERENCES

Bach, W. (1988). Modelling the climatic effects of trace gases: Reduction strategy and options for a low risk policy, Paper prepared for the World Congress "Climate and Development," Hamburg, November 7–10, 1988.

Barnett, T. P. (1984). The estimation of "global" sea level change: A problem of uniqueness, *J. Geophys. Res. 89*, 7980–7988.

Baumgartner, A., and E. Reichel (1975). *The World Water Balance*, Elsevier, Amsterdam.

Bentley, C. R. (1985). Glaciological evidence: The Ross Sea sector, in *Glaciers, Ice Sheets, and Sea Level: Effects of a CO_2-Induced Climatic Change*, Committee on Glaciology, National Research Council, National Academy Press, Washington, D.C., pp. 178–196.

Berner, R. A., A. C. Lasaga, and R. M. Garrels (1983). The carbonate-silicate geochemical cycle and its effect on atmospheric carbon dioxide over the past 100 million years, *Am. J. Sci. 283*, 641–683.

Bond, G. C., M. A. Kominz, and J. P. Grotzinger (1988). Cambro-Ordovician eustasy: Evidence from modelling of subsidence in Cordilleran and Appalachian passive margins, *Frontiers in Sedimentary Geology. New Perspectives in Basin Analysis*, K. L. Kleinspehn and C. Paola, eds., Springer-Verlag, New York, pp. 129–160.

Born, C. H., B. D. Tapley, J. C. Ries, and R. H. Stewart (1986). Accurate measurement of mean sea level change by altimeter satellites, *J. Geophys. Res. 91*(C16), 11,778–11,782.

Broecker, W. S. (1987). Unpleasant surprises in the greenhouse?, *Nature 328*, 123–126.

Bryan, K. S., S. Manabe, and M. J. Spelman (1988). Interhemispheric asymmetry in the transient response of a coupled ocean-atmosphere model to a CO_2 forcing, *J. Phys. Oceanogr. 18*(6), 851–867.

Carter, W. E., D. S. Robertson, T. E. Pyle, and J. Diamante (1986). The application of geodetic radio interferometric surveying to the monitoring of sea-level, *Geophys. J. R. Astron. Soc. 87*, 3–13.

Cess, R. D., *et al.* (1989). Interpretation of cloud-climate feedback as produced by 14 atmospheric general circulation models, *Science 245*, 513–516.

Chappell, J., and N. J. Shackleton (1986). Oxygen isotopes and sea level, *Nature 324*, 137–140.

Christie-Blick, N. (1989). Sequence stratigraphy and sea-level changes in Cretaceous time, in *Cretaceous Resources, Events and Rhythms*, R. N. Ginsburg and B. Beaudoin, eds., NATO Advanced Research Workshop Report, in press.

Christie-Blick, N., J. P. Grotzinger, and C. C. von der Borch (1988). Sequence stratigraphy in Proterozoic successions, *Geology 16*, 100–104.

COHMAP Members (1988). Observations and model simulations, *Science 241*, 1043–1052.

Crowell, J. C. (1982). Continental glaciation through geologic time, in *Climate in Earth History*, Geophysics Study Committee, National Academy Press, Washington, D.C., pp. 77–82.

Frakes, L. A., and J. E. Francis (1988). A guide to Phanerozoic cold polar climates from high-latitude ice-rafting in the Cretaceous, *Nature 333*, 547–549.

Frei, A., M. C. MacCracken, and M. I. Hoffert (1988). Eustatic Sea Level and CO_2, *Northeastern J. Environ. Sci. 7*(1), 91–96.

Goldemberg, J., T. Johansson, A. Reddy, and R. Williams (1987). *Energy for Development*, World Resources Institute, Washington, D.C.

Gornitz, V., S. Lebedeff, and J. Hansen (1982). Global sea level trend in the past century, *Science 215*, 1611–1614.

Haq, B. U., J. Hardenbol, and P. R. Vail (1987). Chronology of fluctuating sea levels since the Triassic, *Science 235*, 1156–1166.

Hekstra, G. P. (1988). Prospects of sea level rise and its policy consequences, discussion paper for Symposium on Controlling and Adapting to Greenhouse Warming, June 14–15, 1988, Resources for the Future, Washington, D.C.

Henderson-Sellars, A., and V. Gornitz (1985). Possible climate impacts of land cover transformations with particular emphasis on tropical deforestation, *Climate Change 6*, 231–257.

Hoffman, J. S., D. Keyes, and J. G. Titus (1983). *Projecting Future Sea Level Rise: Methodology, Estimates to the Year 2150, and Research Needs*, U.S. Environmental Protection Agency, Washington, D.C., 121 pp.

Kominz, M. (1984). Oceanic ridge volumes and sea-level change—An error analysis, in *Interregional Unconformities and Hydrocarbon Accumulation*, J. S. Schlee, ed., American Association of Petroleum Geologists Memoir 36, Tulsa, Okla., pp. 109–127.

Labeyrie, L. D., J. J. Pichon, M. Labracherie, P. Ippolito, J. Duprat, and J. C. Duplessy (1986). Melting history of Antarctica during the past 60,000 years, *Nature 322*, 701–706.

Lorius, C., J. Jouzel, C. Ritz, L. Merlivat, N. I. Barkov, Y. S. Korotkevich, and V. M. Kotlyakov (1985). A 150,000-year climate record from Antarctic ice, *Nature 316*, 591–596.

MacCracken, M. C., M. I. Hoffert, and A. Frei (1989). Rising Sea Level and Warming Climate: Their Dependence on Heat Penetration into the Deep Ocean, unpublished manuscript, Lawrence Livermore National Laboratory and New York University.

Machta, L. (1983). Sensitivity studies using carbon cycle models, in *Changing Climate: Report of the Carbon Dioxide Assessment Committee*, National Research Council, National Academy Press, Washington, D.C., pp. 262–265.

Meier, M. F. (1984). Contribution of small glaciers to global sea level, *Science 226*, 1418–1421.

Mercer, J. H. (1978). West Antarctic ice sheet and CO_2 greenhouse effect: A threat of disaster, *Nature 271*, 321–325.

Micklin, P. P. (1971). An enquiry into the Caspian Sea problem and proposals for its alleviation. Thesis, University of Washington, Seattle.

Mintzer, J. (1987). *A Matter of Degrees: The Potential for Controlling the Greenhouse Effect*, World Resources Institute, Washington, D.C.

Mitchell, J. F. B., C. A. Senior, and W. J. Ingram (1989). CO_2 and climate: A missing feedback? *Nature 341*, 132–134.

NRC (1985). *Glaciers, Ice Sheets, and Sea Level: Effects of a CO_2-Induced Climatic Change*, Committee on Glaciology, National Research Council, National Academy Press, Washington, D.C., 330 pp.

NRC (1987). *Responding to Changes in Sea Level: Engineering Implications*, Committee on Engineering Implications of Changes in Relative Sea Level, Marine Board, National Research Council, National Academy Press, Washington, D.C., 148 pp.

Ostlund, H. G. (1983). Tritium and Radiocarbon: TTO Western North Atlantic Section GEOSECS Re-occupation, Tritium Laboratory Data Release 83-07, Rosenstiel School of Marine and Atmospheric Sciences, Miami, Fla., unpublished data.

Ostlund, H. G., H. G. Dorsey, and C. G. Rooth (1974). GEOSECS North Atlantic radiocarbon and tritium results, *Earth Planet. Sci. Lett. 23*, 69–86.

Patullo, Jr., W. Munk, R. Revelle, and E. Strong (1955). The seasonal oscillation of sea level, *J. Marine Res. 14*(1), 88–155.

PCODF (1981). *TTO Preliminary Hydrographic Data Reports*, Vols. I–IV, Scripps Institution of Oceanography Reports, La Jolla, Calif.

Prentice, M. L., and R. K. Matthews (1988). Cenozoic ice-volume history: Development of a composite oxygen isotope record, *Geology 16* (11), 963–966.

Revelle, R. R. (1983a). Methane hydrates in continental slope sediments and increasing atmospheric carbon dioxide, in *Changing Climate: Report of the Carbon Dioxide Assessment Committee*, National Research Council, National Academy Press, Washington, D.C., pp. 202–267.

Revelle, R. R. (1983b). Probable future changes in sea level resulting from increased atmospheric carbon dioxide, in *Changing Climate: Report of the Carbon Dioxide Assessment Committee*, National Research Council, National Academy Press, Washington, D.C., pp. 433–448.

Robin, G. de Q. (1987). Changing the sea level, projecting the rise in sea level caused by warming the atmosphere, in *The Greenhouse Effect, Climate Change and Ecosystems*, B. Bolin, B. R. Doos, J. Jager, and R. A. Warrick, eds., SCOPE, vol. 29, John Wiley & Sons, New York.

Ronov, A. B. (1982). The Earth's sedimentary shell (quantitative patterns of its structure, composition, and evolution, *Int. Geol. Rev. 24*, 1313–1363, 1365–1388.

Schubert, G., and A. P. S. Reymer (1985). Continental ice volume and freeboard through geologic time, *Nature 316*, 316–319.

Shackleton, N. J., J. Imbrie, and M. A. Hall (1983). Oxygen and carbon isotope record of East Pacific core V19-30: Implications for the formation of deep water in the Late Pleistocene North Atlantic, *Earth Planet. Sci. Lett. 65*, 233–244.

Southam, J. R., and W. W. Hay (1981). Global sedimentary mass balance and sea level changes, in *The Ocean Lithosphere, The Sea 7*, C. Emiliani, ed., John Wiley & Sons, New York, pp. 1617–1684.

Stauffer, B., E. Lochbonner, H. Oeschgar, and J. Schwander (1988). Methane concentration in the glacial atmosphere was only half that of the preindustrial Holocene, *Nature 332*(23), 812–814.

Stewart, R. W. (1989). Sea-level rise or coastal subsidence?, *Atmosphere-Oceans*, (in press).

Thomas, R. H., and C. R. Bentley (1978). A model for Holocene retreat of the West Antarctic ice sheet, *Quat. Res. 10*, 150–170.

Thompson, R. E., and S. Tabata (1987). Steric height trends at ocean station PAPA in the northwest Pacific Ocean, *Mar. Geod. 11*, 103–113.

Topographic Science Working Group (1988). *Topographic Science Working Group Report to the Land Processes Branch, Earth Science and Applications Division, NASA Headquarters*, Lunar and Planetary Institute, Houston, Texas, 64 pp.

Whitehead, J. A. (1989). Giant ocean currents, *Sci. Am. 260* (February), 50–57.

Woods, J. D. (1984). The upper ocean and air-sea interaction in global climate, in *The Global Climate*, J. T. Houghton, ed., Cambridge University Press, Cambridge, England.

Wyrtki, K., and S. Nakahoro (1984). Monthly Maps of Sea Level Anomalies in the Pacific 1975–1981, Hawaii Institute of Geophysics Report HIG-84-3.

THE RECORD

Recent Changes in Sea Level: A Summary

1

TIM P. BARNETT
Scripps Institution of Oceanography

INTRODUCTION

The purpose of this chapter is to review the behavior of sea level over the past century. This review is prompted by recent interest in global warming and the CO_2 problem and the possible increase in ocean level that might accompany such a warming (e.g., Hansen *et al.*, 1981). The change in sea level in such a scenario is due to both an increase in ocean temperature and a decrease (melting) of landbound ice. This change in relative sea level (RSL) will, in the absence of compensating factors, be global in nature (Barnett and Baker, 1981), although the variations need not be uniform over the globe (e.g., Farrell and Clark, 1976; Clark and Lingle, 1977).

The review is organized as follows. The next section summarizes research over the past 20 yr on changes in sea level. Next, the data used to study these changes are examined critically. Two different approaches to estimating RSL change are then reviewed using specific case studies. These sections are intended to point out the nonuniqueness in estimates of RSL change, plus the apparent fact that sea level has not been rising at all points in the world's oceans over the past century. The next section summarizes some of the possible causes of RSL change and thus provides an introduction to other chapters of this volume. A concluding section summarizes briefly the status of research on RSL. Given the review nature of this chapter, it has been convenient to rely heavily on two articles (Barnett, 1983a, 1984a) that will serve as basic reference for the discussion of the two different approaches to estimating RSL change and will offer abundant references for readers interested in delving further into the subject.

SOME PRIOR STUDIES OF "GLOBAL" SEA LEVEL

Sea-level stations are located mainly on coastal margins and a few islands. Over the vast interior of the world's oceans there are no sea-level data. Thus any analysis of mean sea-level (MSL) data that claims to represent a "global" condition is already attended by considerable assumption regardless of the results of the analysis. An effort is made throughout the text to remind the reader of this fact.

At least six different estimates of the "eustatic increase" in mean "global" sea level were made between 1940 and 1961. These estimates are summarized by Lisitzin (1974) and are presented in Table 1.1. Perhaps the most notable aspect of these early analyses is that all values agree remarkably well, even though they were obtained by several different methods and data sets (e.g., sea level or cryological information). All of these studies were se-

TABLE 1.1 Estimates of Mean Global Sea-Level Increase

Author	Rate (mm/yr)	Method
Thorarinsson (1940)	0.5	Cryologic aspects
Gutenberg (1941)	1.1 ± 0.8	Sea level (many stations)
Kuenen (1950)	1.2 to 1.4	Different methods combined
Lisitzin (1958)	1.12 ± 0.36	Sea level (six stations)
Wexler (1961)	1.18	Cryologic estimates
Fairbridge and Krebs (1962)	1.2	Selected sea level stations
Emery (1980)	3.0	Sea level (many stations) and selected stations
Gornitz et al. (1982)[a]	1.2	Sea level (many stations)
Barnett (1983a)	1.51 ± 0.15	Selected sea level stations
Barnett (1984a)	1.43 to 2.27	Sea level (many stations)

[a]The authors attempt a correction for crustal motion and find 1 mm/yr. The 1.2 mm/yr is without this correction.

verely limited by geographic extent, time span, and accuracy. A second notable aspect of the early analysis pointed out by Lisitzin was the existence of an apparent "break" in temporal behavior of sea level. Prior to about 1890 to 1900, she found little or no trend in sea level. After that time, however, sea level begins a slow, monotonic rise.

Fairbridge and Krebs (1962) attempted to create an estimate of "global" sea-level change between 1860 and 1960. They paid careful attention to station location and data quality. In addition, they showed the high correlation that exists between nearby sea-level stations, thus allowing one station to represent a large area. Their final estimates show an increase for RSL of 1.2 mm/yr, although they note the estimate is biased by an overabundance of North Atlantic data. Their world sea-level curve shows, as suggested by Lisitzin (1974), a period between 1890 and 1910 of little change. Unfortunately, they never stated exactly how their "global" curve was obtained.

Emery (1980) used data from selected stations in the worldwide tide gauge net to estimate a rise in "global" sea level of 3 mm/yr over the past 40 yr. (Note: this is roughly three times the rate estimated by earlier workers; see Table 1.1.) He pointed out that his study is hampered by lack of data in the Southern Hemisphere. As noted above, this problem attends all studies of "global" sea level. Emery also found that the rate of sea-level rise between 1970 and 1974 is almost five times the longer-term value.

Unfortunately, a number of problems arise in obtaining these estimates: (1) the erratic distribution and strong clustering of stations, plus the analysis method, make the above number strongly biased to several specific regions and individual stations exhibiting strong MSL increases; (2) the recent estimates of RSL rise have not been cor-

rected for atmospheric effects; simple interannual changes in sea-level pressure (SLP) could easily cause part of the observed change; and (3) the 5-yr averaging time (1970 to 1974) is a time scale typical of short-term climate change and regimes. Changes in ocean temperatures and circulation, for example, associated with these epochs, could also account for much of the observed RSL trend (cf. Fairbridge and Krebs, 1962).

Etkins and Epstein (1982) accepted the 3-mm/yr RSL increase and assumed it applies globally. They then tried to explain the increase in terms of ocean temperature change and discharge from the polar ice sheets. Perhaps their most interesting result relates the observed change in rotation rate of the Earth to the estimated change in the polar ice caps. The apparent agreement so obtained in this comparison would be much less impressive, however, if they had not used the 3-mm/yr RSL rise figure of Emery. They also did not discuss the fact that the rotation rate of the Earth increased between 1900 and 1930, and so the length of day decreased. Sea-level arguments, such as they presented, would dictate an increase in length of day (see Etkins and Epstein, 1982, reference 8). The entire problem of the effect of sea level and ice-cap melting on Earth's rotation characteristics has been described well in the 1950s. Further discussion of this subject is deferred to a later section in this chapter. Additional discussion of the Etkins and Epstein paper is given by Robuck (1983).

Gornitz et al. (1982) also attempted to develop an estimate of "global" MSL change. They developed large area averages of sea level, thus avoiding the clustering problem that plagued Emery. They generally ignored regions that do not show an increasing trend (e.g., Alaska) and eventually ended up with a 1.2-mm/yr increase as a "global" number. They also attempted to remove vertical crustal motion based on geologic data and to find a RSL trend of 1 mm/yr on a "global" basis.

Review of the references supporting the development of their land motion estimate suggests that their corrections may be highly unstable. Indeed, the corrections are generally major, relative to the uncorrected trend, and unstable from millennium to millennium (Newman et al., 1980). It may thus be fortuitous that the two global-trend estimates come out as close as they do. There are other problems in their analysis. (1) Eight of the 13 regions used in the global average border the Atlantic Ocean, a fact that biases their estimates of RSL changes to past variability in that ocean. (2) The regional time series are not temporally homogeneous; thus, for instance, the "global" average between 1880 and 1900 is dictated almost entirely by European data. (3) The use of raw average trends for each region to develop a "global" trend allows a few active regions to dominate the global number (cf. Table 1 of Gornitz et al., 1982).

THE DATA

Sources and Distribution

Sea-level data have been taken routinely since at least 1808 (Brest, France) and even before, e.g., Amsterdam, 1682 (R. Fairbridge, Columbia University, personal communication). These data have been organized by The Permanent Service for Mean Sea Level, Institute of Oceanographic Science, England (Lennon, 1975-1978), and are available in printed volumes and on magnetic tape. The distribution of stations is shown in Figure 1 of Emery (1980). That illustration shows the highest station densities in the United States and western Europe followed by Japan. Long-term records from the Southern Hemisphere, e.g., South America and Africa, are few in number. Long-term records from islands in the open ocean far from land are almost nonexistent. Hence, any attempts to infer "global" changes in sea level from this data set are attended by substantial assumption—the distribution of data is simply not adequate to represent global coverage.

Instruments

Interestingly, little was found in the literature that described the instrumental methods. However, Lennon (1970) provided a good overview of current instruments and their possible problems—and there were many.

Most modern tide data come from instrumental measurement (e.g., a float in a "stilling well") with the measurements continually recorded by a pen and ink device. In recent years the data recording mode has become digital. Before 1900, however, the measurement methods varied considerably. The long-term San Francisco record was obtained in the above manner, even though the actual measurement site was moved four times (S. Hicks, NOAA, personal communication). The old MSL records from India (e.g., Bombay) were obtained by human "tide parties" whose members made frequent visual sightings on a graduated staff inserted in the water (B. Zetler, Scripps Institution of Oceanography, personal communication). Needless to say, the risk associated with the data increases considerably as one reaches back beyond 1900.

Potential Problems

The sea-level data have numerous potential difficulties. Some of these are listed below. [See also Fairbridge and Krebs (1962) for additional problems.]

1. Station locations were subject to position change, for example, as harbors developed through the century. These changes can add discontinuity to the time series and, if undetected, can induce apparent long-term trends.

2. Instrument position changes can be accommodated if the old and new locations are tied to a common geodetic reference point. It is often difficult to discover if this was done. But even the leveling (reference) operation is not without difficulty (see below).

3. Tectonic uplift/subsidence of the land mass on which the gauge sits will induce an apparent sea-level change. In some parts of the world (e.g., Alaska and Scandinavia), this signal is as large or larger than any possible oceanographic change (e.g., Hicks and Shofnos, 1965). [Note: this effect is most pronounced in regions of glacial rebound (high latitudes) and near the mouths of large rivers (e.g., in the Gulf of Mexico) where heavy sedimentation causes subsidence.]

4. Mean sea level is affected by a variety of physical processes (e.g., Lisitzin, 1974; Namias and Huang, 1972), including changes in water temperature and salinity, river runoff, local and nonlocal currents, wind, waves, and atmospheric pressure. All of these can change sea level in conflicting ways, and these changes can be as large as the signal under study. Since all of the variables mentioned above are subject to climatological variations, they characteristically have "red spectra," which means they can contribute to estimates of trends in sea level.

5. Geodetic leveling between sea-level stations apparently cannot be used to remove relative vertical motion of the land, if the stations are more than a few hundred kilometers apart. Sturges (1966) showed that the 500-mm north-south sea-level difference inferred from leveling data to exist along the Atlantic seaboard was inconsistent with oceanographic data. The same situation was found along the west coast of the United States. More shockingly, Balazs and Douglas (1979) reviewed five different attempts between 1969 and 1978 to obtain the relative elevation difference between San Francisco and San Pedro, California (about 700 km apart). According to those surveys, the two cities had a relative vertical displacement rate of 70 mm/yr. The RSL trend between the two was 2.1 mm/yr—the difference is a factor of nearly 35. Thus, the leveling data for both coasts of the United States may well contain serious errors of unknown origin when the separation distance exceeds a few hundred kilometers.

6. The possible instrumental problems are legion. The ability to speculate in this area far exceeds the available information. Lennon (1970) gave a good overview of possible problems with "modern" equipment. Errors of 180 mm or more are possible in sea-level measurement, and they may be virtually impossible to detect in the data. Contrasting this error with an often stated RSL rise of 1 mm/yr gives one an appreciation for some of the potential uncertainties in sea-level studies.

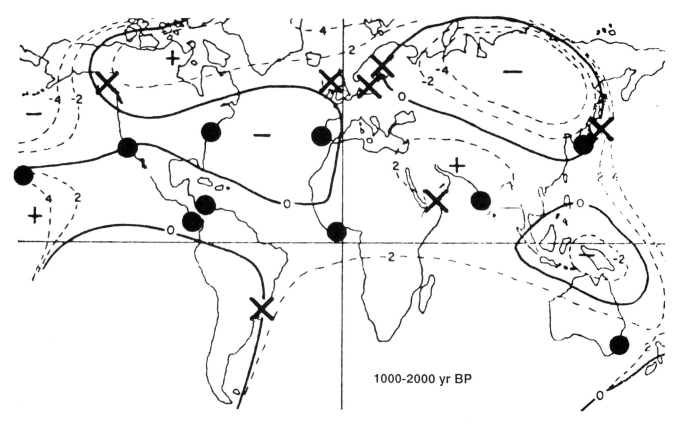

1000-2000 yr BP

FIGURE 1.1 Location of primary (•) and secondary (x) stations versus estimates of vertical crustal motion for the period 1000 to 2000 yrBP. Contours are in millimeters per year. After Newman *et al.* (1980).

ESTIMATES OF RELATIVE SEA-LEVEL CHANGE: KEY STATION APPROACH

Methods

One approach to estimating change in global RSL is to analyze data from *key* stations, arguing that changes at these stations are representative of changes in RSL in large (1000+ km) regions surrounding the stations. Examples of such analyses are found in Table 1.1 (e.g., Fairbridge and Krebs, 1962; Barnett, 1983a). The latter work is summarized here.

The use of the key station approach imposes the following series of constraints on the data to be analyzed.

1. High-quality, continuous measurement that reveals no sudden shifts suggestive of station movements should be used. The records should be as long as possible.

2. Station locations should be away from areas of strong tectonic movement (e.g., separated from areas of major deposition/uplift).

3. Stations should be unaffected by spurious physical processes (e.g., gauge exposed to strong freshwater invasions).

4. The spatial density of stations by oceans should be proportional to the relative areas of the respective oceans, e.g., 2.4/1.5/1.0, for the Pacific/Atlantic/Indian (cf. Sverdrup *et al.*, 1942). The stations should be synchronous in time. Such a distribution will allow equal weight to be given to all oceans and years in the subsequent analysis, thereby avoiding biasing problems.

5. Finally, the selected stations must represent large geographic regions.

The locations of a set of stations that generally satisfy these criteria are shown in Figure 1.1 and are listed in Table 1.2.

A method of inferring, quantitatively, the existence of a signal that is coherent over all portions of a data field using empirical orthogonal function (EOF) analysis (cf. Backus and Preisendorfer, 1978; Barnett, 1978) was applied to the primary key station set and to the secondary station set to check sensitivity of the results to data perturbations. The sea-level data at position i are represented by

$h_i(t)$, $i = 1, 2, \ldots$ NP, where NP is the number of stations in the study. Define

$$\bar{h}_i = \langle h_i(t) \rangle_t,$$

and

$$\sigma^2 = \langle [h_i(t) - \bar{h}_i] \rangle_t, \qquad (1.1)$$

where $\langle \rangle_t$ denotes a time averaging operation. The data were normalized to have unit variance and zero mean, i.e.,

$$h'(t) = \frac{h_i(t) - \bar{h}_i}{\sigma_i}. \qquad (1.2)$$

This is a crucial step for later interpretation of the results, for now each station will have equal weight in the subsequent analysis. This, coupled with equal time and area weighting in the key station selection, ensures a more or less equal geographic weighting for the analysis.

Next form the correlation matrix:

$$C_{ij} = \langle h'_i(t) h'_j(t) \rangle_t. \qquad (1.3)$$

Denote the eigenvectors and eigenvalues of Eq. (1.3) by B_{ni} and A_n, respectively, and the associated principal components:

$$A_n(t) = \sum_i B_{ni} h'_i(t). \qquad (1.4)$$

Finally, it follows that

$$h'(t) = \sum_n A_n(t) B_{ni}. \qquad (1.5)$$

Special properties of the (B_n, A_n) are given in the above references.

The EOF representation of the RSL field is interpreted as follows:

1. The B_{ni} represent average patterns of spatial covariability between the NP-members of the sea-level field. If all members are fluctuating in unison, as would be the case for a global change in RSL, then all components of B_{li}, the first, most energetic eigenvector, will have the same sign.

2. The $A_n(t)$ modulate, in time, the intensity of the spatial patterns of B_{ni}. For instance, a secular trend in the h_i coherent at all stations will be manifested by a secular trend in $A_l(t)$.

3. Using Eq. (1.5) and an average value of the σ_i from Eq. (1.1), it is possible to reconstruct the behavior of global sea-level change—if such a creature exists.

Results

The results of the above analysis for two different time periods, 1903 to 1969 and 1930 to 1975 are shown in

Figures 1.2 through 1.5. For the longer time period, the B_{li} all have the same sign (Figure 1.2), indicating the existence of a coherent pattern of variation in the data set. Note all stations analyzed are not contributing equally to this pattern since all B_{li} are not the same size. Figure 1.3 shows that the coherent change is associated with increasing RSL. Further, the time variation appears well fit by a simple linear trend (1.51 ± 0.15 mm/yr).

Analysis of the more recent period suggests RSL is not rising uniformly within the data set analyzed (Figure 1.4). Indeed the implied decrease in RSL in the eastern Asian region TON is supported by independent hydrographic observations (e.g., White *et al.*, 1979). The implied RSL trend in Figure 1.5 is 1.79 ± 0.22 mm/yr. (Note that neither of the temporal reconstructions suggests an accelerating rate of RSL in recent years.)

Consideration of the results and the data shows that the EOF analysis is being dominated by temporal trends. Stations with positive trends have B_l components of one sign; those with negative trend have B_l components of the opposite sign. The analysis method is extracting a single trend common to all stations that accounts for as much of the total field variance as possible. Extraneous noise, which would normally contaminate study of data from individual stations, is therefore effectively filtered out of data prior to trend estimation.

TABLE 1.2 Sea-Level Stations

Station	Abbreviation	Start	End	Missing Years
PRIMARY				
San Francisco	SFO	1862	1874	—
Honolulu	HNL	1905	1975	—
Tonoura, Japan	TON	1894	1975	1
Balboa, C.Z.	BCZ	1908	1969	—
Sydney	SYD	1897	1977	—
Bombay	BOM	1878	1964	1
Baltimore	BAL	1903	1975	—
Cristobal, C.Z.	CRS	1909	1969	—
Cascais, Port.	CAS	1882	1975	4
Takoradi	TAK	1930	1975	1
SECONDARY				
Ketchikan	—	1919	1975	—
Hosojima	—	1894	1977	—
Aden	—	1879	1960	45
Aberdeen	—	1862	1965	1
Helsinki	—	1879	1970	—
Montevideo	MON	1938	1970	7
Ystad	—	1887	1974	—

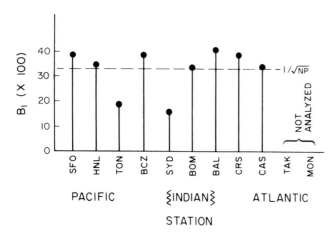

FIGURE 1.2 Mode 1 eigenvector components by station for the analysis period 1903 to 1969. If all stations contribute equally to the pattern, then all eigenvector components would have value $(NP)^{-0.5}$. From Barnett (1983a), *Climate Change* (American Meteorological Society).

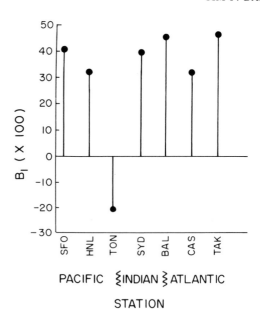

FIGURE 1.4 As in Figure 1.2 but for analysis period 1930 to 1975. From Barnett (1983a), *Climate Change* (American Meteorological Society).

In conclusion, it appears that a large coherent pattern of increasing RSL exists over much of the globe where data are available. The rate of increase is rather uniform spatially ($B_{1i} \cong (NP)^{-0.5}$) except in the western and Indo-Pacific regions in the past 3 to 4 decades. These results are qualitatively unaltered by substantial perturbation of the original set or method of normalization.

ESTIMATES OF RSL CHANGE: AREA AVERAGE APPROACH

Methods

Selection of one or more unrepresentative stations in the key station approach can introduce spurious information into the resulting analysis. An obvious way around this potential problem is to average a large number of

stations within a region to get a representative estimate of RSL in that area. Examples of this approach may be found in Hicks (1978), Emery (1980), Gornitz *et al.* (1982), Barnett (1984a), and Aubrey (1985). Unfortunately, this apparently simple procedure is not without its problems as we shall see below.

One way of objectively obtaining a quantitative area average was proposed by Barnett (1984a). Denote by $h_i(t)$ the time history of RSL at station i in a predetermined region and form C_{ij} as in Eq. (1.3). Again denote the first eigenvector of C_{ij} by B_{1i}. It turns out that for most of the sea-level data studied, this first mode accounts for the

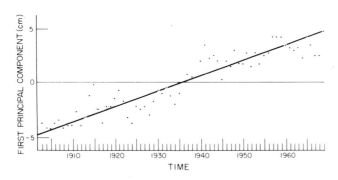

FIGURE 1.3 The principal component for the first eigenvector of the analysis period 1903 to 1969 scaled to give estimated RSL trend in real units (centimeters). The solid line represents a least squares fit to the data. From Barnett (1983a), *Climate Change* (American Meteorological Society).

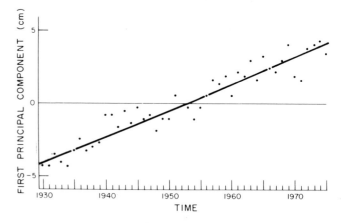

FIGURE 1.5 As in Figure 1.3 but for analysis period 1930 to 1975. From Barnett (1983a), *Climate Change* (American Meteorological Society).

majority of the non-noise variability in the $h_i(t)$. Under these conditions, the B_{li} provide a set of weights that can be used to average the h_i over the domain of analysis. Thus one measure of a regional average would be

$$R_k(t) = \frac{1}{A} \sum_i B_{li} h_i(t),\qquad(1.6)$$

where $A = \sum_i B_{li}$ and R_k is average sea level for the kth region and $i = 1, 2, \ldots$ NP.

Note R_i is comparable to a principal component except for the normalizing factor A. The latter is necessary if the $h_i(t)$ do not all begin and end at the same time. One may use only values of B_{li} with the same sign in Eq. (1.6). This ensures that the signal coherent and in like phase at the NP stations is included in R_k. Alternatively, all elements of B_{li} could be used in Eq. (1.6) to obtain a different version of the area average.

Results

The results of applying the above methods to the six regions shown in Figure 1.6 are given in Figure 1.7. The regional results are much the same as those suggested in Figures 1.2 and 1.4. RSL seems to be increasing generally in most of the study areas. The exception (also seen in Figure 1.4) occurs along the east coast of Asia (Region 4), where RSL is seen to have been decreasing since the mid-1950s. Note also the highly differential behavior of the regional trends. For instance, in Region 1, RSL seems essentially static from 1880 to 1920, after which time it begins to increase. By contrast the east coast of North America (Region 3) shows a rather steady increase since 1900 (the beginning of the data).

For the reasons noted previously, the estimation of "global" sea-level change does not appear possible with present data. However, it is of interest to estimate the average of the data we do have available. With this in mind, the analysis using the area average method was performed on the objectively defined regional averages shown in Figure 1.7. Use of Eq. (1.6), then, gives an "overall average" (R_0), provided the h_i represent the regional averages defined earlier and the B_{li} are the mode-1 eigenvector components of the regional analysis. Note that the distribution of regional averages is roughly area representative of the respective oceans. Prior to the early 1900s, however, the temporal representativeness is biased to the Atlantic and Indian oceans.

The EOF analysis of the regional data (not illustrated)

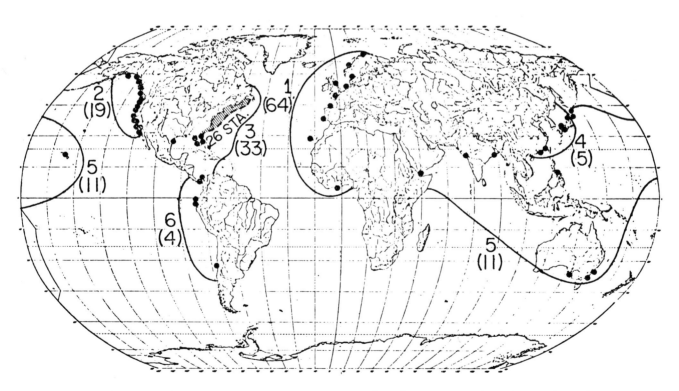

FIGURE 1.6 Station and region location chart. The upper number of each pair refers to region numbers. The numbers in parentheses indicate the number of individual stations in the region. From Barnett (1984a) with permission from the American Geophysical Union.

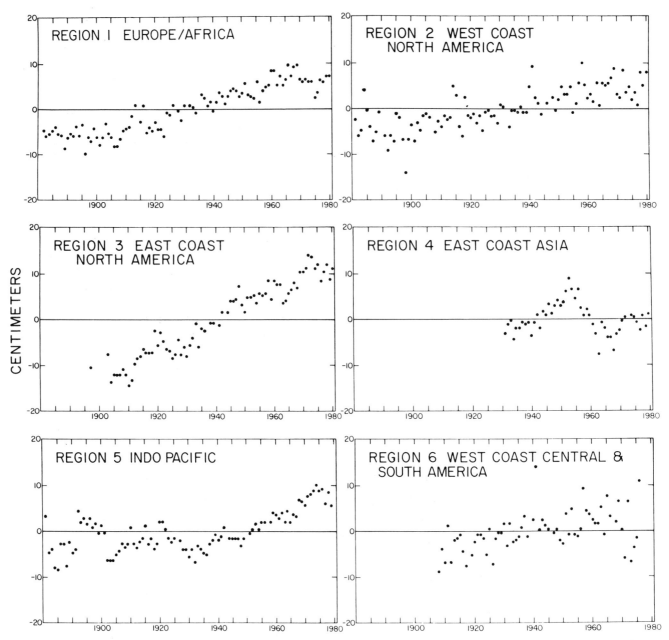

FIGURE 1.7 Regional averages of annual sea-level anomaly. From Barnett (1984a) with permission from the American Geophysical Union.

showed that the first mode components all have the same sign, except for the Southeast Asian region. Thus, to first order, all the regions vary in unison, except the Southeast Asian region. Since the first mode captured approximately 50 percent of the total variance, it contains the dominant signal in the sea-level set. Further, the in-phase eigencomponents were all of approximately equal magnitude, suggesting the uniformity of the relative sea-level variation associated with this mode. However, since there is a well-

defined area that is not in phase with the other regions, one cannot claim the existence of a uniform pattern fluctuation common to all available sea-level data.

The time domain result of the above operations is shown in Figure 1.8 for the periods 1881 to 1980 and 1930 to 1980. The R_0 curve of the 100-yr data set has a slope of the trend that is 1.43 ± 0.14 mm/yr, a value close to that obtained by other workers. This trend accounts for 81 percent of the total variance in R_0. However, it is clear that a simple trend is not the best representation for the entire 100-yr set, as it represents a compromise between times of little change (1881 to 1920) and steady increase (1920 to 1980).

The analysis for the period 1930 to 1980 is well represented by a trend of magnitude 2.27 ± 0.23 mm/yr that accounts for 87 percent of the variance in the R_0 series. The larger slope value for the 1930 to 1980 sea-level set is due to omission of the first 50 years of data (1881 to 1930) when little or no trend was apparent. This result emphasizes the sensitivity of the results to the length of record considered.

The past 50 years of data were exceedingly well fit by a simple linear trend. Would a more exotic function give a better fit to the data? The realistic answer is "No." The residual time series for 1930 to 1980, after removal of the trend, was indistinguishable from white noise (zero mean and a standard deviation of 12 mm), based on analysis of its correlation function. Thus higher-order fits, which appear to capture more variance, are likely fictitious and could lead one to an erroneous view of the data.

Cautions: Nonuniqueness

The selection of the gross regions for EOF analysis was based on geographic considerations. The selection could have been made on the basis of water mass types or relation to specific ocean gyres, for example. Any of these criteria seem reasonable and could be easily rationalized. Using different selection procedures might have changed somewhat the resulting regional definitions. The magnitude of the R_k certainly would have been affected. A final estimate of global RSL obtained by averaging the R_k will also clearly depend on the original selection of the gross regions.

The EOF analysis was performed on the correlation matrix of the $h_i(t)$. Use of the covariance matrix would have given a different set of weight to use in Eq. (1.6). It might also be argued that the weighting in Eq. (1.6) should be proportional to B_{li}^2 since the latter represent pattern variance. Both alternatives would concentrate attention on the sea-level data with the highest variance—in this case, greatest trend—thus guaranteeing a higher value of global RSL change than estimated here.

Finally, one might ask, "Why not compute the EOFs of all the stations together?" This will lead to a result biased by regions of highest station density. The resulting global average would thus be, in reality, no more than the RSL changes associated mainly with one ocean, the Atlantic (50 percent of the stations).

The point of the above discussion is that there are many valid but rather different ways to estimate "global," even regional, RSL changes. Without an adequate global sea-level network, which is unlikely to exist in the near future, the estimate of "global" RSL change will remain an uncertain business. Tests of the different approaches mentioned above to obtain this average suggest that estimates of global change with current data can vary by a factor of 50 percent—a scatter induced solely by the analysis method. One should not become too enamored with a single estimate of "global" RSL change or depend on it for crucial policy decisions.

DISCUSSION: POSSIBLE CAUSES

The results of the previous section indicate that on average there has been a rise in RSL along most of the

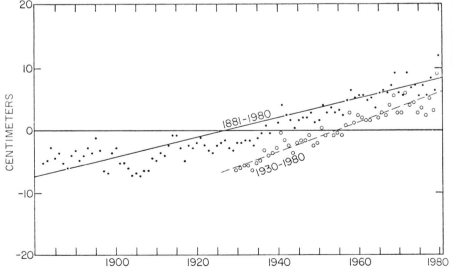

FIGURE 1.8 Estimates of overall average (R_0) of sea-level data for two different time periods. The solid/dashed lines are least squares fits of a linear trend. From Barnett (1984a) with permission from the American Geophysical Union.

Earth's coastal margins and in central ocean regions where data have been gathered. Exceptions to this statement occur in areas of known upward crustal motion and in areas where the sense of the vertical motion is not so well known, e.g., east Asia. It is unknown whether similar changes are occurring over the interior of the southern oceans. At any rate, it is tempting to state the results as follows: "Relative sea level is rising 'globally' at a rate of about 1.5 to 2.0 mm/yr." Subject to the numerous caveats and assumptions stated earlier, let us now take this rate as a working hypothesis. How might this hypothesis be rationalized? This section serves to introduce some of the possible answers to this question and comes largely from Barnett (1983a). Other chapters of this volume will deal with these individual possibilities in more depth.

Subsidence Argument

The average result might simply be due to the fact that all the stations analyzed, plus all the regions they represent, are slowly subsiding because of crustal downthrust and/or sediment compaction. If the Earth is not contracting, then the subsequent compensation could occur at high latitudes (glacial rebound?) or in the ocean basins. The work of Hicks (1972) and Newman et al. (1980) suggests that the value for the RSL increase is not inconsistent with the uplift/subsidence rate alone.

The above argument is not particularly convincing for the following reason. The spatial scale of observed coherent RSL increase, when projected to the regions that the individual stations represent, is substantially larger than the uplift/subsidence spatial scale shown in Figure 1.1. Furthermore, the current stations are rather uniformly distributed among up/down motion areas, at least as they were about 1500 yrBP. In fact, 7 of 11 primary stations are in what were inactive regions. It would be a remarkable coincidence if the current station pattern put all locations in areas of subsidence. Indeed the detailed studies of Clark et al. (1978) suggest this did not happen. However, in this volume, Peltier (Chapter 4) and Bloom and Yonekura (Chapter 6) clearly show the potentially strong contaminating effect that crustal motion has on the RSL data.

Global Warming

Melting the ice caps and warming the oceans due to a global temperature increase would give rise to an increase in global MSL as noted earlier. Assume for the moment that the RSL change is due to this effect. What other evidence might be used to test this hypothesis?

Ocean Warming   One might simply look at the shape of a global or hemispheric temperature curve since 1900 and compare it with the time series of coherent RSL and ocean temperature (Figure 1.9). The air temperature curve rises slowly until around 1940 after which time it drops. The RSL curve shows no such behavior. However, time histories of sea-surface temperature (SST) from the world's oceans show a linear trend (increasing) identical to the RSL curve, although they too show an apparent decrease beginning around 1960 (see also Kukla et al., 1977). The 1°C change in SST since 1900, if distributed in the vertical by some type of mixing process as suggested by Cess and Goldenberg (1981), would raise RSL at the rate of about 1.1 mm/yr—remarkably close to previous and current estimates. Indeed, Gornitz et al. (1982) concluded that much of the observed RSL can be attributed to this thermal expansion.

Roughly one-half of the rise of 1°C since 1900 through 1970 can be ascribed, if so desired, to differing instrumental methods, e.g., bucket versus injection temperature (Barnett, 1984b). Also, it is difficult to conceive of the world's oceans warming by 1°C and the atmosphere not following suit by a like amount. Part of this inconsistency may be due to the general practice of estimating hemispheric temperature from land stations only (Mitchell, NOAA, personal communication). The work by Paltridge and Woodruff (1981) partially overcomes this problem but, in turn, has severe difficulties of its own. At any rate, this "global" temperature curve does not agree with the RSL with regard to phasing, and this is a crucial disparity.

Changes in the vertical density structure of the ocean, such as might be associated with warming, have been investigated over time scales of 40 to 60 yr by Robinson (1960) for the Pacific and Barnett (1983b) for the Northern Hemisphere oceans. Both find no significant change in the density field of the upper 1000 m of the ocean. Both point out the very noisy nature of the hydrographic data they used. Studies of upper ocean variability at time scales of 20 to 30 yr are varied in their conclusion (e.g., Gammelsrod and Holm, 1981; Tabata, 1981). Work by Wunsch (1972), Pocklington (1972, 1978), and Roemmich (Chapter 13, this volume) suggest that, at least near Bermuda, warming of the upper ocean can or cannot explain the local RSL change, the result depending on the length of record analyzed. This is clearly an unsatisfactory situation.

In summary, the connection between global atmospheric warming and the behavior of the ocean over the past 70 to 80 yr, while suggestive, is hardly convincing.

Astronomical Observations   Another possible means of checking indirectly the warming assumption lies in a rich set of scientific work on the theory of the Earth's rotation done in the 1950s. Lambeck (1980) gives an excellent summary to back the discussion, as do Munk and McDonald (1960).

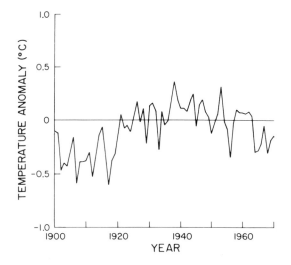

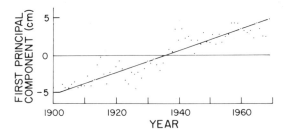

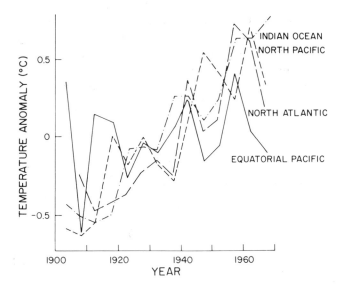

FIGURE 1.9 Time series of key climatic variables for the period 1900 to 1970. Upper panel: Northern Hemisphere temperature anomalies (after Jones *et al.*, 1982); middle panel: "global" sea-level trend reconstruction (Barnett, 1983a); lower panel: sea surface temperature anomalies averaged over large ocean areas by pentad. From Barnett (1983a), *Climate Change* (American Meteorological Society).

Change in the length of day (l.o.d.)—The change in l.o.d. from 1860 to 1970 is shown in Figure 1.10 (courtesy, T. Van Flandern, U.S. Naval Observatory; also see Brouwer, 1952). [The l.o.d. data until 1956 were derived from lunar occultation observations of Ephemeris Time minus Universal Time. After 1956 the data came from U.S. Naval Observatory observations (McCarthy, 1976).] Only after the early 1900s do the data obtain sufficient accuracy to become highly useful (see Lambeck, 1980). Now a rise in sea level of 1 cm, if due to melting on Greenland and/or the Antarctic continent, will increase the length of day by 0.06 ms (Munk and Revelle, 1952a) as the Earth's rotation rate slows to conserve the change in momentum induced by the net mass redistribution. A 1.5-mm/yr change in RSL is equivalent to a 1 ms/century change in l.o.d. On the other hand, if the increase in sea level is due only to a uniform warming of the oceans (thermal expansion), Munk and Revelle (1952b) showed that there will be negligible impact on the l.o.d.

Figure 1.10 shows that the majority of the variance in the l.o.d. must be explained by other processes. Indeed, the rapid decrease in l.o.d. in the early 1900s is exactly opposite to any sea-level induced effect. However, if one considers the secular trend in l.o.d., then the effect of RSL rise due to polar melting cannot be ruled out, but the l.o.d. will also increase as tidal friction slows the Earth's rotation rate. The effect is shown on Figure 1.10 and seems to offer an adequate explanation for the scalar trend in l.o.d. (see Lambeck, 1980 for more in this area). Other reasons for increase in l.o.d. are given by Peltier (Chapter 4, this volume).

Changes in pole position—In a series of papers, Munk and Revelle (1952a,b, 1955) showed that the position of the Earth's instantaneous pole of rotation is highly sensitive to small, differential changes in amount of polar ice and its position. They compute the amount the (north) pole will be displaced toward Greenwich (λ) and toward 90°E of Greenwich (μ) depending on relative changes in ice concentration as reflected in RSL changes.

A crude history of the long-term migration of the rotation pole is shown in Figure 1.11 (Yumi and Yokoyama, 1980; actual data courtesy of D. McCarthy, U.S. Naval Observatory). The quality of data such as these deserves special attention, for they can be attended by numerous sources of error. However, it appears that the general secular trend in the illustration is reasonable, at least with respect to direction and order of magnitude in displacement. For further discussion in this area, see Melchior and Yumi (1972).

Using the theoretical results from Munk and Revelle (1952b, Table I), one finds a 1.5-mm/yr linear increase in RSL, if due to approximate equal melting on both Greenland and Antarctica, will cause the pole to move (linearly) toward 60°W. [Lambeck's (1980) version of these results

FIGURE 1.10 Secular variations in length of day (l.o.d.) in milliseconds. The line denoted "sea-level effect" is the change in l.o.d. expected from the RSL change, if that change was induced by polar melting. The line denoted "TF" is the change in l.o.d. caused by tidal friction. From Barnett (1983a), *Climate Change* (American Meteorological Society).

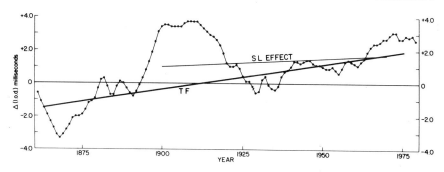

(his Table 9.2) is qualitatively the same but has some quantitative differences.] The heavy line bisecting the data is directed towards 70°W. The observed (linear) travel distance is approximately 730 cm. This contrasts with 180 cm expected on theoretical grounds. Assuming most of the melting occurred on Greenland would change the theoretically expected direction of motion to approximately 37°W and increase displacement to 826 cm. Both possibilities appear clearly within the range and uncertainty of the combined MSL and astronomical data sets.

If melting of the landlocked ice sheets is completely responsible for all the RSL change, then this implies that roughly 45,000 km³ of ice (roughly 0.2 percent of the land-ice volume) has melted since about 1900, i.e., the change is *glacioeustatic*. [Note: this value is about the magnitude that Etkins and Epstein (1982) estimated for the period since 1940.] The astronomical results suggest that this volume would have to come about equally from Greenland and Antarctica or mainly from Greenland alone because the Antarctic ice sheet is not so asymmetric with respect to the mean axis of rotation as Greenland, and hence changes there do not affect the position of the axis as much.

In any event, current knowledge suggests that the ice volume of both Greenland and Antarctica is relatively stable (see Chapter 10, this volume). Unfortunately, the residual in such calculations is enough to explain the increase in RSL (cf. Lambeck, 1980). Lambeck states, however, that estimates of this residual are both positive and negative. Accretion of the polar ice would lower RSL—but then the rotation axis would presumably have a secular motion just opposite to that currently observed.

The assumption of approximately equal melting at both poles seems in reasonable agreement with the astronomical observations; however, the polar wandering can be explained in terms of other processes that need not affect RSL. Continued rebound effects from the Pleistocene deglaciation could explain the polar wandering (Dickman, 1979; Nakiboglu and Lambeck, 1980, 1981), although there is ample room for argument on the details of these theoretical calculations and on the appropriate time scale of the effects (Dickman, 1981; Sabadini *et al.*, 1982). The posi-

tion of the pole also will be affected by continental drift. Given typical drift values (e.g., Minster and Jordan, 1978), the resulting pole displacement will be but 10 to 20 percent of that observed (McCarthy, 1972; Mueller and Schwartz, 1972; Dickman, 1977). These points are considered in depth in Chapter 4, this volume, by Peltier. It is sufficient to say that the astronomical data cannot be used exclusively to explain RSL change.

Glacial Retreat

Mountain glaciers hold roughly 1 percent of the total land ice (cf. Flint, 1971) and at first thought would seem an unimportant source of RSL change. However, mountain glaciers show large variability in time, and one needs only a 20 percent reduction in their volume to obtain the observed RSL increase. Consideration of their retreat rates since 1900 suggests that mountain glaciers could account for a significant fraction of the observed RSL. Because they occur asymmetrically around the Earth at high latitude, their melting will also increase l.o.d. (as

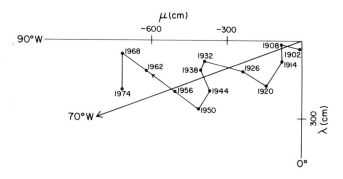

FIGURE 1.11 Observed motion of the mean pole of rotation. Displacements in centimeters toward Greenwich (λ) and 90°E of Greenwich (μ). The solid arrow directed toward 70°W indicates the average direction of pole motion between 1900 and 1975. Each data point represents the average of six annual mean positions thereby minimizing biasing effect of the Chandler wobble. From Barnett (1983a), *Climate Change* (American Meteorological Society).

observed). Their melting should also cause the axis of rotation to wander. Preliminary calculations by Lambeck (1980) suggest that this deglaciation plays a role in the variations of λ and μ, although it does not dominate the observations. Mountain glacial retreat has been studied in more depth by Meier (Chapter 10, this volume), who concludes that a significant fraction of the observed RSL change could be accounted for by this mechanism.

Ocean Circulation Changes

It is well known that RSL is intimately involved with the oceanic circulation systems (e.g., Reid and Mantyla, 1976). Sea level roughly tends to be higher in the central parts of the ocean's gyres because of the general geostrophic balance of the clockwise circulation of the current systems (counterclockwise in the Southern Hemisphere). The reverse situation applies for counterclockwise gyre systems. A unique pattern of reduction/increase in the general circulation of the oceans could give a RSL change at the coastal boundaries as observed and a subsequent change in the central gyres where, generally, we have no information.

The above ideas could be checked through an exhaustive analysis of the historical hydrographic data. The available superficial studies of these data, however, suggest that the effect, if real, may not be important. For instance, the resulting magnitude change in the ocean circulation systems would have been noticed, at least qualitatively. Further, White and Hasanuma (1980) and White *et al.* (1979) showed that coherent dynamic height variations occur throughout the edges and center of the Pacific gyre. If the circulation change idea was correct, the fluctuations in the western edge should have been out of phase with those of the central ocean. Also, Robinson (1960) and Barnett (1983b) showed no statistically significant differences in limited hydrographic data below 100 m depth obtained from nearly identical sites between 1824 and 1958 and between the 1920s and the 1970s, respectively. Finally, it may be noted that circulation changes would have no appreciable effect on the Earth's rotation (Munk and Groves, 1952). Chapters 2 and 3 in this volume by Thompson and Sturges, respectively, deal with the circulation mechanism in more depth.

CONCLUSIONS

Numerous diverse studies over the past four decades have been directed at the problem of sea-level change. These studies have suggested that, on average (over the available data set), RSL is increasing. Notable exceptions to this statement occur around Scandinavia, Alaska, and

parts of eastern Asia. All studies have been attended by serious limitations, which include the following:

1. The lack of data in the Southern Hemisphere and mid-ocean regions means one must accept a substantial risk in interpreting the results in terms of global changes.

2. The effect of vertical crustal motion induces a "signal" in each station's data that cannot reliably be removed. Hence, it is possible, but highly unlikely, that vertical motions of the crust are responsible for almost all of the RSL change. New geophysical measurement systems, e.g., very-long-baseline interferometers, may remove this problem in the future.

3. A coherent global signal can be detected in a relative sense. Attributing a magnitude to it, however, is subject to the problems included in 2 above plus many others.

Given the limitations noted above, the following statements characterize our current knowledge regarding RSL change:

1. The average rate of increase, according to the existing sea-level data set, is roughly 1 to 2 mm/yr. Different methods of analysis alone can cause variations on the order of 50 percent in these estimates.

2. The existence of a "globally" coherent pattern of RSL increase has not been confirmed with existing data and analysis techniques. Neither is the existence of an accelerating rate of RSL increase in recent years.

3. It is not possible, at this time, to explain the reasons why RSL seems to be increasing at many stations. It is possible, but unlikely, that a large part of the RSL change observed is due to vertical crustal motion alone. Fortunately, there are a number of other interesting hypotheses that could explain the RSL change. The current data are not adequate to unambiguously test these ideas. However, many of these possible explanations can be rigorously tested, and the approaches for doing so are discussed in Chapter 14 (this volume).

REFERENCES

Aubrey, D. G. (1985). Recent sea levels from tide gauges: Problems and prognosis, in *Glaciers, Ice Sheets, and Sea Level: Effects of a CO_2-Induced Climate Change*, Polar Research Board, National Academy Press, Washington, D.C., pp. 73–91.

Backus, G., and R. W. Preisendorfer (1978). On the positive components of the first eigenvector of a positive symmetric matrix, Appendix in Barnett (1978).

Balazs, I., and B. Douglas (1979). Geodetic Leveling and the Sea Level Slope along the California Coast, NOAA Tech. Memo., NOS NGS 20, National Geodetic Survey, Rockville, Md.

Barnett, T. P. (1978). Estimating variability of surface air temperature in the Northern Hemisphere, *Mon. Weather Rev. 106*, 1353–1367.

Barnett, T. P. (1983a). Possible changes in global sea level and their causes, *Climate Change 5*(1), 15–38.

Barnett, T. P. (1983b). Long term changes in dynamic heights, *J. Geophys. Res. 88*, 9547–9552.

Barnett, T. P. (1984a). The estimation of "global" sea level change: A problem of uniqueness, *J. Geophys. Res. 89*, 7980–7988.

Barnett, T. P. (1984b). Long term trends in surface temperature over the oceans, *Mon. Weather Rev. 112*, 303–312.

Barnett, T. P., and D. J. Baker, Jr. (1981). Possibilities of detecting CO_2-induced effects: Ocean physics, in *Proceedings of the Workshop on the Needs for Research, Analysis, and Monitoring Efforts for Detecting CO_2 Concentrations in the Atmosphere* (Harpers Ferry, WV), Department of Energy.

Brouwer, D. (1952). A new discussion on the changes in the Earth's rate of rotation, *Proc. Natl. Acad. Sci. 38*, 1–12.

Cess, R. D, and S. D. Goldenberg (1981). The effect of ocean heat capacity upon global warming due to increased atmospheric carbon dioxide, *J. Geophys. Res. 86*, 498.

Clark, J. A., and C. S. Lingle (1977). Future sea-level changes due to West Antarctic ice sheet fluctuations, *Nature 269*, 206–209.

Clark, J. A., W. E. Farrell, and W. R. Peltier (1978). Global changes in postglacial sea level: A numerical calculation, *Quat. Res. 9*, 265–287.

Dickman, S. R. (1977). Secular trend of the Earth's rotation pole: Consideration of motion of the latitude observations, *Geophys. J. Roy. Astron. Soc. 51*, 229–244.

Dickman, S. R. (1979). Continental drift and true polar wanderings, *Geophys. J. Roy. Astron. Soc. 57*, 41–50.

Dickman, S. R. (1981). Investigation of controversial polar motion features using homogeneous International Latitude Service Data, *J. Geophys. Res. 86*, 4904–4912.

Emery, K. O. (1980). Relative sea levels from tide-gauge records, *Proc. Natl. Acad. Sci. 77*, 6968–6972.

Etkins, R., and E. S. Epstein (1982). The rise of global mean sea level as an indication of climate change, *Science 215*, 287–289.

Farrell, W. E., and J. A. Clark (1976). On postglacial sea level, *Geophys. J. Roy. Astron. Soc. 46*, 647–667.

Fairbridge, R., and O. Krebs, Jr. (1962). Sea level and the southern oscillation, *Geophys. J. Roy. Astron. Soc. 6*, 532–545.

Flint, R. (1971). *Glacial and Quaternary Geology*, Wiley, New York.

Gammelsrod, T., and A. Holm (1981). A note on the long-term variations in the upper ocean at Station M (66°N, 2°E), in *Proceedings of the Meeting on Time Series of Ocean Measurements* (Tokyo, Japan), World Meteorological Organization.

Gornitz, V., L. Lebedeff, and J. Hansen (1982). Global sea level trend in the past century, *Science 215*, 1611–1614.

Gutenberg, B. (1941). Changes in sea level, postglacial uplift and mobility of the Earth's interior, *Geol. Soc. Am. Bull. 52*, 721–772.

Hansen, J., D. Johnson, A. Lacis, S. Lebedeff, P. Lee, D. Rind, and G. Russell (1981). Climate impact of increasing atmospheric carbon dioxide, *Science 213*, 957–966.

Hicks, S. D. (1972). Vertical crustal movements from sea level measurements along the east coast of the United States, *J. Geophys. Res. 77*, 5930–3934.

Hicks, S. D. (1978). An average geopotential sea level series for the United States, *J. Geophys. Res. 83*, 1377–1379.

Hicks, S. D., and W. Shofnos (1965). The determination of land emergence from sea level observations in southeast Alaska, *J. Geophys. Res. 70*, 3315–3320.

Jones, P. D., M. L. Wigley, and P. M. Kelly (1982). Variations in surface air temperatures: Part I, Northern Hemisphere 1881 to 1980, *Mon. Weather Rev. 110*, 59–70.

Kuenen, Ph. H. (1950). *Marine Geology*, Wiley, New York, 551 pp.

Kukla, G., J. K. Angell, J. Korshover, H. Dronia, M. Hoshiai, J. Namias, M. Rodenwald, R. Yamamoto, and T. Iwshima (1977). New data on climatic trends, *Nature 270*, 573–580.

Lambeck, K. (1980). *The Earth's Variable Rotation: Geophysical Causes and Consequences*, Cambridge University Press, Cambridge, England, 449 pp.

Lennon, G. W. (1970). Sea level instrumentation, its limitations and the optimization of the performance of conventional gauges in Great Britain, in *Report on the Symposium on Coastal Geodesy*, pp. 181–199.

Lennon, G. W. (1975–1978). *Monthly and Annual Mean Heights of Sea Level*, Permanent Service for Mean Sea Level, Inst. Oceanographic Science, Bedton Observatory, Birkenhead, Merseyside U.K.

Lisitzin, E. (1958). Le niveau moyen de la mer, *Bull. Inf. C.O.E.C. 10*, 254–262.

Lisitzin, E. (1974). *Sea-Level Changes*, Elsevier Oceanography Series 8, Elsevier, Amsterdam, 273 pp.

McCarthy, D. D. (1972). Secular and non-polar variation of Washington latitude, in *Rotation of the Earth*, P. Melchior and S. Yumi, eds., D. Reidel, Dordrecht, Holland, pp. 86–96.

McCarthy, D. D. (1976). The Determination of Universal Time at the U.S. Naval Observatory, Cir. No. 154, U.S. Naval Observatory, Washington, D.C., 11 pp.

Melchior, P., and S. Yumi, eds. (1972). *Rotation of the Earth*, Reidel, Dordrecht, Holland, 242 pp.

Minster, J., and T. Jordan (1978). Present day plate motions, *J. Geophys. Res. 83*, 5331–5354.

Mueller, I. I., and C. R. Schwartz (1972). Separating the secular motion of the pole from continental drift—Where and what to observe, in *Rotation of the Earth*, P. Melchior and S. Yumi, eds., Reidel, Dordrecht, Holland, pp. 68–77.

Munk, W., and G. Groves (1952). The effect of winds and ocean current on the annual variation in latitude, *J. Meteorol. Soc. 9*, 385–396.

Munk, W., and G. J. F. MacDonald (1960). *The Rotation of the Earth*, Cambridge University Press, Cambridge, England, 323 pp.

Munk, W., and R. Revelle (1952a). On the geophysical interpretation of irregularities in the rotation of the Earth, *Mon. Not. Roy. Astron. Soc., Geophys. Suppl. 6*, 331–347.

Munk, W., and R. Revelle (1952b). Sea level and the rotation of the Earth, *Am. J. Sci. 250*, 829–833.

Munk, W., and R. Revelle (1955). Evidence from the rotation of the Earth, *Ann. Geophys. 11*, 104–108.

Nakiboglu, S. M., and K. Lambeck (1980). Deglaciation effects on the rotation of the Earth, *Geophys. J. Roy. Astron. Soc. 62*, 49–58.

Nakiboglu, S. M., and K. Lambeck (1981). Corrigendum, *Geophys. J. Roy. Astron. Soc. 64*, 559.

Namias, J., and H. Huang (1972). Sea level at Southern California: A decadal fluctuation, *Science 177*, 351–353.

Newman, W., L. J. Cinquemani, R. R. Pardi, and L. F. Marcus (1980). Eustasy and deformation of the geoid: 1,000–6,000 radiocarbon years BP, in *Earth Rheology, Isostasy and Eustasy*, N-A. Mörner, ed., Wiley, New York, pp. 555–567.

Paltridge, G., and S. Woodruff (1981). Changes in global surface temperature from 1880 to 1977 derived from historical records of sea surface temperature, *Mon. Weather Rev. 109*, 2427–2434.

Pocklington, R. (1972). Secular changes in the ocean off Bermuda, *J. Geophys. Res. 77*, 6604–6607.

Pocklington, R. (1978). Climatic trends in the North Atlantic, *Nature 273*, 407.

Reid, J. L., and A. W. Mantyla (1976). The effect of the geostrophic flow upon coastal sea elevations in the Northern Pacific Ocean, *J. Geophys. Res. 81*, 3100.

Robuck, A. (1983). Global mean sea level: Indicator of climate change? *Science 219*, 996.

Robinson, M. K. (1960). Statistical evidence indicating no long-term climatic change in the deep waters of the North and South Pacific Oceans, *J. Geophys. Res. 65*, 2097–2116.

Sabadini, R., D. Yuen, and E. Boschi (1982). Polar wandering and the forced responses of a rotating, multilayer viscoelastic planet, *J. Geophys. Res. 87*, 2885–2903.

Sturges, W. (1966). Slope of Sea Level Along U.S. Coasts, Ph.D. thesis, Johns Hopkins University, Baltimore, Maryland, 89 pp.

Sverdrup, J., J. Johnson, and R. Fleming (1942). *The Oceans*, Prentice-Hall, New York, 515 pp.

Tabata, S. (1981). Oceanic time series measurements from station P and along line P in the Northeast Pacific Ocean, in *Proceedings of the Meeting on Time Series of Ocean Measurements* (Tokyo, Japan), World Meteorological Organization.

Thorarinsson, S. (1940). Present glacier shrinkage and eustatic changes in sea level, *Geogr. Ann. 12*, 131–159.

Wexler, H. (1961). Ice budget for Antarctica and changes of sea level, *J. Glaciol. 3*, 867–872.

White, W., and K. Hasanuma (1980). Interannual variability in the baroclinic gyre, *J. Mar. Res. 38*(4), 651–672.

White, W., K. Hasanuma, and G. Meyers (1979). Large-scale secular trend in steric sea level over the Western North Pacific from 1954–1974, *Geodetic Soc. Jpn. 25*, 49–55.

Wunsch, C. (1972). Bermuda sea level in relation to tides, weather, and the baroclinic fluctuations, *Rev. Geophys. Space Phys. 10*, 1–49.

Yumi, S., and K. Yokoyama (1980). Results of International Polar Motion Service in a Homogeneous System 1899.9-1979.0, Publ. Cent. Bureau, International Polar Motion Service, 15 pp.

2 North Atlantic Sea Level and Circulation

KEITH R. THOMPSON
Dalhousie University, Canada

ABSTRACT

Monthly sea levels are examined from 25 North Atlantic tide gauges for the period 1950 to 1975. The influence of local wind forcing is first quantified, using multiple regression techniques, and some of the gains are interpreted in terms of recent theoretical and numerical modeling studies. Three distinct regions of sea-level variability remain after removal of the local meteorological effects, namely, (1) the eastern boundary of the North Atlantic, (2) the western boundary, south of Cape Hatteras, and (3) the western boundary, north of Cape Hatteras. Along the eastern boundary, a statistically significant relationship is obtained between sea level and Ekman pumping of the North Atlantic. It appears that wind-forced changes in ocean circulation can significantly affect the eastern boundary sea level. A similar result could not be found for the western boundary. Examination of the seasonal cycle, however, suggests that the Gulf Stream and an upper-slope boundary current, north of Cape Hatteras, may be important influences.

The value of "correcting" annual sea-level series, in order to detect small changes in their long-term trend, is discussed. An example is given for the eastern boundary of the North Atlantic, where the effect of slow changes in wind-forced circulation is removed from the Newlyn sea-level record. The standard error of the trend estimate is halved, from 0.4 to 0.2 mm/yr, on removal of the meteorological effects.

INTRODUCTION

Interest in the rate of rise of global sea level has been stimulated recently by predictions of a change in air temperature associated with the increasing concentration of atmospheric carbon dioxide. Tide-gauge records have played a key role in determining the sea-level rise this century, mainly because of their length (some exceed 100 yr) and accuracy. Rossiter (1972), for example, showed that an annual mean sea level can be considered accurate to within about 1 mm, certainly less than the variability due to meteorological forcing or steric changes. Hicks (1978) showed that the standard deviation of detrended annual means (σ_a) is about 3 cm along the western boundary of the North Atlantic. This implies that at least 35 yr of data are required to determine a linear trend to within 1 mm/yr, with 95 percent confidence, along this boundary. If σ_a could be halved, by removing the effect of local

52

wind for example, only 22 yr of data would be required to achieve the same accuracy in estimating the linear trend. Clearly this procedure could be applied to short records from other regions and so allow them to usefully contribute to the global picture of sea-level rise. Apart from an improved linear trend estimate, an accelerating sea-level rise could also be detected more readily in long "corrected" series.

Rossiter (1967) was one of the first to correct annual sea level in his study of secular trends on the northwest European shelf. [See Lisitzin (1974) for an historical review of this topic.] Rossiter used linear combinations of air pressures to implicitly represent the joint effect of air pressure and wind forcing over shelf seas and the North Atlantic. Multiple regression techniques are also employed in this chapter to model meteorological and density effects on North Atlantic monthly sea level. One major difference between this study and Rossiter's is in the choice of independent variables for the regression model; only those variables that correspond to a direct physical influence (e.g., local wind stress, wind-forced ocean circulation) are used here (see also Thompson, 1986). The advantages are twofold. First, physically motivated regression models contain useful oceanographic information on shelf/ocean circulation. Second, it is important to know what is being removed when forming the corrected annual series; many geophysical time series are dominated by low-frequency variations that could mistakenly be interpreted as the cause of the sea-level trend when in fact there is no physical connection.

In the next section, the main features of North Atlantic sea-level variability (e.g., variance, power spectra, seasonal cycle, and space scales) are described. In following sections, the relevant forcing functions and the regression analysis (with physical interpretation) are discussed. The correction of long annual series is then finally illustrated with an example from the eastern boundary of the North Atlantic.

OBSERVED SEA-LEVEL VARIABILITY

The monthly sea levels used in this study were recorded by 25 North Atlantic gauges over subperiods of 1950 to 1975 (Table 2.1). The data were obtained from the Permanent Service for Mean Sea Level. An earlier empirical orthogonal function (EOF) analysis of a more extensive array showed that there are three distinct groupings of tide gauges in the North Atlantic (Thompson, 1981, 1986). One group is found along the eastern boundary; apart from the seasonal cycle, the coherence is weak between sea-level changes on opposite sides of the North Atlantic. The other two groups are found along the western boundary and are separated by Cape Hatteras. The split at Cape

TABLE 2.1 Positions of the Tide Gauges and the Number of Monthly Sea Levels Used in This Study

Station	Latitude (°N)	Longitude (°W)	Months	σ_1 (cm)	σ_2 (cm)
GULF OF MAINE AND SCOTIAN SHELF					
Halifax	44.7	63.6	300	4.3	3.3
Yarmouth	43.8	66.1	95		
Bar Harbour	44.4	68.2	307	4.1	3.9
Portland	43.7	70.3	312	5.5	5.0
Portsmouth	43.1	70.8	228		
Boston	42.4	71.1	312	4.3	3.5
Cape Cod	41.8	70.5	235	4.4	3.4
MID-ATLANTIC BIGHT					
Nantucket	41.3	70.1	125		
Woods Hole	41.5	70.7	272		
Buzzards Bay	41.7	70.6	229	3.6	3.0
Newport	41.5	71.3	302	4.0	3.3
New London	41.4	72.1	300	3.9	3.0
Montauk	41.1	72.0	283	4.1	3.1
Sandy Hook	40.5	74.0	312	5.0	3.8
Atlantic City	39.4	74.4	294		
Cape May	39.0	75.0	84		
Lewes Harbour	38.8	75.1	276	6.0	4.7
Kiptopeke Beach	37.2	76.0	280	5.2	4.0
Virginia Beach	36.9	76.0	102		
SOUTH ATLANTIC BIGHT					
Morehead City	34.7	76.7	110		
Charleston	32.8	79.9	312	6.0	5.1
Fort Pulaski	32.0	80.9	288	5.8	4.8
Mayport	30.4	81.4	300	5.9	5.1
Miami	25.8	80.1	276	4.6	4.5
EASTERN BOUNDARY OF NORTH ATLANTIC					
Newlyn	50.1	5.6	312	7.0	3.1

Note: All data was for the period 1950 to 1975; the exact coverage can be obtained from the data publications of the Permanent Service for Mean Sea Level.

The last two columns are the standard deviation of deseasonalized monthly sea level (σ_1) and residual from the multiple regression model (σ_2), both based on data from the common period 1961 to 1970.

Hatteras immediately suggests that the circulation of the North Atlantic may be affecting the coastal sea levels. [Blaha (1984) inferred variations in the strength of the Gulf Stream from the South Atlantic Bight data.]

Western Boundary

To avoid swamping the text with statistics (e.g., EOFs and cross spectra), three typical series are shown in

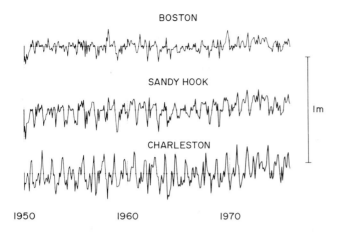

BOSTON

SANDY HOOK

CHARLESTON

1m

1950   1960   1970

FIGURE 2.1 Typical monthly mean sea-level series for the South Atlantic Bight (Charleston), mid-Atlantic Bight (Sandy Hook), and Gulf of Maine (Boston), 1950 to 1975.

Figure 2.1 to illustrate some of the main features of the sea-level variability. [A more detailed description is given by Thompson (1986).] There is a clear seasonal cycle at Charleston (~10 cm, South Atlantic Bight) that is attenuated as one moves north to Sandy Hook (~5 cm, mid-Atlantic Bight) and Boston (~2 cm, Gulf of Maine). These series are regionally representative, as confirmed by the

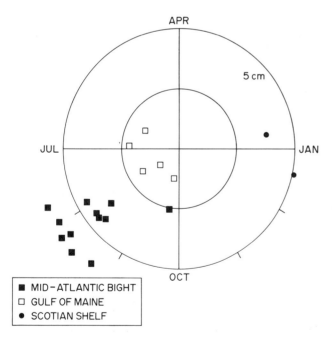

APR

5 cm

JUL

JAN

OCT

■ MID-ATLANTIC BIGHT
□ GULF OF MAINE
● SCOTIAN SHELF

FIGURE 2.2 Amplitude and phase of the annual cycle in observed sea level, north of Cape Hatteras. Tick mark corresponds to a maximum on the fifteenth of each month. The tide-gauge positions are listed in Table 2.1. The two Scotian Shelf gauges are Halifax and Yarmouth; the former has the largest amplitude.

amplitude/phase plot of the annual cycle of sea level in Figure 2.2. [See Blaha (1984) for a description of the ~7 cm seasonal oscillation in the South Atlantic Bight.] The coastal sea-level gradient between the South Atlantic Bight and the Gulf of Maine varies over the year by at least 2×10 cm/1500 km ($=1.4 \times 10^{-7}$). Also from Figure 2.2, the sea-surface slope varies by at least 2×6 cm/200 km ($=6 \times 10^{-7}$) between the Scotian Shelf and Gulf of Maine. These are dynamically significant slopes of the sea surface; Csanady (1976), for example, had to impose a long-shelf gradient of 1.4×10^{-7} to correctly model the mean circulation of the mid-Atlantic Bight.

The standard deviations of the deseasonalized monthly series (σ_1, Table 2.1) show that the most energetic stations are found in the South Atlantic Bight ($\sigma_1 \sim 6$ cm); further north, in the mid-Atlantic Bight and Gulf of Maine, $\sigma_1 \sim 4$ cm. Changes of monthly sea level are similar at Boston and Sandy Hook but distinct from those at Charleston (Figure 2.1), in agreement with the EOF analysis described above. Note, for example, the anomalous 10- to 20-cm drop in both the mid-Atlantic Bight and the Gulf of Maine in early 1950s. This change was not recorded by the South Atlantic Bight gauges. An upward trend of sea level is evident in all three series in Figure 2.1. Note, for example, the 10-cm change in the mean level from 1950 to 1960 through 1970 to 1975 at Sandy Hook. This corresponds to a mean rate of rise of 5 mm/yr, considerably larger than the global average of 1.5 mm/yr obtained by Barnett (1983a). [Hicks (1978) suggested that the gauge at Sandy Hook may be subject to localized subsidence.] The (typical) spectrum of Boston sea level shows that half of the energy in this record is at periods between 7 months and 11 yr (Figure 2.3). There is also a sharp spectral peak at 6 months and a broad peak at 12 to 15 months that will be related in part to the pole tide in a later section.

Eastern Boundary

A detailed description of the sea level along the eastern boundary is given by Thompson (1986). The main point to note here is that, in contrast to the western boundary, the standard deviation of the deseasonalized series (σ_1) increases poleward. This coincides with an increase in the variance of wind and air pressure at the more northerly stations and suggests that local meteorological forcing of sea level may be important.

FORCING FUNCTIONS

A very brief description of some of the more important influences on sea level is given below in order to motivate the regression analysis and aid in its physical interpretation.

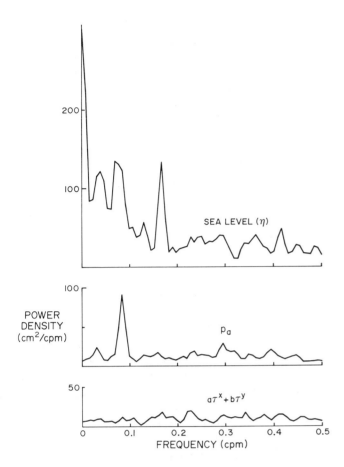

FIGURE 2.3 Power spectra of Boston monthly sea level, air pressure (1 mbar is equivalent to 1 cm), and contribution of local wind stress according to Eq. (2.8), 1950 to 1975. The spectra have been smoothed by "Hamming," and there are 12 degrees of freedom to each spectral estimate.

Air Pressure

The well-known inverse barometer law relates local air pressure (p_a) and sea level η according to the relationship

$$\rho g \eta = \bar{p}_a - p_a,\qquad(2.1)$$

where $\bar{p}_a$ is the average pressure over the world's oceans. Pattullo *et al.* (1955) showed that p_a has a surprisingly large mean annual range of 2.1 mbar and included it in their study of the seasonal oscillation of sea level. It is unlikely however that p_a has a significant trend and its effect can probably be ignored on a decadal time scale. [For example, Figure 3 of Bunker (1980) shows that the average air pressure over the Atlantic (1948 to 1972, 40°S to 70°N) has a trend of only 0.01 mbar/yr.] The time taken for a shelf sea to adjust to changes of p_a is complicated by stratification and topography. However the spin-down time under bottom friction is probably a controlling factor

on the wide, tidally energetic shelf north of Cape Hatteras. This implies a response time of several days and suggests that Eq. (2.1) is valid for monthly means.

The (typical) air pressure spectrum for Boston shows that Eq. (2.1) cannot account for much of the low-frequency sea-level variability, although it is an important contributor at the annual and shorter periods (Figure 2.3). Assuming a typical standard deviation of 5 mbar for monthly p_a in mid-latitudes, a white p_a spectrum implies that the standard deviation of annual and decadal means of p_a would be 1.4 and 0.5 mbar, respectively.

Wind Stress

Both observation and theory confirm that wind stress acting over the shelf can have an important effect on sea level. Csanady (1982) described some simple analytical models and showed that the coastal sea-level response to a steady longshore wind stress (τ^y) can be written in the form

$$\rho g \eta / \tau^y = fL/r,\qquad(2.2)$$

where L is the cross-shelf scale of the wind-driven coastal boundary current. This scale is a function of the Coriolis parameter (f), linear bottom friction coefficient (r), bottom slope, and the spatial structure of τ^y. The time taken to achieve a steady state is again complicated by stratification and topography, but it is probably less than the present averaging period of 1 month. [Wright *et al.* (1986) calculated an e-folding time of 20 hr for the spin-up of their barotropic model of the Gulf of Maine.] The response has therefore been assumed quasi-steady on a monthly time scale in this chapter, and the empirically determined gains of sea level on longshore stress have been used to obtain estimates of L. This is described in the next section. The combined contribution of longshore and cross-shore winds at Boston is shown in Figure 2.3 to illustrate the magnitude of the wind effect. (The results of the regression analysis have been anticipated in order to define the gains.) Wind stress effect is similar in magnitude to that of air pressure at Boston; it is not a major contributor to the low-frequency changes of sea level.

Wind-Forced Ocean Circulation

Recent theoretical and numerical modeling studies (e.g., Anderson *et al.*, 1979) show that the initial response of a mid-latitude, baroclinic ocean to an imposed wind stress is essentially barotropic. Away from the western boundary, the quasi-steady barotropic response can be approximated by the bottom-modified Sverdrup relationship, i.e.,

$$J(\psi, f/h) = k \cdot \nabla x(\tau/\rho h),\qquad(2.3)$$

where ψ is the stream function and J denotes Jacobian. The associated sea-level slopes are given implicitly by

$$gJ(h/f, \eta) = w_e , \qquad (2.4)$$

where g is acceleration due to gravity and w_e is the Ekman pumping [i.e., $k \cdot \nabla x(\tau/\rho h)$]. The sea-surface topography can be determined, up to an arbitrary constant, by integrating Eq. (2.4) from the eastern boundary along f/h contours. To calculate the arbitrary constant of integration, the ocean is assumed closed and conservation of mass is applied, i.e.,

$$\bar{\eta} = 0 , \qquad (2.5)$$

where the overbar denotes a basinwide average and η is measured relative to the undisturbed level. (Note that the contribution to η from the western boundary region is assumed relatively small and ignored.) The longshore momentum equation for the eastern boundary and Eqs. (2.4) and (2.5) then give the interior change in sea level. If it is assumed for simplicity that h is a constant and that wind setup is negligible along the eastern boundary, then the sea level along the eastern boundary is

$$\eta^e = \frac{f^2}{gh\beta} (xw_e) , \qquad (2.6)$$

where $x = 0$ on the western boundary and increases eastward.

If all the dissipation is assumed to occur in a narrow western boundary current, then the sea-level head along the shoreward edge of the western boundary current (but still in deep water) is approximately given by

$$\frac{\delta\eta^W}{\delta y} = \frac{\tau^y}{\rho gh} + \int_0^W k \cdot \nabla x \left(\frac{\tau}{\rho gh}\right) dx , \qquad (2.7)$$

where W is the width of the ocean. (Again h is assumed constant, but the results can be readily generalized to include bottom topography.)

How big are the sea-level changes predicted by Eqs. (2.6) and (2.7)? Clearly the results depend on h. If variations in τ are slow compared to the time taken for a baroclinic Rossby wave to cross the ocean (i.e., decades), then h can be approximated by the mean thermocline depth and bottom topography plays no part (Anderson *et al.*, 1979). For shorter periods (i.e., months), h is the ocean depth. The sea-level response therefore depends on the frequency of wind forcing. Power spectra of monthly w_e were calculated for 55°N, 35°W and 35°N, 35°W in the manner outlined by Thompson and Hazen (1983). The spectra were white, apart from an annual peak. The standard deviations of the monthly changes were 15×10^{-7} and 8×10^{-7} m/s at 55°N and 35°N, respectively. If typical values are assumed of $w_e = 5 \times 10^{-7}$ m/s and $W = 5000$ km,

then $\eta^e = 2$ cm (for $h = 4$ km, initial barotropic) and 8 cm (for $h = 1$ km, final baroclinic). Thus the large-scale wind field becomes increasingly important on longer time scales. Similarly, sea-surface slopes along the western boundary are 0.6×10^{-8} (initial barotropic) and 2×10^{-8} (final baroclinic) if the same values for W and w_e are taken and τ^y is assumed to be 0.1 Pa.

Thermohaline Changes

Fluctuations of the local heat and salt content of the top 200 m of the ocean are responsible for a pronounced seasonal oscillation in sea level (Figure 2.4). The amplitude of this oscillation in the deep water adjacent to the mid-Atlantic Bight and Scotian Shelf is about 8 cm (Figure 2.4). Csanady (1979) extrapolated the deep-ocean steric field to the coast of North America under the assumption that the geostrophic velocity is zero at the seafloor (Figure 2.5). His topography for spring shows a well-defined surface geostrophic flow along the 1000-m isobath; the difference between summer and spring shows that this current has a strong seasonal variation. This seasonal difference in sea level (Figure 2.5b) is consistent with Figure 2.4 in deep water but shows that the deep-ocean amplitude ($\sim$10 cm) is attenuated at the coast ($\sim$2 cm, mid-Atlantic Bight; 0 cm, Gulf of Maine).

The influence of thermohaline changes is not limited to the annual period. Roemmich and Wunsch (1984) identified decadal changes in the large scale temperature field of

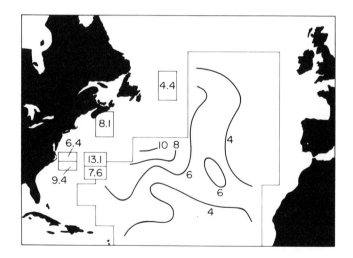

FIGURE 2.4 Co-range lines of the annual cycle of North Atlantic sea level (centimeters) as calculated by Gill and Niiler (1973); an annual wave was fitted to the sum of contributions given in their Figure 3. The individual boxed values are the amplitudes (centimeters) of the annual steric oscillation calculated by Pattullo *et al.* (1955). The maximum sea level generally occurs in August throughout the North Atlantic.

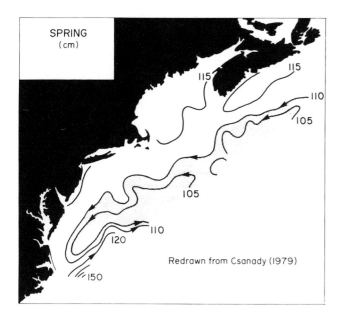

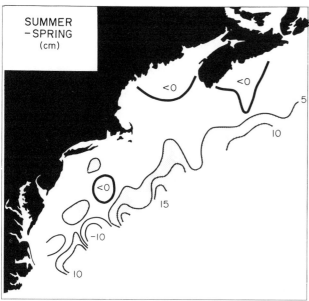

FIGURE 2.5 (a) Sea-surface topography during spring (April through June) calculated by Csanady (1979, Figure 4), under the assumption that the bottom geostrophic velocity is zero. Values are in centimeters. (b) Differences in the summer (July through September) and spring sea-surface topographies, calculated by Csanady (1979, Figures 4 and 7).

the North Atlantic. The observed warming of the ocean between 700 to 3000 dbar, across 24°N and 36°N, results in a thermal expansion of several centimeters. Roemmich (Chapter 13, this volume) shows that Bermuda sea level does reflect such changes in the density field. On a larger spatial scale, Barnett (1983b) examined slow changes of dynamic height in the major oceans (0 to 1000 dbar, early

1900s to date), but he did not find a significant global trend.

MULTIPLE REGRESSION ANALYSIS

In this section, multiple regression models based on the above forcing functions are used to explain some of the observed features of sea-level variability. The following model has been fitted to each series in order to quantify the effect of local meteorology and seasonal changes in density:

$$\rho g \eta + p_a = a \tau^x + b \tau^y + c_1 \cos(\omega_1 t + \phi_1)$$
$$+ c_2 \cos(\omega_2 t + \phi_2) + \varepsilon , \qquad (2.8)$$

where $\omega_1 = 1/12$ cpm and $\omega_2 = 1/6$ cpm. The influence of air pressure has been assumed to follow Eq. (2.1) and has been removed by adding the local air pressure and sea level to obtain the total pressure ($\rho g \eta + p_a$). Local air pressure could of course be included as a forcing term in Eq. (2.8). However, p_a could alias the influence of wind stress, and this would complicate our physical interpretation of the model's coefficients. The influence of local wind stress (τ^x, τ^y) has been modeled by $a \tau^x + b \tau^y$, where a and b are regression coefficients to be determined. This form assumes a quasi-steady response to monthly mean winds (no lags) and a sinusoidal dependence on wind direction, i.e., the direction of maximum sea-level response is given by $\tan^{-1}(b/a)$, and there is no response to winds perpendicular to this direction. [In contrast to the results of Noble and Butman (1979), I found no evidence in the monthly sea levels of an asymmetrical dependence on the direction of wind forcing.] The influence of seasonal changes of density has been modeled by the periodic terms in Eq. (2.8). Unfortunately, there were insufficient hydrographic data to improve on this representation.

Eastern Boundary

There is insufficient space in this chapter to discuss the seasonal cycles and wind gains for the eastern boundary [see Thompson (1986) for a detailed discussion]. One of the most interesting results from the regression analysis, however, was that the residual series (ε) were still correlated, i.e., the coherent sea-level signal along the eastern boundary could not be explained by local air pressure and wind forcing. The influence of North Atlantic circulation was therefore examined by first calculating a time series of η^e using Eq. (2.6) and the 3-month mean Ekman upwelling fields of Thompson and Hazen (1983). (The ocean was assumed closed at 30°N, 60°N; the depth was taken to be 4 km.) The coherence and gain between η^e and Newlyn residuals (ε) are shown in Figure 2.6. (The posi-

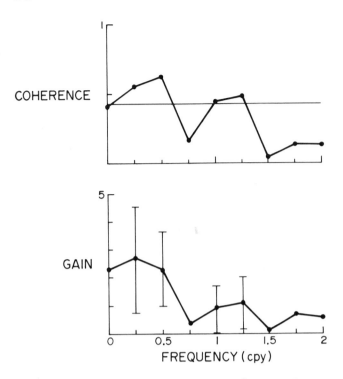

FIGURE 2.6 Coherence and gain between η^e and Newlyn residuals (ε), 1950 to 1975. Seasonal mean values were used. Confidence intervals are at the 95 percent level. The horizontal line in the coherence plot is that coherence that is significantly different from zero at the 0.05 level.

tion of the Newlyn tide gauge is given in Table 2.1; this series was chosen because it was the longest available from the eastern boundary.) The gain increases with decreasing frequency as expected from the above discussion of the response of a baroclinic ocean to Ekman pumping. The coherence also increases with decreasing frequency; the slight reduction at the lowest frequencies may be due to the quasi-linear trend in the Newlyn record, due to eustatic changes and land movement (Rossiter, 1967), which is not in η^e.

Thus it appears that the North Atlantic circulation does influence sea level along the eastern boundary. Further, the gain (Figure 2.6) will transform a white w_e spectrum into a "redder" η^e spectrum and so allow the meteorology to make a significant contribution to the interannual changes of sea level. This point is discussed further in the next section.

Western Boundary

Blaha (1984) recently presented a thorough analysis of the monthly sea-level variability observed in the South Atlantic Bight. The following discussion focuses therefore on the stations north of Cape Hatteras.

Local Wind Effect Wind gains from the regression model (*a*, *b*) are shown in Figure 2.7. To illustrate the type of information that can be extracted from this figure, consider the Scotian Shelf, which is relatively straight and to which Csanady's idealized models are relevant. The longshore gain at Halifax implies that the cross-shelf scale of the wind-forced coastal boundary current is about 16 km [*L*, see Eq. (2.2)]. A typical value of 5×10^{-4} m/s was used for *r* (see Csanady, 1982). This width is in reasonable agreement with the value of 23 km obtained from Csanady's "box-car" forcing model (Csanady, 1982, Eq. 6.53), if we assume (1) *r* is the same as above; (2) the longshore wind forcing starts at the deep Laurentian Channel, the natural "upstream" boundary; and (3) the bottom slope is 5×10^{-3}, a value that is representative of the inshore bottom topography felt by the boundary current.

The gains for the Gulf of Maine have a stronger onshore component than the Scotian Shelf gains, presumably the result of enhanced wind setup in this wide, semienclosed sea. Recent results from a numerical modeling study of the Gulf of Maine agree favorably with Figure 2.7 [see Wright *et al.* (1986) for a detailed comparison].

Seasonal Cycle The coefficients c_1 and c_2 define the annual cycle of sea level that is not forced by local wind or air pressure. This cycle is more regionally coherent than

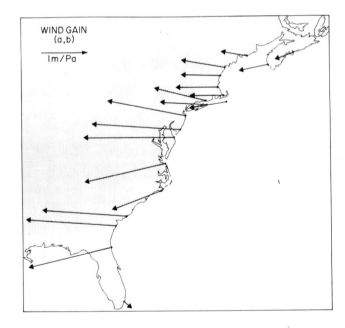

FIGURE 2.7 Wind gains (*a*, *b*) of sea level on local wind stress from Eq. (2.8). The tide-gauge positions are given in Table 2.1. Several of the gains for the mid-Atlantic Bight have been omitted to avoid cluttering the figure, but they conform to the overall pattern. All the gains are significantly different from zero at the 0.05 level except Miami.

TABLE 2.2 Mean Correlation of Residual Series (ε), Both Within and Between Tide-Gauge Groups for the Common Period 1961 to 1970

	GMSS	MAB	SAB
GMSS	0.64		
MAB	0.63	0.74	
SAB	0.26	0.28	0.70

Note: The groupings are SAB (Miami, Mayport, Fort Pulaski, Charleston); MAB (Kiptopeke Beach, Lewes, Sandy Hook, Montauk, New London, Newport, and Buzzards Bay); GMSS (Cape Cod, Boston, Portland, Bar Harbour, and Halifax).

the annual cycle in observed sea levels and has a September maximum in both the mid-Atlantic Bight and the Gulf of Maine (compare Figures 2.2 and 2.8). The amplitude is about 4 cm in the mid-Atlantic Bight and about 2 cm in the Gulf of Maine. These weak seasonal cycles are in favorable agreement with the change in coastal sea level, from spring to summer, predicted by Csanady (1979), i.e., 2 cm in the mid-Atlantic Bight and 0 cm in the Gulf of Maine. Both sets of results agree on an attenuated amplitude at the coast. Thus our sea-level data provide some evidence for the existence of Csanady's upper slope current that was calculated under the major assumption that the geostrophic bottom velocity was zero. The attenuation and phase propagation of the annual cycle along the Scotian Shelf (Halifax-Yarmouth, see Figure 2.8) reflect the seasonal freshwater discharge from the Gulf of St. Lawrence (Drinkwater *et al.*, 1979). The maximum westward flow in winter would correspond (geostrophically) to an increased coastal sea level as observed. (The influence of the freshwater discharge is also evident in the Csanady's spring topography shown in Figure 2.5.)

Residuals Empirical orthogonal function analysis of the residuals (ε) showed that the large-scale modes of sea-level variability remained after removal of the seasonal cycle and the influence of local meteorology. The results of the EOF analysis are confirmed by the overall correlation structure of the residuals given in Table 2.2. (The correlations have been averaged according to the tide-gauge groupings suggested by the EOF analysis.) The average correlation is high (~0.7) for station pairs on the same side of Cape Hatteras. The average correlation between station pairs on different sides of Cape Hatteras is much lower (~0.3). Three typical residual series are shown in Figure 2.9. Note the similarity of the Boston and Sandy Hook records (both north of Hatteras). Apparently, the

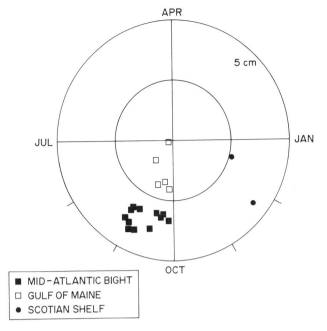

FIGURE 2.8 As in Figure 2.2 but for the annual cycle in the regression model, i.e., c_1 and c_2, the annual cycle not forced by local air pressure or wind. The largest amplitude, Scotian Shelf station, is Halifax.

10-cm anomaly in the mid-Atlantic Bight and Gulf of Maine in early 1950 was not due to local air pressure or wind (compare Figures 2.1 and 2.9).

The standard deviations of the residuals (σ_2, Table 2.1) show that the most energetic stations are in the South Atlantic Bight, even though the influence of local meteorology has been removed ($\sigma_2 \sim 5$ cm). Further north, σ_2 is

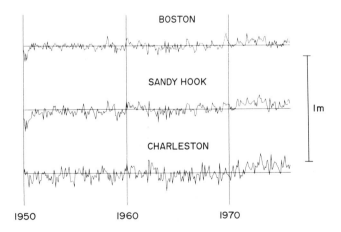

FIGURE 2.9 Typical residual (ε) series for the South Atlantic Bight (Charleston), mid-Atlantic Bight (Sandy Hook), and Gulf of Maine (Boston), calculated from Eq. (2.8) for the period 1950 to 1975.

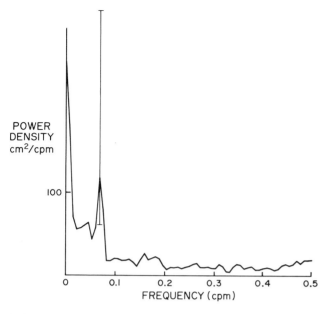

POWER
DENSITY
cm²/cpm

100

FREQUENCY (cpm)

FIGURE 2.10 As in Figure 2.3 but for Boston residual series, 1950 to 1975. The 95 percent confidence interval is given for the pole tide peak.

about 4 cm. The proportion of sea-level variance that can be accounted for by local air pressure and wind $(1 - \sigma_2^2/\sigma_1^2)$ is everywhere less than 42 percent along the western boundary, in contrast to 80 percent at Newlyn on the eastern boundary (Table 2.1). The power spectrum of the Boston residuals (Figure 2.10) shows that the regression model has been able to account for about half of the energy at periods less than 1 yr but is noticeably less successful at lower frequencies (compare Figures 2.3 and 2.10). The pronounced peak in the residual spectrum at 14.7 months is presumably due to the pole tide. Miller and Wunsch (1973) also detected a weak pole tide in the Boston monthly sea-level record but did not attempt to reduce the background noise by removing the variations coherent with the meteorology. This analysis suggests that such a procedure would significantly improve the chances of detecting such a small signal.

What causes the large-scale residual variations north and south of Cape Hatteras? Given the EOF split at Cape Hatteras and the higher residual variance in the South Atlantic Bight, one obvious possibility is the Gulf Stream. I attempted to relate the residuals to monthly fluctuations in the Sverdrup transport across f/h contours (i.e., transports were calculated from Eq. (2.3) using North Atlantic bathymetry). No significant relationships were found. It was also impossible to explain the difference in sea level between the mid-Atlantic Bight and South Atlantic Bight using the pressure head from Eq. (2.7). In short, no relationship could be found between the residual variations along the western boundary and the large-scale wind field

over the North Atlantic. It is however still likely that fluctuations in the surface current of the Gulf Stream contribute significantly to the sea-level variability in the South Atlantic Bight, particularly as its influence has been so clearly demonstrated at Miami on shorter time scales (Maul et al., 1985). If this is indeed true, then monthly changes in the surface current are not dominated by the bottom modified Sverdrup transport of the North Atlantic or by the local wind field.

Some statistically significant relationships were obtained between residuals and changes in shelf hydrography north of Hatteras. (Temperature and salinity data from U.S. lightships were used from the well-mixed time of year, October to March, 1956 to 1971.) The correlations, however, are so low (~0.3) that it seems unlikely that local hydrographic changes are the main cause of the coherent month to month variations in the residuals. Comparison of the time series of sea level and salinity did suggest, however, that density may be important on time scales exceeding 1 yr.

SECULAR CHANGES OF SEA LEVEL

We have seen that wind stress and air pressure are only small contributors to the interannual changes of sea level along the western boundary of the North Atlantic. Shelf salinity and baroclinic boundary currents may be important, but further work is required to quantify their effect on sea level. Thus it has not been possible to correct the western boundary records and so obtain a "cleaner" signal for detecting a change in the rate of rise of sea level.

Along the eastern boundary, however, the large-scale wind field does appear to exert a significant influence on the low-frequency changes of sea level. The combined influence of local wind, air pressure, and η^e has been subtracted from an extended annual series for Newlyn by means of a multiple regression model. The marked reduction in the variability about the trend after "correction" is clear from Figure 2.11. The trends in the observed and residual records are 1.0 ± 0.5 and 1.4 ± 0.2 mm/yr, respectively. The standard errors clearly indicate the increased confidence that can be placed in the latter estimate. Perhaps more important than the reduced standard error is the possibility of detecting a change in the trend more readily in the residual, rather than observed, series (Figure 2.11). There is no evidence for an increasing rate of rise in the Newlyn record.

DISCUSSION

What has been learned about the sea level and circulation of the North Atlantic? Ekman pumping of the North Atlantic may be causing significant changes in sea level

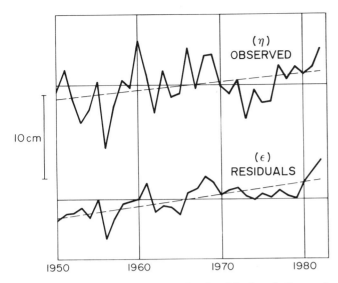

FIGURE 2.11 Annual mean sea level at Newlyn, before and after removal of the effect of local p_a, τ, and η^e with a multiple regression model. The linear trends ($\pm$ standard error) are 1.0 ± 0.5 and 1.4 ± 0.2 mm/yr before and after correction, respectively.

along the eastern boundary. Clearly, more work is required to check this hypothesis because of the arbitrary closing of the ocean at 30°N. One check would be to compare Ekman pumping data for the North Atlantic with changes in the observed density field. Sea-level differences between island stations, notably, Bermuda, and the eastern boundary could also be compared with Ekman pumping. It would be worthwhile, using the Panulirus data, to first remove the effect of density changes below the main thermocline (and so probably not directly wind forced) from the Bermuda sea-level record.

Local wind stress is a significant contributor to sea-level variability at all tide gauges except Miami, although it does not explain the EOF split at Hatteras. Along the Scotian Shelf, the longshore wind gain implies that the quasi-steady, wind-forced coastal boundary current is trapped to within about 16 km of the coast. This value is in good agreement with the inclined beach model of Csanady (1982) if the longshore wind forcing is assumed to start at the deep Laurentian Channel. In the Gulf of Maine, the gains are in favorable agreement with a numerical modeling study (Wright *et al.*, 1986), thereby adding credibility to the gains further south. The most useful oceanographic application of these empirical analyses is probably the provision of such checks on numerical and analytical models. Ideally, the gains should be based on hourly data and made frequency dependent. This can lead to estimates of the spin-down time of shelf circulation (Garrett *et al.*, 1985) and possibly the influence of nonlocal winds.

Changes in the intensity of the surface Gulf Stream are believed to dominate the seasonal oscillation of Miami sea level; it is also probable that the surface Gulf Stream makes a significant contribution to the aperiodic sea-level variability at Miami and the South Atlantic Bight as a whole. If coastal sea levels are assumed to be a measure of the intensity of the surface Gulf Stream, then my failure to relate the sea-level residuals to bottom-modified Sverdrup transport requires some other forcing mechanism for monthly fluctuations in the surface Gulf Stream. North of Cape Hatteras, the seasonal oscillation of sea level suggests the importance of baroclinic current variations, specifically the upper-slope current identified by Csanady (1979) from hydrographic data. Smith and Petrie (1982) recently showed that the surface position of the shelf/slope water boundary along the Scotian slope exhibits large-scale, submonthly onshore translations that appear related to changes in the longshore current in deep water and at the shelf break. This suggests that aperiodic variations in the upper-slope current may make a significant contribution to the monthly sea-level variability north of Hatteras. A comparison of coastal sea level with some of the long-term current records now available for the shelf break and slope is required.

Long and precise tide-gauge records will probably continue to play a key role in determining the rate of rise of global sea level. One of the objectives of this chapter has been to show how more reliable trends can be obtained by first removing meteorological effects from the tide-gauge records. We have seen that meteorological forcing is relatively unimportant to the interannual changes of sea level along the western boundary. Salinity variations may be important, but further work is required to quantify their effect. Fluctuations in the Gulf Stream, and perhaps the upper-slope current, probably contribute to the sea-level variability, but their effect is difficult to quantify. Thus in the absence of any effective independent variables for the regression analysis, it has not been possible to "correct" the western boundary sea-level series. [Meade and Emery (1971) showed that river discharge can account for only 7 to 13 percent of the annual sea-level variance.]

However, the variance of the annual sea-level series from Newlyn (eastern boundary) was reduced significantly by removing the influence of local wind, air pressure, and Ekman pumping of the North Atlantic. The standard error of the trend in the record was thus halved from 0.4 to 0.2 mm/yr. Perhaps more important than the reduced error bars on the trend is the possibility of detecting changes in the trend more readily by using the corrected series. Given the present concern about an accelerating rise of sea level, and the shortage of long series from important areas of the globe, it appears worthwhile to develop similar regression models for other strategic locations.

ACKNOWLEDGMENTS

I would first like to thank the Permanent Service for Mean Sea Level for providing all the sea-level data used in this study. Both Chris Garrett and Adrian Gill made some useful suggestions during the course of this work for which I am grateful. Thanks also to Sara Bennett for reviewing a draft version of this chapter.

REFERENCES

Anderson, D. L. T., K. Bryan, A. E. Gill, and R. C. Pacanowski (1979). Transient response of the North Atlantic: Some model studies, *J. Geophys. Res. 84*, 4795–4815.

Barnett, T. P. (1983a). Recent changes in sea level and their possible causes, *Climatic Change 5*, 15–38.

Barnett, T. P. (1983b). Long-term changes in dynamic height, *J. Geophys. Res. 88*, 9547–9552.

Blaha, J. P. (1984). Fluctuations of monthly sea level as related to the intensity of the Gulf Stream from Key West to Norfolk, *J. Geophys. Res. 89*, 8033–8042.

Bunker, A. F. (1980). Trends of variables and energy fluxes over the North Atlantic Ocean from 1948 to 1972, *Mon. Weather Rev. 108*, 720–732.

Csanady, G. T. (1976). Mean circulation in shallow seas, *J. Geophys. Res. 81*, 5389–5399.

Csanady, G. T. (1979). The pressure field along the western margin of North America, *J. Geophys. Res. 84*, 4905–4915.

Csanady, G. T. (1982). *Circulation in the Coastal Ocean*, D. Reidel, New York, 279 pp.

Drinkwater, K., B. Petrie, and W. H. Sutcliffe (1979). Seasonal geostrophic volume transports along the Scotian Shelf, *Estuarine Coast. Mar. Sci. 9*, 17–27.

Garrett, G. J. R., F. Majaess, and B. Toulany (1985). Sea level response at Nain, Labrador, to atmospheric pressure and wind, *Atmosphere-Ocean 23*(2), 95–117.

Gill, A. E., and P. P. Niiler (1973). The theory of the seasonal variability in the ocean, *Deep-Sea Res. 20*, 141–177.

Hicks, S. D. (1978). An average geopotential sea level series for the United States, *J. Geophys. Res. 83*, 1377–1379.

Lisitzin, E. (1974). *Sea-Level Changes*, Elsevier Oceanography Series 8, Elsevier, Amsterdam, 273 pp.

Maul, G. A., F. Chew, M. Bushnell, and D. A. Mayer (1985). Sea level variation as an indicator of Florida Current volume transport: Comparison with direct measurements, *Science 227*, 304–307.

Meade, R. H., and K. O. Emery (1971). Sea level as affected by river runoff, eastern United States, *Science 173*, 425–428.

Miller, S. P., and C. Wunsch (1973). The pole tide, *Nature 246*, 98–102.

Noble, M., and B. Butman (1979). Low-frequency wind-induced sea level oscillations along the east coast of North America, *J. Geophys. Res. 84*, 3227–3236.

Pattullo, J., W. Munk, R. Revelle, and E. Strong (1955). The seasonal oscillation in sea level, *J. Mar. Res. 14*, 88–155.

Roemmich, D., and C. Wunsch (1984). Apparent changes in the climatic state of the deep North Atlantic Ocean, *Nature 307*, 447–450.

Rossiter, J. R. (1967). An analysis of annual sea level variations in European waters, *Geophys. J. Roy. Astron. Soc. 12*, 259–299.

Rossiter, J. R. (1972). Sea level observations and their secular variation, *Phil. Trans. Roy. Soc. London, A 272*, 131–139.

Smith, P. C., and B. Petrie (1982). Low-frequency circulation at the edge of the Scotian Shelf, *J. Phys. Oceanogr. 12*, 28–46.

Thompson, K. R. (1981). Monthly changes of sea level and the circulation of the North Atlantic, *Ocean Modelling 41*, 6–9.

Thompson, K. R. (1986). North Atlantic sea level and circulation, *Geophys. J. Roy. Astron. Soc. 87*, 15–32.

Thompson, K. R., and M. G. Hazen (1983). Interseasonal changes of wind stress and Ekman upwelling: North Atlantic, 1950–80, Canadian Technical Report of Fisheries and Aquatic Sciences, No. 1214.

Wright, D. G., D. Greenberg, J. Loder, and P. C. Smith (1986). The steady state response of the Gulf of Maine to a surface wind stress, *J. Phys. Oceanogr. 15*, 947–966.

Large-Scale Coherence of Sea Level at Very Low Frequencies

3

W. STURGES
Florida State University

ABSTRACT

The coherence of sea level has been examined between a number of widely distributed stations having the longest data sets, such as at San Francisco, where the data begin in 1855. The sea-level signals in the eastern Pacific appear to be dominated by propagating Rossby waves, so that the variability, having periods of 5 to 8 yr (e.g., between San Francisco and Honolulu) is coherent but out of phase by several years. Sea level is coherent on opposite sides of the Atlantic at periods of about 6 yr, but this may be the result of direct atmospheric forcing rather than of wave propagation. From the Pacific to the Atlantic, coherence is also found at the most energetic periods (6 to 8 yr). At the longest periods detectable—40 to 50 yr—the sea-level signals have amplitudes of 5 to 15 cm and are "visually coherent" between the west coasts of the United States and Europe. The amplitude of these extremely long-period signals is the same as the apparent "rise of sea level over the past century," although the rate of rise from these fluctuations is larger. Because there is so much variability at extremely long periods, the sea-level data must be treated carefully in space as well as in time to avoid contaminating the "sea-level rise" signal by propagating signals. If the data were adjusted, or "corrected" for these signals, the signal-to-noise ratio might be substantially improved, allowing better estimates of the observed rise of sea level, but the forcing mechanisms are not well known at the longer periods. Until the data are so corrected, changes in the rate of rise of sea level, on time scales of about 10 to 50 yr, cannot be distinguished from the background "noise."

INTRODUCTION

One of the difficulties in addressing the sea-level rise problem by using tide-gauge data is that the signals are noisy. Many geophysical processes contribute vigorous time-dependent signals to the data; yet, we wish to extract the very small background sea-level trend in the midst of this variability. The purpose of this chapter is to use the longest tide-gauge records to see if there is hope for improving the signal-to-noise ratio. The answer is "yes."

Barnett (1984) found that the observed rise of sea level, averaged over middle and low latitudes, has been approximately 10 cm in the last century. Lambeck and Nakiboglu (1984) conclude that roughly half of this rise is the result of continuing, long-term response to glacial unloading, i.e., with no addition of water from (high-latitude) glaciers. A major concern, however, comes from the possibility that much of this rise has taken place preferentially in the past 50 yr. If it should turn out that the rise is caused by recent CO_2 addition to the atmosphere, the rate of rise could increase because of increasing input of additional CO_2 and other gases.

It is extremely difficult to determine whether the average sea-level rise is caused by the much-publicized "carbon dioxide effect," the long-term response of the Earth to glacial unloading, or other causes. If the primary cause of sea-level rise is the CO_2 increase, we would expect to see an increasing rate of sea-level rise. However, if the observed rise is primarily the result of continuing adjustment since the last deglaciation, we would expect the sea-level rise to be a simple linear trend when viewed on appropriate time scales. Local tectonic effects, such as the differences between the east and west coasts of the United States, are not addressed here.

Because it seems important to try to understand this problem, it is natural that many workers will examine sea-level data, particularly over the past few—and the next few—decades, in an attempt to determine whether the rate of rise of sea level is increasing. Such analyses will be difficult, as the data have "red spectra." That is, the spectral energy keeps rising at low frequencies. Moreover, the variability at the lowest frequencies we can resolve is the same order as—or larger than—the rise of sea-level "signal."

The point of view taken in this chapter is that, although there is substantial regional variability (as found by Barnett and others), we should expect the effects of a nearly global sea-level rise to be coherent and in phase over the appropriate regions of the globe. It seems useful, therefore, to examine the longest available tide-gauge records to ask: At what frequencies are widely separated gauges coherent, and at what frequencies are they not? Such results might prove useful in telling us what frequencies are (or are not) appropriate for examining the basic problem of "sea-level rise." If there are coherent, large-scale signals, the data can be adjusted for such effects. Most studies take the point of view that the analysis is limited by the time base over which a near-global data set is available. Here, by contrast, we maximize the length of the record available, and deal with a correspondingly much smaller set of stations. Before we explore these low-frequency data, however, it should be noted that a study by Sturges (1987) was made of sources of errors in the usual data.

One hopes, of course, to learn something about the ocean in a study such as this. The primary aim of this chapter, therefore, is twofold. First, sea-level studies obviously must contend with noisy data. The analysis can be improved if as much of this "noise" as possible can be understood, and treated appropriately as signal. It is found here that substantial parts of the low-frequency signals are coherent over basin-wide scales. By averaging many such stations, therefore, one may not necessarily reduce the noise—although by understanding the physical mechanisms involved one can adjust the data for such effects. A second point is that in the longest data sets (beginning in the middle to late 1800s) fluctuations rather like the "rise of sea level since the 1930s" appear to be part of the normal *background variability* of sea level—but at periods too long to resolve with statistical reliability. The problem, then, is not one so much of measuring the present rate of rise as of separating it (if possible) from the low-frequency background variability.

SEA LEVEL ALONG A SINGLE COAST

Useful and convenient summaries of tide-gauge data around U.S. coastlines have been prepared by Hicks and Crosby (1974) and Hicks *et al.* (1983). Their figures show both the yearly values and the lower frequency trends. Simply by lining up the highs and lows "by eye," it is apparent that, on these time scales, the major features of sea-level variability are coherent along a single coast line. These effects have been studied elsewhere at great length (e.g., Brooks, 1979; Enfield and Allen, 1980; Chelton and Davis, 1982). The large scale of this coherence may be partly the result of the large scale of the forcing, but is primarily the result of the efficiency of wave propagation along coastlines. This coherence along the U.S. coasts is widely known. The Hicks and Crosby figures show that the major highs and lows are coherent along the east coast, just as along the west coast. The details vary, but from Miami and Key West to Atlantic City and New York, the major features at these low frequencies are remarkably similar, and they extend into the Gulf of Mexico. On the Pacific coast, the El Niño events are a major forcing phenomenon, whereas variations in the large-scale wind field and the Gulf Stream are Atlantic coast counterparts (e.g., Blaha, 1984; Thompson, Chapter 2, this volume). There is some question as to whether the oceanic response to large-scale atmospheric forcing is deterministic. That is, the response may be found only in an averaged, spectral sense. See, for example, the discussion by White (1985).

Because the lowest frequencies are coherent along the entire coastline, it seems plausible that, at the longest periods, a few gauges having the longest records can be examined in detail and assumed to be representative of the whole coast.

SEA LEVEL AT SAN FRANCISCO AND CASCAIS

From the results of previous studies we know that the rate of sea-level rise appears to increase after about 1930. One is naturally curious about what the longest available data sets show. Figure 3.1 is a "background" figure. It was prepared to show the frequency bands where the

energetic variations occur in nature. In the lower part of the figure, the central peak—a result of wind forcing—is seen to be at periods less than a month. Figure 3.2 shows yearly mean data at San Francisco and Cascais, Portugal. The data have been heavily filtered to suppress the energetic fluctuations at periods less than about 10 yr. The slope of the long-term trend at both stations, 12 cm/cen-

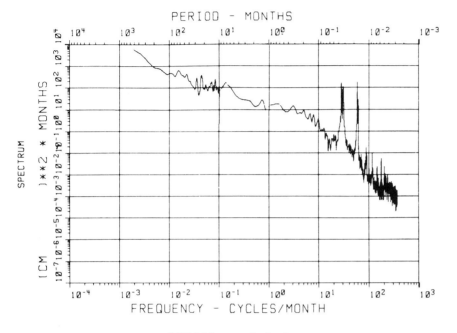

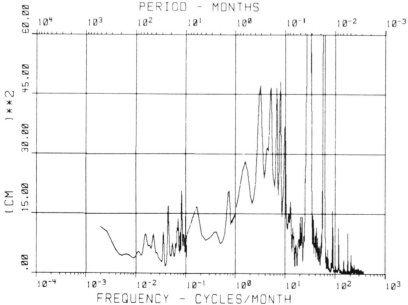

FIGURE 3.1 A composite sea-level spectrum based on data processed in a variety of ways, shown in the normal method above and the variance-preserving form below. Highest frequencies are computed from hourly data at St. Petersburg, Florida, smoothed with three Hanning passes. The intermediate periods, from 1 month to 10 months, are based on 5-hourly filtered data, smoothed with three Hanning passes. Monthly data at San Francisco are used for periods longer than 10 months and are smoothed with five Hanning passes. From Sturges (1987), *J. Physical Oceanography* (American Meteorological Society).

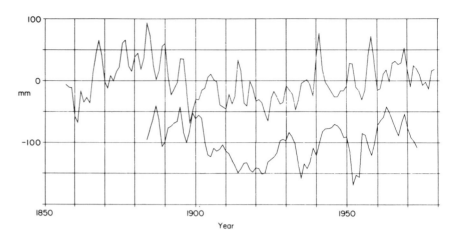

FIGURE 3.2 Sea level at San Francisco (upper curve) and at Cascais, Portugal (lower curve). The means have been subtracted; the curve at Cascais is offset by 100 mm; and a linear trend (12 cm/century) has been removed from each. The data have been filtered to suppress the energetic signals at 5 to 8 yr; half the signal remains at periods of about 6.5 yr (less than 10 percent at 4 yr, over 90 percent at 16 yr). From Sturges (1987), *J. Physical Oceanography* (American Meteorological Society).

tury, has been removed. The second digit is scarcely significant, however, as changing the length of the record by about 10 yr can change the mean slope by about 1 cm/century.

In the San Francisco data a noticeable, qualitative feature is an "event" between approximately 1860 and 1900. In other words, there is a "signal" in this record having a period of approximately 40 yr, and an amplitude essentially the same as that observed in the "mean rise of sea level" since about 1930. Likewise, sea level at Cascais was high in the late 1800s, low around 1920, and rose again after then, showing a pattern similar to, but slightly out of phase with, the San Francisco record.

Clearly, these records are not long enough for statistically significant calculations at periods of 40 to 50 yr. Nevertheless, one cannot help but notice the obvious "visual correlation" between the longest-period variability in the two records—from different continents and different oceans. The essential idea is that the data contain a substantial amount of energy at periods of many decades. The "observed rise of sea level" since 1930 takes place in a frequency band that has energy from many sources. In terms of what the data actually show, the "mean sea-level rise" could fairly be described as low-frequency variability having a peak-to-peak signal of at least 10 cm and a period of at least 100 yr.

On the basis of the data in Figure 3.2, one finds it difficult to argue for any significant change in sea-level behavior beginning in the 1920s. Given this variability, assigning a "beginning date" at which a linear trend should be fit to the data is quite difficult, unless the records could be prefiltered to remove the energetic 40- to 50-yr variability.

These two sea-level records are coherent, in a statistically significant sense, at periods on the order of years to decades. Before we discuss this result, however, it seems appropriate to discuss the idea that sea levels are coherent, on large scales, within a single ocean.

SEA-LEVEL SIGNALS FROM PROPAGATING OCEAN WAVES

In an attempt to be quantitative about the longest possible periods, I have computed cross spectra between sea level at a number of locations, using mean monthly data for some of the calculations and mean yearly data for others. For the details and some hand wringing, see Sturges (1987).

Pacific Ocean

Figure 3.3 shows cross spectra between sea levels at San Francisco and Honolulu. The principal results are twofold. First, for periods of about 5 yr and longer, where the power is found, the records are highly coherent. Second, there is a substantial phase difference; San Francisco leads. While this is clearly a straightforward result in terms of Rossby wave propagation, it has important implications for the "sea-level rise problem," as discussed in a later section. The gain factor (or frequency-response function) shows that the (coherent portion of the) amplitude at Honolulu is essentially the same as that at San Francisco.

It is standard practice to introduce an artificial delay into one of a pair of records to allow for the energy travel time between the two. Such a technique enhances the signal-to-noise ratio, and is also known to improve the accuracy of the phase estimates (e.g., Jenkins and Watts, 1968, p. 396ff). Because it is known that the Rossby wave energy takes several years to travel these distances, an artificial delay was induced between the records for the calculations shown in Figure 3.3. The maximum coherence was found with delays of 2 to 3 yr (a reasonable result, as shown below).

The propagation of low-frequency waves across the ocean has been studied extensively. White (1985) reported agreement between observations and models of wind-forced, baroclinic Rossby waves. Price and Magaard (1983) found a peak in energy in thermocline motions in the Pacific associated with Rossby waves at periods consistent with the peak shown in Figure 3.3. He found that the energy is almost all in the first baroclinic mode; so we expect to see it in sea-level fluctuations.

Mysack (1983) studied Rossby waves at the annual period (an energetic band suppressed in the present work) traveling away from the U.S. west coast. In his model, the waves are generated by fluctuations in coastal currents rather than directly by wind. The important feature to note, for this application, is that the group velocity is toward the southwest (toward the Hawaiian Islands). The phase, however, propagates toward the northwest. At the annual period the waves are still dispersive. Using Mysack's values (his Eq. 3.10), but for a 7-yr period, long enough to be nondispersive, the group velocity is estimated to be about 2.5 cm/s. Thus, in the 4500 km between San Francisco and Honolulu, the travel time for energy would be about 6 yr.

We compared Balboa (Canal Zone) with San Francisco. It is well known (e.g., Enfield and Allen, 1980) that sea-level signals propagate to the north along the west coast of North America. Chelton and Enfield (1986) showed typical amplitudes greater than 10 cm. Chelton and Davis (1982) found that the dominant signal [their first empirical orthogonal function (EOF)] accounted for about 40 percent of the energy and represented a fairly uniform rise

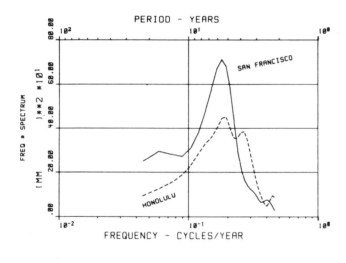

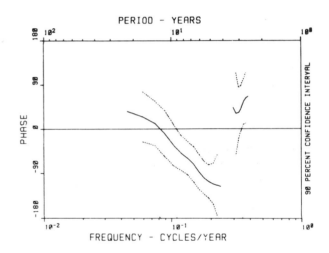

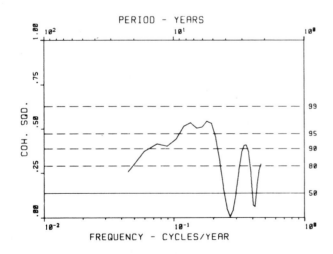

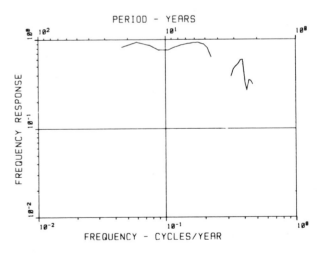

FIGURE 3.3 Cross spectra between sea level at San Francisco and Honolulu: yearly data are given for 1905 through 1971 at San Francisco, and beginning in 1907 at Honolulu. The Honolulu data are delayed by 2 yr for increased coherence (see text). The spectral amplitudes, upper left, are shown in the variance-preserving form; coherence is shown in lower left. Frequency-response curve, lower right, shows the ratio of the coherent power at Honolulu to that at San Francisco. On the phase plot, upper right, the 90 percent confidence intervals are shown. Smoothing is by five Hanning passes. From Sturges (1987), *J. Physical Oceanography* (American Meteorological Society).

and fall of sea level along the whole coast. They found that low-frequency signals traveled up the coast at about 40 cm/s. Hence, at low frequencies, the signals should be essentially in phase between Balboa and San Francisco.

Most of the low-frequency signals at Balboa lead those at San Francisco. At periods near 6 yr, however, the computed result is in the other direction, although the energy level at Balboa is low. The phase error bars (at 90 percent confidence limits) reach to zero; thus, in a statistical sense, this result may not be repeatable—but for the present data the phase calculation is correct. From this we conclude that the phase difference along the coast must be at least partly the result of the distribution of forcing, and not simply of wave propagation.

Atlantic Ocean

A similar set of calculations was done between sea-level records in the Atlantic. On the basis of results in the Pacific, we anticipate that wave energy will propagate generally toward the southwest, suggesting comparison between Cascais and Mayport, Florida. Figure 3.4 shows that coherence is high for the band of energy near 5 to 6 yr. These data overlap, after filtering, for only 38 yr. There is a data segment from 1897 to 1924 at Fernandina, Florida (near Mayport). Because there is not a reliable benchmark connection between the Fernandina and Mayport data, and because there is a 5-yr gap between the two data sets, it seemed risky to try to form a single, long, interpolated data set from the two. It is possible, however, to compare Cascais with Fernandina data, and on the basis of 25 yr of overlapping data, the results were consistent with the Cascais-Mayport calculations.

A most surprising result, however, is found in these comparisons. The coherence is about 90 percent confidence (periods of 6 to 7 yr) when the data are adjusted to introduce a 2- or 3-yr lead into the Cascais data—to be consistent with ocean wave propagation. The coherence is highest, however, when a 3- to 4-yr lead is introduced into the Mayport data—which is in the "wrong direction." The most plausible explanation is that direct atmospheric forcing is more important in these records than is ocean wave propagation. This lag direction for maximum coherence is also supported by the comparisons between San Francisco and Cascais, as discussed in the next section.

It is likely that the mid-Atlantic ridge system interferes with the oceanic wave propagation, making the coherence lower here than between the U.S. west coast and Hawaii. As in the Pacific, we find significant phase shifts. At periods near 6 to 7 yr, the observed phase shift is about 135°, and for periods near 19 yr the phase difference is about 45° on the basis of records beginning in 1929. We should note that, while the statistical reliability of coher-

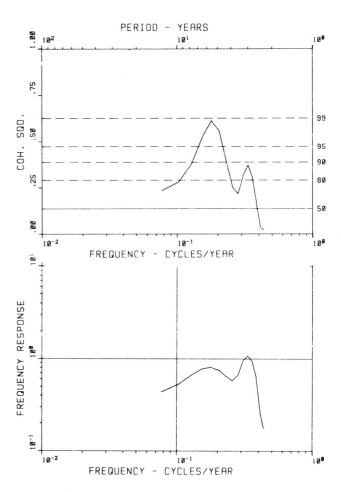

FIGURE 3.4 Coherence (upper) between sea level at Mayport, Florida, and Cascais, Portugal, and gain (lower), based on 38 yr of yearly mean data beginning in 1938. An artificial delay of 4 yr was introduced into the Cascais data. From Sturges (1987), *J. Physical Oceanography* (American Meteorological Society).

ence of periods this long is beyond the limits of the data, the observed phase difference, as calculated, is a valid result and is potentially important in the context of the present work.

Sturges and Summy (1982) found large-scale coherent waves in the central Atlantic thermocline whose energy propagated toward the west southwest. On the basis of their results for group velocity we expect that it should take the energy about 3 yr to propagate across the Atlantic Ocean. It was found that introducing an artificial delay of 3 yr into the records, in the sense of the direction of energy propagation, resulted in increased coherencies.

Thus we find that the lagged coherence has maxima for an induced delay of 2 to 3 yr in one direction or 3 to 4 yr in the other. This is consistent with the idea that the 6- to 7-yr period energy has highest coherence. One cannot tell (simply from these calculations) whether the secondary maximum, consistent with ocean wave propagation, is merely the effect of atmospheric forcing offset by one period. Further work is required to distinguish between the two mechanisms.

On the east coast of the United States, we compared New York with Mayport. The long-period fluctuations are marginally coherent (about 80 percent or higher) for periods longer than about 4 yr, with New York leading, as expected. The much higher coherence typical of the U.S. west coast, however, is not evident.

On the eastern side of the Atlantic, the coherencies were surprisingly low. Other than Cascais the longest records seem to be at Brest, France, and Takoradi (5°N, 2°W) on the coast of Ghana. For periods longer than a few years coherence was low between Cascais and Brest. At the energetic periods found at Cascais (about 7 to 8 yr) there is little power in the Brest record.

Similar lack of coherence was found between Takoradi and Cascais. While there is significant energy in the near-6-yr band at Takoradi, it was marginally coherent (not quite 80 percent confidence level) with that at Cascais. Energy having periods near 3.5 yr, however, was barely coherent between the two stations. The direction of phase propagation for this signal, however, was from Cascais to Takoradi, suggesting that the phase difference is a result of the forcing, rather than simple wave propagation (or that this result is from aliasing).

SAN FRANCISCO TO CASCAIS

San Francisco and Cascais are located at similar latitudes and positions in the Pacific and Atlantic. Because it is known that the sea-level signals on the U.S. coasts are strongly coherent with atmospheric forcing, and because atmospheric signals at low frequencies have very large scales, it seemed reasonable to calculate coherence between sea level observed at San Francisco and Cascais. That is, in the example of San Francisco and Honolulu, the observed coherence results from oceanic wave propagation. Any coherence between San Francisco and Cascais, however, would be attributed presumably to coherence in atmospheric forcing—if the coherence is found in a frequency range where atmospheric signals are present. Figure 3.5 shows, first, that there is high coherence at resolvable periods. Second, the energetic signals in the two oceans have peaks that are at slightly different periods, but the differences are small in a spectral sense (i.e., within 90 percent confidence limits). Third, the most energetic fluctuations are out of phase by several years. At the longest periods resolvable (about 30 yr) the records are also out of phase by about 3 yr.

There is some evidence that atmospheric fluctuations at low frequencies tend to advance from south to north, at

periods of order about 2 yr, so the east-west phase difference may be a result of the slope of phase lines, rather than simple east-west propagation. So far as we are aware, however, comparisons of this type at the longer periods have not been done with the atmospheric data (J. M. Wallace, University of Washington, personal communication). It was not clear whether an easterly or westerly phase lead was more likely, and so the phases were computed with lags in both directions. Assigning the lead to Cascais gives a slightly broader band of coherence at periods from 5 to 8 yr. Assigning the lead to San Francisco gives a slightly higher coherence, but the results are not very sensitive to the choice.

DISCUSSION

Several people have suggested a possible connection between the low-frequency sea-level perturbations shown in Figure 3.2 and the length-of-day (l.o.d.) record shown in Figure 3.6. The l.o.d. variations show striking correlation with atmospheric angular momentum (e.g., Morgan *et al.*, 1985). The l.o.d. variations shown by Barnett (1983) and by Morgan *et al.* have the same amplitudes but are at enormously different periods. Brosche and Sondermann (1985) examined the transfer of angular momentum between the solid earth and the ocean; they find that variability in the Antarctic Circumpolar Current can account for the observed discrepancy in the semiannual angular momentum budget of the Earth and atmosphere, and that the amplitudes are comparable. Thus there is reason to suspect that at very long periods, there may be a rational connection between oceanic circulation and the angular momentum of the Earth—hence the l.o.d. signal. Because the sea-level record at Brest is rather unlike the record at Cascais at the longest periods, one suspects that these variations have to do with changes in the gyre-scale circulation.

Although it is perhaps speculative to suggest a connection between variations in the l.o.d. and sea level at periods of many decades, it may be important to point out such a possibility in the context of this chapter. That is to the extent that the variability in these records can be ascribed to known forcing mechanisms, the uncertainty in determining the rate of any real, global sea-level rise may be improved significantly.

There is a natural temptation to invoke the climate variability of the 1800s in an attempt to explain the lowest frequency variability of Figure 3.2. Stommel and Stommel (1983) showed a temperature record from England beginning in 1740, in which the January temperatures show the most variability at periods of roughly 40 to 50 yr. To "prove" a causality between the forcing mechanism and the observations, however, one must overcome three in-

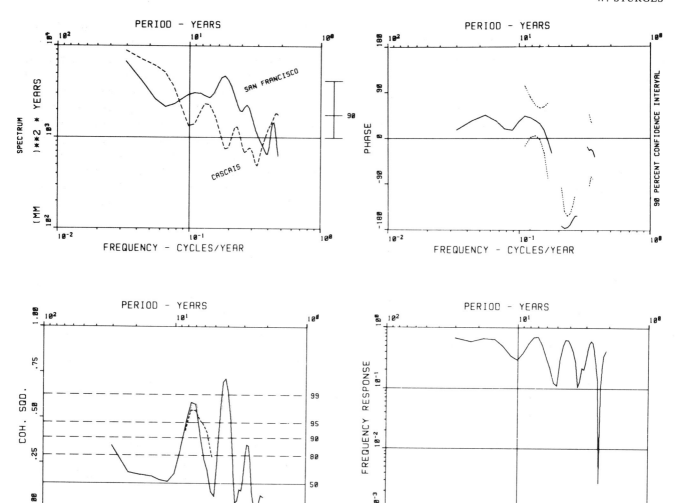

FIGURE 3.5 Cross spectra between sea level at San Francisco and Cascais using 90 yr of data beginning in 1884. The phases are computed using data that are delayed by 4 yr at Cascais. The dashed curve on the coherence plot is for data delayed by 2 yr at San Francisco (see text). Smoothing is by five Hanning passes. From Sturges (1987), *J. Physical Oceanography* (American Meteorological Society).

timidating hurdles. The records are not very long, compared with the periods involved; the events took place during a time that was "historical but prethermometric" (in the words of Stommel and Stommel); and, finally, as Chelton and Davis (1982) pointed out, on scales as large as the subtropical gyres, almost all the observed signals are coherent, making it difficult to distinguish between a variety of reasonable hypotheses.

CONCLUSIONS

The dominant sea-level signals at periods of 5 to 10 yr and longer are coherent between the U.S. Pacific coast and Hawaii, and on both sides of the Atlantic. The amplitudes, at long periods, are about 5 to 15 cm. These signals are out of phase, with delays of several years, and are consistent in part with baroclinic wave propagation across the ocean. At the longest periods detectable in the records, "signals" having periods of 40 to 50 yr are found with amplitudes of about 10 cm. There is "visual correlation" between records on the U.S. Pacific coast and the west coast of Europe, at the same latitude. Similar periodicities are found in English temperature records, but no causal mechanisms have been established.

Because these changes in sea level are coherent on ocean-wide scales, inferences about the "sea-level rise problem" must be treated with this in mind. That is, the longest records appear to be "contaminated" by ocean-

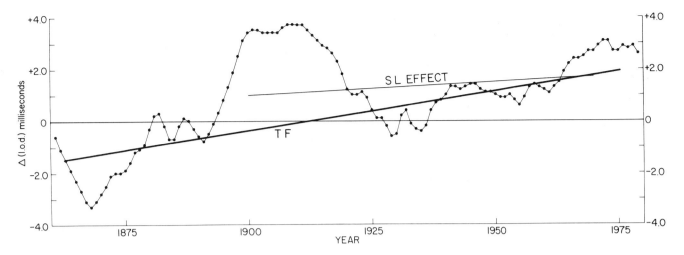

FIGURE 3.6 Secular variation in length of day (l.o.d.) in milliseconds. The line denoted "SL effect" is the change in l.o.d. expected due to the relative sea-level change estimated by Barnett (1983), if that change was due to polar melting. The line denoted "TF" is the change in l.o.d. due to tidal friction. After Barnett (1983), *Climate Change,* (American Meteorological Society).

wave propagation, as well as by regional tectonic motions and other processes. A wave of 10-yr period and 5-cm amplitude has a rate of rise of order 2 cm/yr, which is an order of magnitude larger than the presumed sea-level rise of 10 cm in the past 50 yr. The 40- to 50-yr variations have slopes of about 30 cm/century. Records of shorter duration have a much better geographical distribution, and modern records may have higher precision, but they hide the extent of the low-frequency variability.

These conclusions do not mean that there is no "rise of sea level" problem. On the U.S. east coast, the present rise of sea level (about 26 cm/century) is clearly observable. Of the observed mean global low-latitude rise, roughly half is consistent with long-term glacial rebound. The remaining about 6 cm/century is thoroughly buried in a background of low-frequency variability. If we are to make significant comparisons between the rise of sea level and forcing mechanisms, such as increases in atmospheric gases, in order to have any predictive capability, further effort will be required to extract the signal from the normal background variability associated with all geophysical variables.

ACKNOWLEDGMENTS

During the course of this work, I have benefited from discussions with many people as well as comments from readers of a previous version of this manuscript, including Tim Barnett, John Blaha, Allan Clarke, Nelson Hogg, Jim McCullough, and others. The spectral analysis programs were written by Chris Evans. Data were very kindly provided by Elaine Spencer, Permanent Service for Mean Sea Level, Bidston Observatory. I also thank Pat Klein for cheerfully revising many versions of the manuscript. This work was partially supported by NSF grant OCE 8416458 and by Florida State University.

REFERENCES

Barnett, T. P. (1983). Possible changes in global sea level and their causes, *Climate Change* 5(1), 15–38.

Barnett, T. P. (1984). The estimation of "global" sea level change: A problem of uniqueness, *J. Geophys. Res. 89,* 7980–7988.

Blaha, J. P. (1984). Fluctuations of monthly sea level as related to the intensity of the Gulf Stream from Key West to Norfolk, *J. Geophys. Res. 89*(C5), 8033–8042.

Brooks, D. A. (1979). Coupling of the Middle and South Atlantic Bights by forced sea level oscillations, *J. Phys. Oceanogr. 9,* 1304–1311.

Brosche, P., and J. Sondermann (1985). The Antarctic Circumpolar Current and its influence on the Earth's rotation (abs.), paper presented at IAMAP/IAPSO Joint Assembly, August 5–16, 1985, Honolulu, Hawaii.

Chelton, D. G., and R. E. Davis (1982). Monthly mean sea level variability along the west coast of North America, *J. Phys. Oceanogr. 12,* 757–784.

Chelton, D. G., and D. B. Enfield (1986). Ocean signals in tide gauge records, *J. Geophys. Res. 91,* 9081–9098.

Enfield, D. B., and J. S. Allen (1980). On the structure and dynamics of monthly mean sea level anomalies along the Pacific coast of North and South America, *J. Phys. Oceanogr 10,* 557–578.

Hicks, S. D., and J. E. Crosby (1974). *Trends and Variability of Yearly Mean Sea Level 1893–1972,* U.S. Dept. Commerce NOAA, National Ocean Service, 14 pp.

Hicks, S. D., H. A., Debaugh, and L. E. Hickman (1983). *Sea Level Variations for the United States, 1855–1980*, U.S. Dept. Commerce, NOAA, National Ocean Service, 170 pp.

Jenkins, G. M., and D. G. Watts (1968). *Spectral Analysis and Its Applications*, Holden Day.

Lambeck, K., and S. M. Nakiboglu (1984). Recent global changes in sea level, *Geophys. Res. Lett. 11*, 959–961.

Morgan, P. J., R. W. King, and I. I. Shapiro (1985). Length of day and atmospheric angular momentum: A comparison for 1981–1983, *J. Geophys. Res. 90* (B14), 12645–12652.

Mysack, L. A. (1983). Generation of annual Rossby waves in the north Pacific, *J. Phys. Oceanogr. 13*, 1908–1923.

Price, J. M., and L. Magaard (1983). Rossby wave analyses of subsurface temperature fluctuations along the Honolulu-San Francisco great circle, *J. Phys. Oceanogr. 13*, 258–268.

Stommel, H., and E. Stommel (1983). *Volcano Weather*, Seven Seas Press, 177 pp.

Sturges, W. (1987). Large-scale coherence of sea level at very low frequencies, *J. Phys. Oceanogr. 17*, 2084–2094.

Sturges, W., and A. Summy (1982). Low-frequency temperature fluctuations between Ocean Station Echo and Bermuda, *J. Mar. Res. 40* (Suppl.), 727–746.

White, W. B. (1985). The resonant response of interannual baroclinic Rossby waves to wind forcing in the eastern mid-latitude North Pacific, *J. Phys. Oceanogr. 15*, 403–415.

Glacial Isostatic Adjustment and Relative Sea-Level Change

4

W. RICHARD PELTIER
University of Toronto

ABSTRACT

Secular trends of relative sea level revealed on tide-gauge records have recently been interpreted as requiring some eustatic increase of sea level to have occurred during the past century. Although allowance in interpreting these data is usually made for a contribution due to thermal expansion of ocean volume, it is not generally recognized that the continuing influence of glacial isostatic disequilibrium also contributes significantly to the observed secular trends. The eastern seaboard of the continental United States is a geographical region in which this source of "contamination" is especially large. This fact is discussed in illustration of a global model of deglaciation-induced relative sea-level change, which may be employed to filter the isostatic signal from the tide-gauge data. The same geophysical model has also been successfully employed to explain certain anomalies of the Earth's rotation that have been previously invoked to support the notion that a eustatic increase of sea level must be occurring at present. The anomalies consist of the ongoing wander of the rotation pole toward Hudson Bay at a rate near 1° per million years, and the so-called nontidal component of the Earth's acceleration of rotation that has recently been observed with the LAGEOS satellite. Both anomalies are explicable as memories of the most recent deglaciation event of the current ice age.

INTRODUCTION

The purpose of this chapter is to review results of research conducted over the past decade concerning the nature of relative sea-level (RSL) variations forced by disintegration of the huge continental ice sheets that covered Canada, Northwestern Europe, and West Antarctica at Würm-Wisconsin maximum 18,000 years ago (18 ka). The melting of these ice sheets was essentially complete by 7 ka and resulted in a net rise of mean sea level over the global ocean of about 130 m. The rate of eustatic rise during the glacial-interglacial transition was therefore approximately 10 mm/yr. In spite of the fact that melting has been complete for so many millennia, RSL continues to vary in response to this cause due to the extremely high value of the effective viscosity of the planetary mantle that governs the rate of return to isostatic (gravitational) equilibrium. To fix orders of magnitude it is useful to recall that pres-

FIGURE 4.1 (a) This photograph shows a flight of raised beaches that are located in the Richmond Gulf on the southeast shore of Hudson Bay near the center of postglacial rebound. (b) The relative sea-level (RSL) curve from the Richmond Gulf beaches shown in (a). The age of individual horizons is determined by ^{14}C dating of relict beach material.

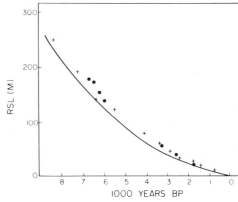

ent-day rates of RSL fall in the three regions that were previously ice covered are approximately 1 cm/yr. The main observational data that have been employed to demonstrate this fact consist of ^{14}C-controlled RSL histories in the age range 0 to 12 ka. These data are obtained from flights of raised beaches such as those shown in Figure 4.1a, which are located in the Richmond Gulf on the southeast shore of Hudson Bay near the center of postglacial

rebound. Individual relict beaches in the sequence may be dated using the radiocarbon method and their heights above present-day sea level plotted as a function of their age to obtain a local history of relative sea level such as that shown in Figure 4.1b for the Richmond Gulf sequence. Such data make up the primary geophysical database for the inference of mantle viscosity, a physical parameter that is fundamental to the understanding of continental drift and many other geodynamic processes. A recent review of solid-earth geophysical applications of RSL data is provided in Peltier (1982).

Although radiocarbon data on raised beaches, such as those shown in Figure 4.1, demonstrate that the surface of the solid earth has been rising monotonically with respect to the geoid since glaciation was complete at sites that were once ice covered, the opposite sense of RSL variation has persisted in the regions immediately peripheral to the margins of the major ice sheets. That this should be the case is simply understandable on the basis of the principle of conservation of mass. At glacial maximum, the Earth is depressed under the weight of the ice, approximately by the amount required for the buoyant restoring force due to the surface deflection (the Archimedes force) to balance the weight of the load. The material that must be removed from beneath the ice sheet to accommodate the sinking of the Earth flows viscously into the immediately peripheral region where the land is elevated. This region will subsequently be referred to as the peripheral bulge. When the ice sheet disintegrates, this process reverses and the land rises with respect to sea level in places that were once ice covered, and beaches such as those shown on Figure 4.1 are cut into the land where it contacts the sea, thus constituting a history of the recovery process. In the peripheral region, however, the land sinks with respect to sea level and relict beaches are drowned. An excellent example of a drowned coast produced by glacial peripheral bulge collapse is provided by the east coast of the continental United States, all of which is peripheral to the huge Laurentide ice mass that covered Canada 18 ka. An illustrative set of ^{14}C-controlled RSL data from sites along this coast is shown in Figure 4.2. Inspection of these data demonstrates that the rate of sea-level rise at some sites along the coast (e.g., Barnstable, Massachusetts) is in excess of 1 mm/yr. Although this rate is an order of magnitude less than the rate of sea-level fall that obtains within the central Laurentide depression, it is nevertheless a rate that is rather significant from several points of view.

Most important for our purposes is the fact that this rate of RSL rise is close to that which has been inferred on the basis of tide-gauge observations of secular sea-level trends and interpreted in terms of implications for climate change (e.g., Hansen et al., 1981). Recent analyses of these data (reviewed by Barnett, Chapter 1, this volume) reinforce the conclusions of earlier workers to the effect that on a global basis the data suggest that RSL is currently rising at a rate of 1 to 2 mm/yr. Gornitz et al. (1982), for example, concluded that the tide-gauge data imply a global increase of RSL at a rate near 1 mm/yr. The most straightforward interpretation of this observation, assuming that the claim made for its global validity is justified, is that the volume of water in the global ocean is currently increasing. Although this could, in principle, be caused by thermal expansion of the water column (a process discussed in detail by Roemmich, Chapter 13, this volume), the estimate by Gornitz et al. (1982) of the possible contribution owing to this effect suggests that it is able to explain no more than 50 percent of the observed rate of RSL rise. If this is a valid upper bound, it implies that the mass of water in the oceans must also be increasing and, therefore, that water "normally" stored in continental ice complexes is being returned to the oceanic reservoir, this being the only conceivable source for the amounts of water required to balance the observation. Hansen et al. (1981) suggested that this implied wastage of continental ice masses could be the first indication of the global climate amelioration that is expected as a result of the increasing atmospheric concentrations of carbon dioxide and other "greenhouse" gases. Meier (1984) recently discussed the compatibility of this hypothesis of a current retreat of continental ice with the available glaciological evidence. His conclusion (also discussed in Chapter 10, this volume) is that although there is no evidence supporting the possibility of a significant ongoing reduction of either Greenland or Antarctic ice, there is strong evidence to the effect that a reduction in volume of the small ice sheets and glaciers of the world is occurring, which could very well provide the mass required to balance the RSL analyses of Gornitz et al. (1982).

The possible climatological implications of the secular variation of RSL, discussed above, that has been extracted from the tide-gauge data, are clearly important. They do hinge crucially, however, on the assumption that the secular trend of 1 to 2 mm/yr is a number that is globally representative. Unfortunately, and as discussed at length by Barnett (Chapter 1, this volume), the existing network

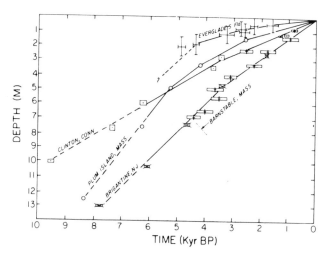

FIGURE 4.2 Relative sea-level data from a number of sites along the east coast of the United States. Adapted from Bloom (1967).

of tide gauges does not provide uniform coverage over the globe so that a severe sampling problem exists. Furthermore, many of the most heavily instrumented coastlines of the world, such as the west coast of the North American continent and coastal Japan, are strongly influenced by local tectonic processes and, therefore, are unlikely to provide any reliable information on eustatic water rise. At these locations, the vertical motions of the solid earth associated with the dominantly strike-slip earthquakes that mark the San Andreas Fault and the dip-slip earthquakes associated with active subduction in the Japan trench are expected to obscure the eustatic signal. Other heavily instrumented coastlines such as the passive continental margins of the east coast of North America and northwestern Europe might be considered more likely regions to provide a signal that is free of such tectonic contamination. These coastlines are however located, respectively, on the forebulges of the ancient Laurentian and Fennoscandian ice sheets so that, as discussed above, sea level in these regions will appear to be rising as a consequence of the sinking of the land with respect to the geoid that accompanies the isostatic readjustment of the surface following deglaciation. If it were possible to correct these passive margin data for the influence of glacial isostatic disequilibrium, the residual trends so obtained might be particularly useful as constraints on the current rate of eustatic water rise (fall) cause by steric and ice-sheet disintegration effects.

Over the past decade a geophysical model has been developed that is ideally suited to the task of filtering from the RSL record, at any tide-gauge station on the Earth's surface, the contribution to the secular change of RSL resulting from current glacial isostatic disequilibrium. This model is based on the mathematical analysis of the deformation of a viscoelastic Earth produced by surface mass loads originally presented in Peltier (1974). It was first employed by Peltier and Andrews (1976) to predict variations of RSL forced by Pleistocene deglaciation; their calculations were based on the assumption that the water produced by ice-sheet melting could be assumed to be distributed uniformly over the ocean basins. Farrell and Clark (1976) showed how this mathematical analysis could be extended to determine the way in which the meltwater must be distributed within and among the ocean basins such that the instantaneous geoid remained an equipotential surface at all times during and subsequent to the ice-sheet disintegration event. The first predictions of expected RSL variations with the complete theory were described in Clark et al. (1978) and Peltier et al. (1978). By comparing these predictions with ^{14}C-controlled RSL observations such as those illustrated in Figure 4.1, it was demonstrated that a large fraction of the variance in the global RSL record over the age range 0 to 18 ka could be

explained. In particular the previously enigmatic observation of raised beaches at sites in the southern ocean at times subsequent to 6 ka (e.g., Russell, 1961) was shown to be a natural consequence of the global-scale viscoelastic adjustment process following a deglaciation that ended about 7 ka.

Although one other global model of RSL change has also been produced (Cathles, 1975), it suffers from two crucial flaws. First, it is not gravitationally self-consistent and so does not accurately predict far-field RSL data; second, it does not incorporate a surface lithosphere and so cannot accurately predict RSL data in the near field. All of the discussion to be presented here will therefore be based on the more accurate model.

In the past few years several additional geophysical observations have been shown to be explicable in terms of extensions of the same theory developed in the papers cited above to explain deglaciation-induced RSL change. The large-scale negative free-air gravity anomalies observed over the three main centers of postglacial rebound have been shown to be compatible with the same viscoelastic model that fits the RSL data (Peltier, 1980, 1981; Peltier and Wu, 1982; Wu and Peltier, 1983). Also, the large misfits between predicted and observed RSL data (Peltier et al., 1978; Clark et al., 1978) at sites in the peripheral bulge regions have been shown to be due to the neglect of the influence of the surface lithosphere (Peltier, 1984b). Finally, two rather important and previously unexplained anomalies in the Earth's rotation, namely, the true wander of the rotation pole toward Hudson Bay at a rate near $1°/10^6$ yr, which is revealed in the International Latitude Service path, and the so-called nontidal component of the acceleration of the axial rate of rotation, which has recently been observed using the LAGEOS satellite, have also been shown to be consistent with the same viscoelastic Earth structure (Sabadini and Peltier, 1981; Peltier, 1982, 1983, 1984a, 1985a; Peltier and Wu, 1983; Wu and Peltier, 1984). That the latter effect might be due to glacial isostatic adjustment was first suggested by Dicke (1969), who suggested that it could be employed to constrain mantle viscosity—a notion that was later investigated in more quantitative fashion by O'Connell (1971). The former observation is particularly important as it was originally invoked by Munk and Revelle (1952) in support of the notion that some current retreat of the Earth's major ice sheets must be occurring to account for it. Their argument has been resurrected in the recent literature on the secular variation of RSL seen on tide gauges, in support of the same idea (e.g., Barnett, 1983). The analysis first reported in Peltier (1982) demonstrated that the theoretical argument employed by Munk and Revelle is incorrect, however. When the radial variations of viscoelastic structure of the real Earth are properly taken into account, it is shown that the observed polar wander toward Hudson Bay can be explained entirely as an ongoing effect due to the deglaciation event that ended about 7 ka. No current wastage of continental ice is therefore required to understand this observation, and the data may in fact be construed as arguing against this possibility.

The remainder of this chapter provides a more detailed discussion of the evidence cited above to the effect that this viscoelastic model does in fact fit the wide range of geophysical data. Since the model does fit, in particular, the observed RSL record in the age range 0 to 18 ka rather well, it may reasonably be employed to predict the present-day rate of RSL variation that should be occurring at any site at which a tide gauge is located. This is the secular change that a tide gauge should record if glacial isostatic disequilibrium were the only source of RSL change at the site in question. The viscoelastic model is then employed to predict the present-day rate of vertical motion with respect to the geoid of the surface of the land everywhere on the North American continent. A map of this field is presented. This demonstrates that present-day rates of RSL rise along the U.S. east coast caused by glacial isostatic disequilibrium may be as high as 2 mm/yr. The field theory is then employed to filter this effect from the secular trends observed on all tide gauges located along both the east and west coasts of the continental United States. The average of the residual trends at U.S. East Coast sites is still significantly different from zero and has a value near 1 mm/yr. However, variations about the mean are extreme on rather small spatial scales and indicate contributions to the residual from sources other than eustatic water rise. In conclusion, a brief summary and discussion of future prospects for further applications of the isostatic adjustment model of RSL change reviewed here is presented.

THE GLOBAL MODEL OF GLACIAL ISOSTASY

The mathematical structure of the global model of glacial isostatic adjustment has been reviewed in Peltier (1982), to which the interested reader is referred for details and for a much more complete review of the history of this subject than could be presented here. In this model the planetary interior is assumed to be radially stratified with an elastic structure fixed by observations of the frequencies of the elastic gravitational normal modes of free oscillation as described, for example, by Gilbert and Dziewonski (1975). The rheology of the interior is assumed to be linearly viscoelastic and of the "Maxwell" type, in which the initial response to an applied shear stress is Hookean elastic, but the final response is Newtonian viscous. The depth-dependent viscosity in the model is therefore the only parameter that one can vary in order to fit RSL data, such as

those shown in Figure 4.1. In the actual prediction of such RSL variations, one assumes a melting history for all of the surface ice loads that exist at glacial maximum and then computes, in a gravitationally self-consistent fashion, the manner in which the meltwater must be distributed over the global ocean in order to ensure that the instantaneous surface of the new ocean is maintained as an equipotential surface at all times. This operation requires inversion of an integral equation at every instant during and subsequent to the deglaciation, and results in a direct prediction of the time-dependent separation of the geoid and the surface of the solid earth at any point on the Earth's surface where ocean and land meet. In the model the geography of oceans and continents is realistically described and the full effects of gravitation are accounted for including the gravitational attraction of the water by the ice and the self-attraction of the water. The integral equation is inverted using a finite-element discretization of the surface, and a Green's function formalism is employed to describe the gravitational interactions between the aquasphere, cryosphere, and solid-earth components of the model that are fundamental to the determination of sea-level change. It is important to be able to make a reasonably accurate *a priori* estimate of the deglaciation history subsequent to 18 ka since we would not otherwise be in a good position to invert the predictions of the model to recover the radial variation of mantle viscosity. Peltier and Andrews (1976) described the way in which ^{14}C-age-controlled terminal moraine data can be combined with far-field observations of RSL and ice-mechanical considerations to develop first-order models of the glacial chronology. Their initial model, called ICE-1, has since been improved by Wu and Peltier (1983), who have called the new model ICE-2. This model is illustrated in Figure 4.3. The forward problem for RSL prediction takes as input this glaciation history and a model of the planet's radial viscoelastic structure and produces as output a prediction of the RSL variation that should be observed at any site of interest. These predictions are illustrated in the following subsections.

Postglacial Variations of Relative Sea Level

Typical examples of observed and predicted RSL variations at a number of sites on the North American continent are illustrated on Figure 4.4 for six locations, three at sites that were once ice covered (a, b, and c) and three for sites along the east coast of the continental United States in the peripheral region of monotonic subsidence (d, e, and f). The Earth model employed to make these predictions has elastic structure 1066B of Gilbert and Dziewonski (1975), an upper mantle viscosity of 10^{21} Pa s, and a lower mantle viscosity beneath 670 km of 2×10^{21} Pa s. On

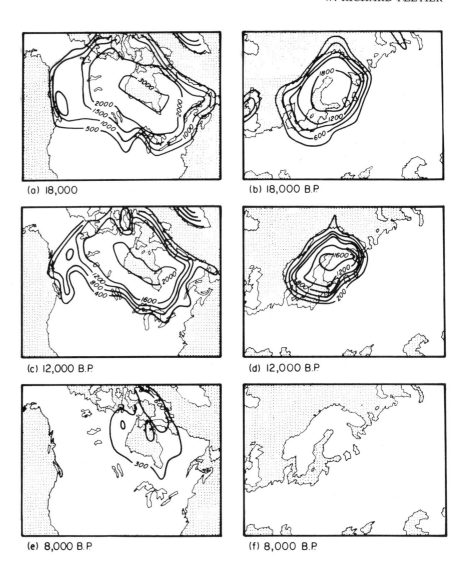

(a) 18,000

(b) 18,000 B.P.

(c) 12,000 B.P.

(d) 12,000 B.P.

(e) 8,000 B.P.

(f) 8,000 B.P.

FIGURE 4.3 Time slices through the ICE-2 deglaciation chronology of Wu and Peltier (1983). Ice thickness is shown at three times for both the Laurentian and Fennoscandian ice sheets.

Figure 4.4, comparisons are shown for three different models that differ from one another only in terms of their lithospheric thicknesses. Inspection of these comparisons shows that the RSL data at sites inside the ice margin are reasonably well fit by the theoretical model and that the RSL variations at these sites are rather insensitive to changes of lithospheric thickness. This is entirely expected as the spatial scale of the Laurentian ice sheet (Figure 4.3) is so large that the lithosphere is transparent to the response at locations that were once under the ice-sheet center. At sites in the peripheral region, on the other hand, the response is extremely sensitive to lithospheric thickness as the deformation at such sites is significantly affected by relatively short horizontal wavelengths that see the lithosphere clearly. This sensitivity was first exploited (Peltier, 1984b) to measure lithospheric thickness, and a relatively high value in excess of 200 km was obtained. As reviewed in Peltier (1982), the totality of ^{14}C-controlled RSL data also requires an almost uniform profile of mantle viscosity with little variation between the upper and lower mantles. Weertman (1978) commented on this result from the point of view of theoretical ideas concerning the microphysical basis of solid-state creep in the Earth and suggested that it might be taken to imply that the relaxation of the lower mantle that occurs in postglacial rebound is controlled by transient creep rather than the steady-state creep, which is assumed in the Maxwell analogue. This possibility may be tested using the simple Burger's body rheology derived in Peltier et al. (1981), which includes the transient component of the response via a single Debye peak governed by two additional physical constants. An analysis of the asymptotic properties of the Burger's body rheology, however, demonstrates that when the elastic defect is large and the short- and long-time-scale viscosities sufficiently different, the Burger body rheology again behaves like a Maxwell solid but with a viscosity equal to that which governs the short-time-scale transient response. Under these circumstances the lower mantle viscosity inferred by analysis of rebound data based on the Maxwell analogue would be the transient viscosity as originally suggested by Weertman (1978).

The Free-Air Gravity Anomaly over Centers of Post-Glacial Rebound

Figure 4.5 shows maps of the free-air gravity anomalies over the present-day centers of postglacial rebound in Canada [Figure 4.5(a)] and Fennoscandia [Figure 4.5(b)]. Comparison of these maps with those for ice thickness at

glacial maximum, shown previously in Figure 4.3, demonstrates a high degree of correlation between these two fields and provides strong support for the hypothesis that the observed free-air anomalies are to be interpreted as measures of the currently existing degree of isostatic disequilibrium of these two regions. Figure 4.6 shows a comparison of observed and predicted present-day peak free-air anomalies for Laurentia and Fennoscandia for a number of Earth models having fixed lithospheric thickness of 120.7 km, and an upper mantle viscosity of 10^{21} Pa s as required by the sea-level data discussed previously. The Earth models employed differ from one another only in terms of their elastic structures, and lower mantle viscosity is varied through the same sequence of values for all models. As described in the figure caption, four of the

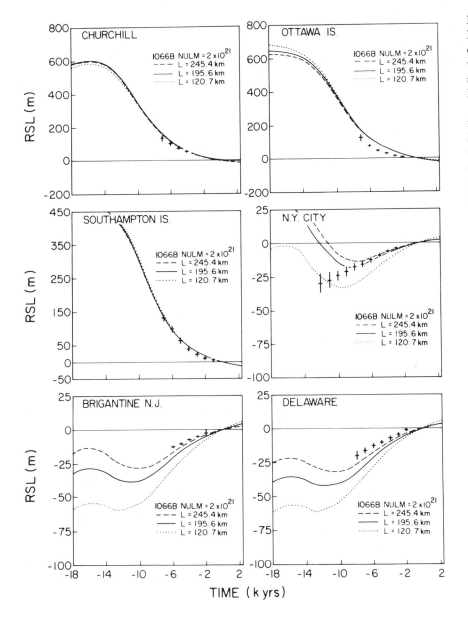

FIGURE 4.4 Observed RSL histories at 6 North American sites and predictions based on the ICE-2 melting chronology coupled with an Earth model having 1066B elastic structure, an upper-mantle viscosity of 10^{21} Pa s, and a lower-mantle viscosity of 2×10^{21} Pa s. Theoretical predictions are shown for models that differ only in their lithospheric thicknesses. The response is sensitive to this parameter only at the last three sites (d, e, and f), which are located in the peripheral bulge region south of the ice sheet margin along the U.S. east coast.

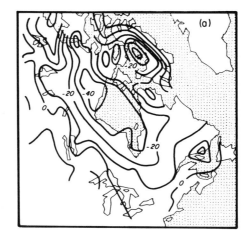
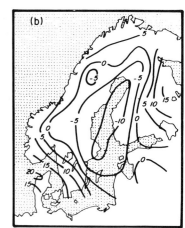

FIGURE 4.5 Observed free-air gravity anomalies over (a) Canada and (b) Fennoscandia. The locations of these anomalies should be compared with the ice-sheet topographic maxima shown on Figure 4.3.

models are flat, homogeneous, layered approximations to the 1066B elastic structure that have either one or two internal discontinuities of elastic parameters within the mantle at 420 km and/or 670 km depth corresponding to the depths to the olivine → spinel and spinel → post-spinel phase boundary horizons. The remaining two curves

are for the seismically realistic models 1066B of Gilbert and Dziewonski (1975) and PREM of Dziewonski and Anderson (1981). In these models, all of the radial variation of density is treated as though it were nonadiabatic. Inspection of these results shows that fitting the observed free-air gravity anomalies with a model with weak radial

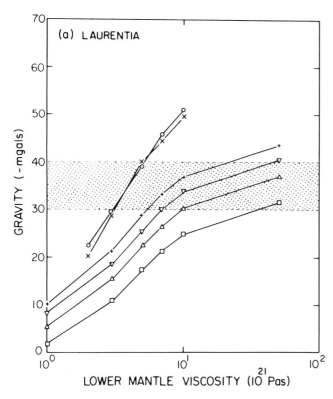
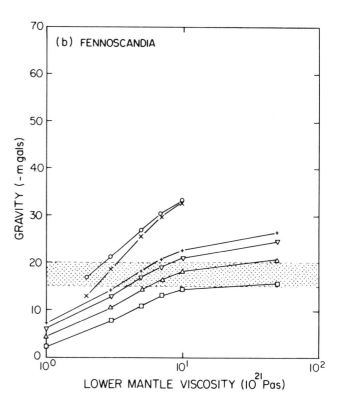

FIGURE 4.6 Observed (hatched regions) and predicted peak free-air gravity anomalies for both Laurentia and Fennoscandia. Predictions of the peak free-air anomaly are shown as a function of lower mantle viscosity with the upper mantle value held fixed at 10^{21} Pa s and the lithospheric thickness fixed at 120 km. The individual curves are predictions for models that differ only in their elastic structures: (×) 1066B; (○) PREM; (+) flat

incompressible approximation to 1066B with a single internal density discontinuity with 12 percent increase at 670 km depth: (∇) same as (+) but having a 6 percent increase at 670 km depth and a 3 percent increase at 420 km depth; (△) same as (+) but having only a 6 percent increase of density at 670 km depth; and (□) same as (+) but having no internal density discontinuities in the mantle. See text for discussion.

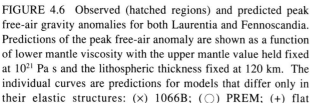

variation of viscosity requires the presence in the model of significant internal buoyancy associated with a density structure that behaves nonadiabatically on the time scales of glacial rebound. This requires at least that the density variations across the phase boundaries at 420 and 670 km depth behave in this fashion, which is possible only to the extent that these transitions may be considered univariant (e.g., O'Connell, 1976; Mareschal and Gangi, 1977). Although such behavior is not inconceivable, it may also prove possible to reconcile these data by appealing to other physical effects. This is clearly an extremely important issue insofar as the problem of mantle convection is concerned (Peltier, 1985b). In any event, the demonstration that the incorporation of internal buoyancy of the mantle allows the essentially isoviscous model preferred by the sea-level data to simultaneously reconcile the large observed free-air anomalies does rather strongly undermine the claim of Walcott (1970, 1980) that these anomalies require that the viscosity of the lower mantle be high.

PLEISTOCENE DEGLACIATION AND EARTH ROTATION

From about 1900 until 1982 the location of the Earth's north pole of rotation was carefully monitored by the ILS using a global network of photo-zenith tubes equipped observatories. Since 1982 this observing system has been replaced by the much more accurate VLBI-based network of the IRIS Earth-orientation monitoring system that routinely determines the pole position at 5-day intervals with a verified accuracy of 2 milliseconds of arc (Carter and Robertson, 1985). Although these new data will quickly replace the old as the industry standard, the duration of the time series is still sufficiently short that the ILS data remain the best source of information on the secular motion of the pole. These data are shown in Figure 4.7 as x and y components of the displacement relative to the axes shown on the inset polar projection. The origin of the coordinate system corresponds to the Conventional International Origin (CIO). Inspection of these data, which are based on the reduction by Vincente and Yumi (1969, 1970), demonstrates that the dominant oscillatory signal, which consists of a 7-yr periodic beat generated by the interference between the 14-month Chandler and 12-month annual wobbles, is superimposed on a secular drift at the rate of $0.95° \pm 0.15°/10^6$ yr toward Hudson Bay. The direction of this drift is shown by the arrow on the inset polar projection of Figure 4.7.

In 1952, Munk and Revelle interpreted this observed secular drift of the rotation pole as requiring some present-day variation of surface mass load and suggested that the cause of the apparently required variation might be found in the melting of ice on Greenland and/or Antarctica. Their inference that such an effect was required to explain the

data was, however, based on a dynamical model in which it was assumed that the Earth could be treated as a homogeneous viscoelastic sphere insofar as its rotational response to surface loading was concerned. To the extent that this approximation is valid, the inference of Munk and Revelle is completely correct since the theory then shows that the pole must be fixed at any instant of time in which the surface load is steady. As demonstrated by Peltier (1982), this holds true for homogeneous Earth models, since in this limit the isostatic adjustment and rotational contributions to the rotational forcing exactly annihilate one another. For radially stratified models, however, the dynamical symmetry that underlies this cancellation is broken and polar wander can occur even at a time when the surface load is steady. It therefore becomes plausible that the secular drift of the rotation pole shown on Figure 4.7 could simply be an effect due to the influence of planetary deglaciation that began 18 ka and ended about 7 ka. In Peltier (1982), Peltier and Wu (1983), Peltier (1984a), and Wu and Peltier (1984), it was in fact demonstrated that both the observed rate and direction of drift were just as expected if the Earth has the viscoelastic stratification required by the previously discussed RSL and free-air gravity data, and if the only forcing to which the system has been subject is that due to a glaciation-deglaciation cycle that ended about 7 ka.

Figure 4.8 illustrates the nature of the fit to the observed polar-wander speed as a function of the viscoelastic model employed to make the prediction. Observations of oxygen-isotope composition in deep-sea sedimentary cores (e.g., Peltier, 1982) are employed to constrain the cyclic variation of planetary ice cover that has occurred over the past 10^6 yr. These data demonstrate that the major continental ice masses have appeared and disappeared in a highly periodic fashion with a time interval of 10^5 yr separating successive interglacials. Individual glaciation pulses in the sequence are each observed to have a sawtooth form with a slow glaciation period lasting about 9×10^4 yr followed by a fast collapse lasting 10^4 yr. The calculations illustrated in Figure 4.8 are based on the assumption that seven such cycles have occurred and the observed polar-wander speed of about $1°/10^6$ yr is shown at a time 6000 yr following the last 10^4-yr disintegration event. Again this choice gives a best fit of the simple glaciation history model to the oxygen isotope data. Polar-wander speed predictions are shown on this figure for six different viscoelastic models of the interior, each of which has the same lithospheric thickness $\overline{L} = 120$ km. The prediction denoted by L0F (no core) is for the homogeneous model and verifies that the predicted speed following the end of the last glaciation phase of the load cycle is essentially identically zero in accord with the theory of Munk and Revelle (1952). However, as radial structure is added to the model, the symmetry that enforces the null response in

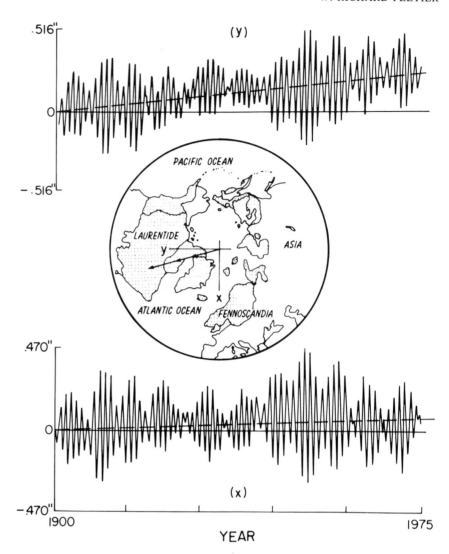

FIGURE 4.7 Position of the rotation pole relative to the CIO since 1900. See text for discussion.

the homogeneous model is broken, and the speeds predicted for times subsequent to the last glacial–deglacial pulse differ from zero. The effect of adding an inviscid high-density core to the model is illustrated by the calculation denoted L0F for which the elastic structure of the mantle is taken to be the average of model 1066B, and the mantle viscosity is assumed to have the value 10^{21} Pa s. Model L1F includes, in addition, the influence of the density jump at 670 km depth in the Earth based on the assumption that this discontinuity is capable of inducing a buoyant restoring force when it is displaced from equilibrium by the applied surface loads. The effect of this internal buoyancy in the mantle is to further increase the speed prediction in the model with 10^{21} Pa s uniform mantle viscosity. Adding a second density discontinuity at 420 km depth (model L2F with uniform viscosity) does not produce a significant further increase in the speed prediction, however. The final calculation illustrated on Figure

4.8, denoted L2F ($v_{LM} = 3 \times 10^{21}$ Pa s), demonstrates that the predicted speed can be reduced to the observed speed simply by elevating the viscosity of the lower mantle to the same value required by the free-air gravity and RSL data discussed previously. As discussed in Wu and Peltier (1984), this model also correctly predicts the observed direction of polar wander. These results establish that the data shown in Figure 4.7 cannot be construed as requiring any currently ongoing variation in surface load due to ice-sheet disintegration on Greenland and Antarctica. Rather, they are entirely explicable as a memory of the planet of the last deglaciation event of the current ice age, an event that was complete by about 7 ka.

A second effect of deglaciation on Earth's rotation addressed here concerns its influence on rotation rate and, therefore, on the so-called length of day (l.o.d.). Although the most important ongoing change in l.o.d. is the deceleration of the rate of rotation due to the gravitational interaction between Earth and Moon through the agency of the lunar semidiurnal tide in the oceans, it has been known for some time that at least one other source of l.o.d. variation must exist. Analysis of ancient eclipse data (e.g., Müller and Stephenson, 1975) demonstrated that if one assumes that the lunar-tidal torque has been constant over the past few thousands of years and uses this assumption to predict the times and locations in the past of total eclipses of the Sun and Moon, then as one goes further back into the past, one makes an increasingly large systematic error in the predictions, which is such as to imply the action of a nontidal acceleration of rotation that is working in opposition to the tidal effect. Lambeck (1980) summarized such estimates and concluded that the best available from such data is the number $(\dot{\omega}/\omega)_{NT} = (6.9 \pm 2.6) \times 10^{-11}$/yr. A recent confirmation of this number has been provided through analysis of 5.5 yr of laser-ranging data to the LAGEOS satellite that has delivered an estimated of $(\dot{\omega}/\omega)_{NT} = (7.1 \pm 0.6) \times 10^{-11}$/yr (Yoder *et al.*, 1983). Peltier (1982, 1983) demonstrated that this observed rate of nontidal acceleration was also predicted as an effect of the last deglaciation event of the current ice age. In this analysis, exactly the same model as that employed to fit the secular drift of the pole revealed in the ILS data was employed to predict $(\dot{\omega}/\omega)_{NT}$, and the results are shown on Figure 4.9. The three calculations illustrated

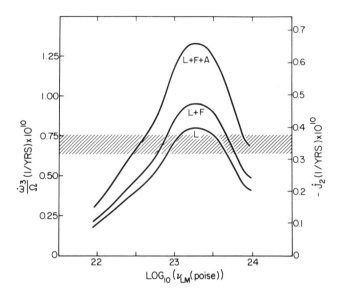

FIGURE 4.9 Observed nontidal acceleration of rotation, $(\dot{\omega}/\omega)_{NT}$, and equivalent $\dot{J}_2$ obtained from the analysis of laser-ranging data to the LAGEOS satellite. Predictions are shown as a function of lower mantle viscosity for models with 1066B elastic structure and for models that include only one ice sheet (L), two ice sheets (L + F), and three ice sheets (L + F + A). See text.

on this figure are for loading models that include only the Laurentian ice sheet (L), one that also includes the effect of the Fennoscandian ice sheet (L + F), and one that adds the effect of the West Antarctic ice sheet melting to that of the other two (L + F + A). The influence of the latter is clearly much more important as a source of forcing for rotation rate than as a source of forcing for polar wander. All speed predictions are shown as a function of lower mantle viscosity v_{LM} with the upper mantle value held fixed at 10^{21} Pa s. Again, the preferred value of v_{LM} is near 3×10^{21} Pa s. As mentioned previously, the lower-mantle viscosity could very well be associated with the transient component of the response rather than with the steady-state component that would govern the long-time scale thermal circulation. The manner in which the interpretations of both of these rotational data are influenced by small ice sheets and glaciers has been discussed in detail by Peltier (1988).

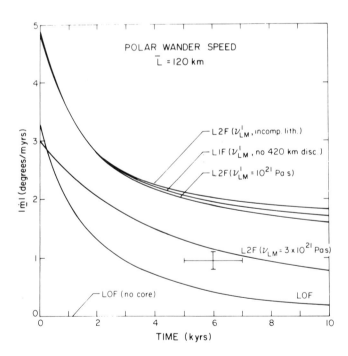

FIGURE 4.8 Predictions of polar-wander speed for several different viscoelastic models discussed in the text. The observed speed obtained from the secular rate of drift visible on Figure 4.7 is shown as the cross.

SECULAR VARIATIONS OF RELATIVE SEA LEVEL WITH GLACIAL ISOSTASY REMOVED

Given that the global model of glacial isostasy, described above, is able to explain much of the observed variability in the record of RSL change over the past 10^4 yr, it is natural to employ it to filter from the recent historical record of tide-gauge observations of secular sea-level change that component which is due to this cause.

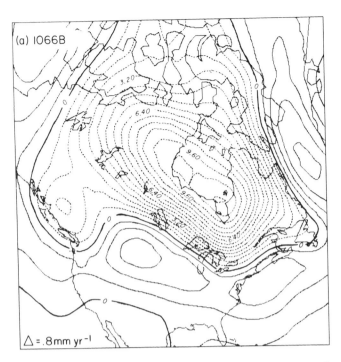

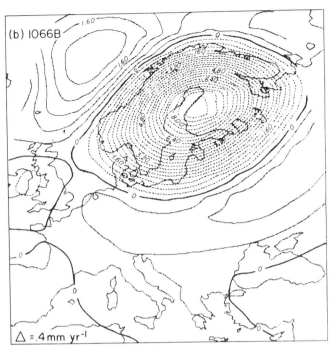

FIGURE 4.10 Continent scale maps of the predicted rate of vertical motion of the surface of the solid earth relative to the geoid using a "best" viscoelastic model with parameters fixed by fitting to the data of postglacial rebound and the ICE-2 deglaciation history shown on Figure 4.3: (a) is for North America and (b) for Europe. Dashed contours are in regions in which the land is currently rising out of the sea and solid contours are in regions that are currently submerging. Rates of vertical motion are in units of millimeters per year and the contour intervals are shown on the individual maps.

One simply employs the data in the long-time scale records of RSL history (controlled by ^{14}C dating) to constrain the viscous component of mantle rheology. One then employs this Earth structure to predict the present-day rate of sea-level rise and fall that should be observed at any location at which a tide gauge is installed, subtracts this prediction from the secular trend observed on the tide gauge, and analyzes the filtered data so produced.

As a preliminary to this procedure, it is useful to first illustrate the continent-scale variability of present-day RSL variation that is predicted by the isostatic adjustment model. Figure 4.10 shows the present-day rate of sea-level rise and fall predicted for North America and northwestern Europe using an Earth model with 1066B elastic structure. The viscous component of the model is one that has a lithospheric thickness of 200 km, an upper-mantle viscosity of 10^{21} Pa s, and a lower-mantle viscosity of 2×10^{21} Pa s. This model provides a reasonably good fit to the ^{14}C record of RSL rise along the U.S. east coast when employed in conjunction with the ICE-2 deglaciation history of Wu and Peltier (1983). The rates shown in the continental interior (where no ocean exists) represent the rates of separation between the surface of the solid earth and the geoid at such locations. (The geoid is an imaginary surface continued inland from the oceans, on which the gravitational poten-

tial has the same value as that obtained on the sea surface.) Notable on these maps is the fact, mentioned above, that the present-day maximum rates of RSL fall in the regions that were once ice covered are near 1 cm/yr. Surrounding each of the two main Northern Hemisphere centers of postglacial rebound, however, are ring-shaped regions in which RSL is predicted to be rising at rates that may be as high as 2 mm/yr, a maximum calculated along the passive continental margin of the U.S. east coast. The variation with position along the coast is fairly extreme, however, with very low calculated values obtaining both to the north and to the south of the maximum. Notable also on this map is the fact that the predicted rates of glacial isostatic submergence along the U.S. west coast are rather different from those on the east coast. The former region is much further distant from the main Laurentian ice mass and is also quite strongly influenced by the separate Cordilleran ice sheet that existed west of the Canadian Rocky Mountains. As a consequence of these effects, the predicted variations of the rate of present-day RSL change along this active continental margin are more complex.

Examination of Figure 4.10(b) illustrates a similar degree of complexity of the pattern of present-day RSL change predicted by the model for northwestern Europe. The maximum present-day rates of RSL fall near the center of uplift in the Gulf of Bothnia are again near 1 cm/yr. Surrounding this central region of uplift is the region of peripheral submergence. Again the latter region is very strongly asymmetric with respect to the former, just as in North America, a consequence of the geometric complexity of the distribution of water and land. One important feature of these results is the fact that rates of present-day sea-level rise are predicted to be much lower along the coast of France, which extends to the southwest away from the center of uplift in Fennoscandia, than those along the east coast of the United States, which is similarly located with respect to the larger Laurentian ice mass.

Perhaps the most interesting aspect of the predictions discussed above of the rates of present-day RSL variation induced by the most recent deglaciation, is that they may be compared directly to tide-gauge observations of these rates at any point on the Earth's surface that may be of interest. Figure 4.11 summarizes the comparisons. The tide-gauge observations employed to construct this figure consist of the secular trends extracted from the individual

time series of observations at each gauge over the time interval 1940 to 1980 as published in the recent catalogue of the National Ocean Service (1983). Figure 4.11(a) shows the results of the analysis for sites along the U.S. east coast. Each cross represents the secular trend at a specific gauge corrected for the secular trend expected from the influence of glacial isostatic adjustment. These corrected data are plotted as a function of the distance (in radians) of each gauge from the station at Key West, Florida, which is the southernmost station along the coast. The northernmost data point is for the gauge at Eastport, Maine. Also shown on this figure is the secular drift that has been subtracted from the raw data to make the correction for isostatic disequilibrium. This correction attains a maximum near 2 mm/yr about midway along the coast with very small rates obtaining at sites in Florida and Maine to the south and north of the maximum, respectively. To give some indication of the error in the model predicted rates, the predictions are shown not only for the present (18,000 yr after glacial maximum) but also for

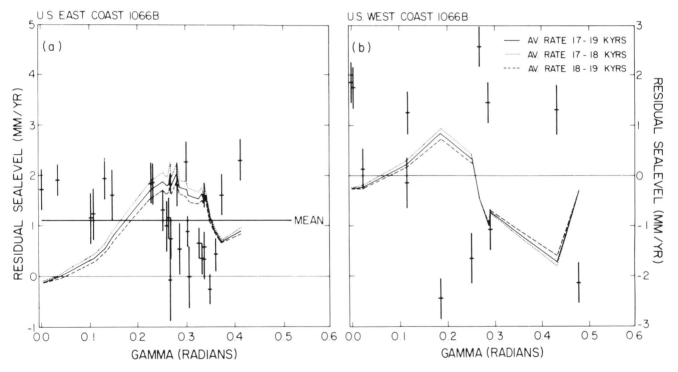

FIGURE 4.11 Tide-gauge secular trends of RSL rise along the (a) U.S. east coast and (b) west coast that have been corrected for the glacial isostatic motions shown on Figure 4.10 are shown as individual data points with the cross through the points indicating the error assigned to these data by the National Ocean Service (1983). The magnitude of the correction that was applied to the raw data is shown as a function of distance along each coast measured positive to the north of the southernmost tide gauge, which on the east coast is at Key West and on the west coast is

at San Diego. Some indication of the error that might be involved in making the correction is provided by the three estimates of the isostatic rates computed at 500-yr separation through model "present." The solid line through the corrected data on (a) is the average secular rate of rise that remains after glacial isostatic contamination has been removed. No similar average was computed for the west coast data as the scatter is so extreme as to imply that the average would be meaningless physically.

17.5 ka and 18.5 ka. Noticeable on these predicted curves are the several locations at which the predicted rates deviate from the smooth variation of rate with distance that otherwise obtains. These are sites at which errors due to the finite-element discretization occur (see Wu and Peltier, 1983, for a detailed discussion of the finite-element discretization employed to solve the sea-level equation).

The main point to note by inspection of the data shown on Figure 4.11(a) is that the correction, due to the influence of glacial isostasy, of the secular rate of RSL rise along the U.S. east coast accounts for as much as 100 percent of the observed rate of rise at some sites. That is, at some sites along the U.S. east coast, all of the observed secular rates of rise are explicable in terms of the influence of glacial isostatic adjustment. In general, however, there is a systematic misfit between the predicted and the observed rates such that the observed rate of rise exceeds that predicted by the glacial isostatic adjustment model that fits the long-time scale, ^{14}C-controlled RSL histories. At no tide gauge along the U.S. east coast is the observed rate of rise significantly slower than that predicted by the adjustment model. The average of the reduced rates of RSL rise at sites in this region is also shown on Figure 4.11(a) and is near 1.1 mm/yr. The interpretation of this average in terms of a eustatic increase of sea level (either of steric or nonsteric origin) is clearly made rather difficult by virtue of the fact that the residual trends vary erratically as a function of distance along the coast. Other physical processes must be contributing substantially to the secular variations of RSL that individual gauges are recording.

An identical treatment of the tide-gauge data from the U.S. west coast is shown on Figure 4.11(b). Here, the corrected tide-gauge-observed rates of RSL rise are shown as a function of position, measured positive, north from the southernmost gauge, which in this case is the one located at San Diego. The observations along this active continental margin are much more erratically scattered than those along the passive east coast continental margin, demonstrating the severe contamination of this record (for our present purposes) that is presumably associated with local tectonic activity. Furthermore, the corrected data are scattered about zero and show no systematic bias toward positive rates as is evident for east coast sites. It would clearly be unreasonable to employ these data to make any inference whatsoever concerning eustatic sea-level variations.

CONCLUSIONS

The analysis of RSL variations presented in the previous sections of this chapter demonstrate that effects due to the most recent deglaciation of the current ice age continue in the modern record of sea-level change even though

Würm-Wisconsin ice had disappeared from the continents by about 7 ka. In regions that were once ice covered, RSL is currently falling at a rate near 1 cm/yr due to this cause. In the immediately peripheral region of the collapsing forebulge, sea level would appear to be rising at rates in excess of 1 mm/yr due to the same process of glacial isostatic adjustment, if this were the only effect operative. The east coast of the continental United States is perhaps the best location illustrating this "drowning" effect, which is responsible for many of the unique features of its near-shore environment, including the extensive occurrence of salt marshes. A global model of glacial isostatic adjustment was described that has been employed to filter from the tide-gauge records of RSL change the influence of this effect. Although it is especially useful along the U.S. east coast, the model may also be applied to filter the global data base before these data are employed to draw conclusions concerning eustatic sea-level variations. An initial attempt to perform such a global analysis has been presented by Peltier and Tushingham (1989), whose analysis of all of the records in the data base (Permanent Service for Mean Sea Level, Bidston Observatory, Birkenhead, Mercyside, England) has led them to conclude that correcting the observed secular sea-level trends for the influence of glacial isostasy reveals a global eustatic signal with much increased coherence of strength of 2.4 ± 0.9 mm/yr. It is noteworthy that the secular drift of the rotation pole evident in the ILS pole path, which was previously interpreted by Munk and Revelle (1952) as requiring some ongoing decrease in surface ice load and thus a present-day increase in eustatic sea level, is also fully explicable as a rotational memory of the Earth of the most recent deglaciation of the current ice age, an event that ended about 7000 ka.

REFERENCES

Barnett, T. P. (1983). Possible changes in global sea level and their causes, *Climate Change 5*, 15–38.

Bloom, A. L. (1967). Pleistocene shorelines: A new test of isostasy, *Geol. Soc. Am. Bull. 78*, 1477–1493.

Carter, W. E., and D. S. Robertson (1985). Earth rotation from VLBI measurements, in *Space Geodesy and Geodynamics*, A. J. Anderson and A. Cazanave, eds., Academic Press.

Cathles, L. M. (1975). *The Viscosity of the Earth's Mantle*, Princeton University Press, Princeton, N.J.

Clark, J. A., W. E. Farrell, and W. R. Peltier (1978). Global changes in postglacial sea level: A numerical calculation, *Quat. Res. 9*, 265–287.

Dicke, R. H. (1969). Average acceleration of the Earth's rotation and the viscosity of the deep mantle, *J. Geophys. Res. 74*, 5895–5902.

Dziewonski, A. M., and D. L. Anderson (1981). Preliminary reference Earth model, *Phys. Earth Planet. Int. 25*, 297–356.

Farrell, W. E., and J. A. Clark (1976). On postglacial sea level, *Geophys. J. Roy. Astron. Soc. 46*, 647–667.

Gilbert, F., and A. M. Dziewonski (1975). An application of normal mode theory to the retrieval of structural parameters and source mechanisms from seismic spectra, *Phil. Trans. Roy. Soc. Ser. A*, 187–269.

Gornitz, V., L. Lebedeff, and J. Hansen (1982). Global sea level trend in the past century, *Science 215*, 1611–1614.

Hansen, J., D. Johnson, A. Lacis, S. Lebedeff, P. Lee, D. Reid, and G. Russell (1981). Climate impact of increasing atmospheric carbon dioxide, *Science 213*, 957–966.

Lambeck, K. (1980). *The Earth's Variable Rotation: Geophysical Causes and Consequences*, Cambridge University Press, Cambridge, England, 449 pp.

Mareschal, J.-C., and A. F. Gangi (1977). Equilibrium position of phase boundary under horizontally varying surface loads, *Geophys. J. Roy. Astron. Soc. 49*, 757–772.

Meier, M. F. (1984). Contribution of small glaciers to global sea level, *Science 226*, 1418–1421.

Müller, P. M., and F. R. Stephenson (1975). The acceleration of the Earth and Moon from early astronomical observations, in *Growth Rhythms and History of the Earth's Rotation*, G. D. Rosenberg and S. K. Runcorn, eds., John Wiley and Sons, New York, pp. 459–534.

Munk, W. H., and R. Revelle (1952). On the geophysical interpretation of irregularities in the rotation of the Earth, *Mon. Not. Roy. Astron. Soc. Geophys. Suppl. 6*, 331–347.

National Ocean Service (1983). *Sea-Level Variations for the United States 1855–1980*, U.S. Dept. of Commerce, National Oceanic and Atmospheric Administration, Rockville, Maryland.

O'Connell, R. J. (1971). Pleistocene glaciation and the viscosity of the lower mantle, *Geophys. J. Roy. Astron. Soc. 23*, 299–327.

O'Connell, R. J. (1976). The effects of mantle phase changes on postglacial rebound, *J. Geophys. Res. 81*, 971–974.

Peltier, W. R. (1974). The impulse response of a Maxwell Earth, *Rev. Geophys. Space Phys. 12*, 649–669.

Peltier, W. R. (1980). Mantle convection and viscosity, in *Physics of the Earth's Interior*, A. M. Dziewonski and E. Boschi, eds., North Holland, Amsterdam, pp. 362–431.

Peltier, W. R. (1981). Ice age geodynamics, *Ann. Rev. Earth Planet Sci. 9*, 199–225.

Peltier, W. R. (1982). Dynamics of the ice age Earth, *Adv. Geophys. 24*, 1–146.

Peltier, W. R. (1983). Constraint on deep mantle viscosity from LAGEOS acceleration data, *Nature 304*, 434–436.

Peltier, W. R. (1984a). The rheology of the planetary interior, *Rheology 28*, 665–697.

Peltier, W. R. (1984b). The thickness of the continental lithosphere, *J. Geophys. Res. 89*, 11303–11316.

Peltier, W. R. (1985a). The LAGEOS constraint on deep mantle viscosity: Results from a new normal mode method for the inversion of viscoelastic relaxation spectra, *J. Geophys. Res. 90*, 9411–9421.

Peltier, W. R. (1985b). Mantle convection and viscoelasticity, *Ann. Rev. Fluid Mech. 17*, 561–608.

Peltier, W. R. (1988). Global sea level rise and Earth rotation, *Science 240*, 895–901.

Peltier, W. R., and J. T. Andrews (1976). Glacial isostatic adjustment I: The forward problem, *Geophys. J. Roy. Astron. Soc. 46*, 605–646.

Peltier, W. R., and A. M. Tushingham (1989). Global sea level rise and the greenhouse effect: Might they be connected? *Science 244*, 806–810.

Peltier, W. R., and P. Wu (1982). Mantle phase transitions and the free air gravity anomalies over Fennoscandia and Laurentia, *Geophys. Res. Lett. 9*, 731–734.

Peltier, W. R., and P. Wu (1983). Continental lithospheric thickness and glaciation induced true polar wander, *Geophys. Res. Lett. 10*, 181–184.

Peltier, W. R., W. E. Farrell, and J. A. Clark (1978). Glacial isostasy and relative sea level: A global finite element model, *Tectonophysics 50*, 81–110.

Peltier, W. R., P. Wu, and D. A. Yuen (1981). The viscosities of the planetary mantle, in *Anelasticity in the Earth*, F. D. Stacey, A. Nicholas, and M. S. Paterson, eds., American Geophysical Union, Washington, D.C.

Russell, R. J., ed., (1961). Pacific Island Terraces: Eustatic? *Zeit. Geomorph. Suppl. 3*, 106 pp.

Sabadini, R., and W. R. Peltier (1981). Pleistocene deglaciation and the Earth's rotation: Implications for mantle viscosity, *Geophys. J. Roy. Astron. Soc. 66*, 552–578.

Vincente, R. O., and S. Yumi (1969, 1970). Co-ordinates of the pole (1899–1968), returned to the conventional international origin, *Publ. Int. Latitude Observ. Mizusawa 7*, 41–50.

Walcott, R. I. (1970). Isostatic response to loading of the crust in Canada, *Can. J. Earth Sci. 7*, 716–727.

Walcott, R. I. (1980). Rheological models and observations of glacio-isostatic rebound, in *Earth Rheology, Isostasy and Eustasy*, N.-A. Mörner, ed., John Wiley and Sons, New York, pp. 3–10.

Weertman, J. (1978). Creep laws for the mantle of the Earth, *Phil. Trans. Roy. Soc. London A288*, 9–26.

Wu, P., and W. R. Peltier (1983). Glacial isostatic adjustment and the free–air gravity anomaly as a constraint on deep mantle viscosity, *Geophys. J. Roy. Astron. Soc. 74*, 377–449.

Wu, P., and W. R. Peltier (1984). Pleistocene deglaciation and the Earth's rotation: A new analysis, *Geophys. J. Roy. Astron. Soc. 76*, 753–792.

Yoder, C. F., J. G. Williams, J. O. Dickey, B. E. Schutz, R. J. Eanes, and B. D. Tapley (1983). J_2 from LAGEOS and the nontidal acceleration of Earth rotation, *Nature 303*, 757–762.

5

Quaternary Sea-Level Change

ROBLEY K. MATTHEWS
Brown University

ABSTRACT

Attempting to understand Quaternary sea-level history provides a vigorous intellectual workout. After negotiating a long path through data and concepts of mixed quality, one finds global ice volume has fluctuated by tens of meters sea-level equivalent at rates that are difficult to resolve. Satellite monitoring of global ice volume appears only prudent.

INTRODUCTION

From a geologist's point of view, the entire history of modern civilization is written within the context of an anomalously high stand of sea level. While 10,000 yr seems like an extremely long time in the history of man, it is an extremely brief time in the history of glacio-eustatic sea-level fluctuations. While late Quaternary sea-level events revealed by the geologic record are not likely to influence short-term decision making, they do serve as an awesome reminder of the dynamic geologic context of mankind. We will not spend vast sums planning for an event that is perhaps thousands of years into the future. However, it is equally certain that we shall not simply acknowledge the imminent demise of the human race. Knowledge must come first; planning can come later.

Acquisition of knowledge in sedimentary geology is an interesting subject in itself. Historically, this field is driven by the inductive mode of investigation; more recently, the deductive mode is taking on new importance. In the in-

ductive mode, one is driven by necessity or by curiosity. The task is begun with virtually zero understanding, but with faith that some new truth will emerge if one works at it hard enough. The emphasis is on data-gathering first, followed by ad hoc explanation of the data later. Conversely, in the deductive mode one attempts to begin with a general *a priori* model and then add sparse amounts of carefully chosen new data to solve problems where the model is in obvious need of refinement.

Intellectual conflict arises when the scientist in a deductive mode balks at "explaining" vast amounts of substantially irrelevant information compiled by the scientist operating in an inductive mode. The inductive scientist does not like being told that his study area is too complicated to be fit to a general model; the deductive scientist does not wish to make his model unnecessarily complicated simply to fit every random scrap of data.

By way of example, classic inductive investigation of late Cenozoic sedimentary geology is rooted in the winter vacations of wealthy Englishmen. Nevermind that the

Mediterranean region is an extremely complicated tectonic area; it was a great place to spend the cold winter months. The local fossiliferous strata constituted a very interesting hobby. Surely if one described enough stratigraphic sequences in enough detail, some new truth would emerge. To this day, a "classic locality syndrome" remains in force. Because much is known about these strata, some workers continue to return to them to learn still more, regardless of the fact that we do not understand their regional tectonic history.

More recently, we find a lot of inductive geologic investigation driven by government edict. The geologic mapping of a nation is substantially an inductive process. One does it because the government is paying one to do it. More to the point at hand, one cannot fault a government for inductive investigation of coastal stratigraphy and neotectonic history with regard to the siting of nuclear power stations, for example. However, the fact that these inductive investigations needed to be carried out by no means ensures that the resultant data will be worthy of incorporation into *a priori* models of Quaternary sea-level history.

This chapter takes a deductive approach to the question of Quaternary sea-level change. The $\delta^{18}O$ record in deep-sea cores provides the best opportunity to obtain a continuous record of Quaternary (perhaps even Cenozoic) sea-level change. There is elegant simplicity to the statement that $\delta^{18}O$ variation in deep-sea cores reflects changes in the isotopic composition of sea water and thereby reflects changes in continental ice volume. The major focus of this chapter will be to add simple refinements to this *a priori* model. The prospects for future ice-volume variation and the prospects for a more predictive approach to stratigraphy will then be examined.

APPROPRIATE (AND INAPPROPRIATE) TECHNOLOGY

There are several technologies that are appropriate to attacking problems in the late Quaternary. These are reviewed below. Further, the inductive mode (and the need to come to some conclusion concerning a practical problem) often leads the geologist to use inappropriate technology. One's mind becomes set on gathering all the data one possibly can; one loses sight of the fact that much of the data may be seriously flawed. These situations are likewise reviewed below.

Stratigraphic Context of Study Materials

To obtain a local sea-level history on the 10^3- to 10^4-yr time scale, one must have a stratigraphy that can be related

to sea level and one must be able to date events within that stratigraphy. There are lots of good types of data to work with and there are lots of pitfalls.

Morphostratigraphic Units To work with the stratigraphic record of sea-level events, it is highly desirable to establish formal morphostratigraphic units. Coral-reef terraces and former strand-line complexes in clastic sediments are examples of morphostratigraphic units. They can be mapped as a physical topographic or bathymetric feature and exist as stratigraphic units regardless of age or age relationships among units. A reasonably formal definition of morphostratigraphic units has been followed with regard to coral-reef terrace sequences on Barbados (Bender *et al.*, 1979), New Guinea (Bloom *et al.*, 1974), and Haiti (Dodge *et al.*, 1983). To a lesser extent, a similar formalism has been applied to fossil strand-line deposits in clastic environments of the U.S. Gulf coast and east coast (see Cronin, 1983, for an excellent review).

Layer-Cake Stratigraphy In the absence of morphostratigraphic units, geologists often fall back to physical stratigraphic relationships defined in vertical profile such as a core or borehole. In some cases, solid inference can be derived from age dating within such a physical stratigraphy. A good example of this is the dating of Holocene peat deposits that immediately overlie late Pleistocene subaerial exposure surfaces (e.g., Nelson and Bray, 1970). Bad examples of this practice exist especially where geologists correlate on the basis of questionable radiometric dates rather than on the basis of physical characteristics or stratigraphic continuity.

Deep-sea cores from many portions of the world ocean sample an extremely reliable "layer-cake stratigraphy." Sedimentation rates commonly can be assumed to be nearly constant for long periods of time. Magnetic stratigraphy and $\delta^{18}O$ variation can be correlated globally (e.g., Imbrie *et al.*, 1984).

Peats Holocene brackish water peat deposits in close proximity to late Pleistocene subaerial exposure surfaces provide quite good stratigraphic indicators of paleosea level, which can be easily dated using the ^{14}C method. Importantly, it is difficult to move peats around as sea level moves up and down. The peats would simply disintegrate if moved at all. Thus, to find peat resting on former subaerial exposure surfaces provides good indication of the time at which sea level first inundated the former subaerial landscape.

Care must be taken in interpreting the sea-level significance of peat dates taken from within thick sections of peat. Time lines within peat accumulation may be hard to define. Correction of present stratigraphic elevation of the sample with regard to compaction of peat below may prove difficult. In the discussion that follows, only peat samples from close proximity to subaerial exposure surfaces below are used to establish sea-level history.

Mollusc Shells Mollusc shells from dredge hauls are by far the most abused "datable material." If mollusc samples are to have stratigraphic integrity, one must be able to demonstrate that the shells have not been transported by rising sea level, subsequent current activity, seagulls, or man. Most commonly, this means recovering articulated pelecypod valves (e.g., Curray, 1960). Numerous dates in the literature undoubtedly represent shells that were transported landward by rising sea level (see MacIntyre *et al.*, 1978, for a good discussion). Still other mollusc shells dredged from bathymetric lineations are species that are known to live in tens of meters water depth. Thus, old shells are washed upwards to shallower depth than where they actually grew and younger shells represent deep subtidal environments where offshore molluscs thrive.

A further problem, rampant in the ^{14}C dating of mollusc shell, is the exposure of sea-bottom samples to Holocene diagenetic processes, such as boring and infilling by microorganisms. A small amount of contamination by Holocene calcium carbonate can make a very old mollusc sample look like 20,000- or 30,000-yr-old material when dated by [14]C.

Corals Reef-crest corals (*A. palmata* in the Atlantic) can be outstanding indicators of paleosea level. Such corals live within a few meters of sea level and are not subject to extensive transport. They provide excellent material for ^{14}C dating or for [230]Th dating (e.g., Lighty *et al.*, 1978, 1982; Mesolella *et al.*, 1969). Virtually all corals other than the reef-crest species have the obnoxious habit of living either in the shallow back-reef or the deep fore-reef environment. Thus, while they do contain datable calcium carbonate, their relationship to past sea level must be inferred from paleoenvironmental reconstruction. Clearly, the coral did not live above sea level; but beyond that, things get complicated.

"Beach Rock" Some of those who work on submerged late Pleistocene and Holocene deposits take cementation of sandy sediments to be an indication of deposition and cementation in the beach environment. None of these studies offer sufficient documentation of the beach environment as the source of cementation in these materials. Submarine cementation is equally capable of producing a lithified sandstone (e.g., Land and Moore, 1980). The claim that "beach rock" identifies a former intertidal environment must be held to close scrutiny. Further, all of the datable materials to be found in such sediments are subject to the problems noted above.

Geochronology/Chronostratigraphy

Given good stratigraphic context of study materials, there are several dating methods available for the study of Quaternary sea-level change.

[14]C Dating Back to the last glacial maximum (approximately 18,000 yr before present (18 kaBP)), the geologist can virtually "date at will" with regard to ice margin advance on the land, with regard to sea-level fluctuations at the shoreline and with regard to downcore stratigraphy in deep-sea cores. In the best of worlds, one would hope that all the material ^{14}C dated would have met selection criteria noted above. In fact, this is not the case, and much of the existing data must simply be regarded as irrelevant to the questions under consideration.

Uranium-Series Dating [230]Th and [231]Pa dating of aragonite corals constitutes the backbone of late Pleistocene reef-terrace geochronology. Unfortunately, a large amount of literature provides only one or two dates per terrace. When one takes the time to make multiple analyses on multiple samples from the same terrace, one learns that there is a considerable spread to these data. Presumably, part of this spread is due to the complicated nature of the analysis (e.g., Harmon *et al.*, 1979). Undoubtedly, some of the spread is due to slight diagenetic alteration of sample in the subaerial environment (Bender *et al.*, 1979). Even more outrageous numbers are obtained when one tries to date materials from within the present-day phreatic lense. Uranium mobility is rampant, and such dates have no geochronologic significance.

The summary data of Harmon *et al.* (1979) with regard to three particular samples exemplify the problem faced with regard to geochronologic significance of [230]Th dates within the time interval of the last interglacial. On the basis of impeccable geologic arguments, Rendezvous Hill (Barbados III) terrace, the Curacao +6 m terrace, and the Key Largo limestone of South Florida all represent deep-sea isotope stage 5e (event 5.5 in the SPECMAP terminology; taken to be centered around 122 kaBP; Imbrie *et al.*, 1984). However, on the basis of 13 analyses of each sample, the [230]Th coral-dating community came up with a pooled age estimate of 118 ± 9, 124 ± 12, and 139 ± 19 kaBP for these samples, respectively (see Harmon *et al.*, 1979).

Two points are worthy of note. First, there appear to be systematic differences among these three samples as reflected by the respective means. Further, Barbados is significantly uplifted whereas Curacao and Key Largo limestone samples occur at an elevation presumed to represent the actual high stand of the sea. If anything, the Barbados sample should appear somewhat older than the other two because an uplifting island would experience a null point with rising sea level earlier in the history of sea-level rise and subsequent high stand of the sea. In fact, the reverse is true; the Barbados date is the youngest of the three examples. Very likely, some small amount of postde-positional alteration has affected at least two of three of these samples. Indeed, x-ray data indicate considerable calcite in the Key Largo sample; but the other samples appear quite satisfactory.

Next, consider precision of these age determinations. Even with 13 analyses of aliquots of the same sample, precision (1σ) remains in the range of 10,000 yr. This is quite large in comparison with the chronostratigraphic precision attainable from deep-sea cores by stacking and spectral tuning of the $\delta^{18}O$ record (discussed below).

Thus, even under the best of circumstances—uranium-series dating in the range of 125 kaBP (an age range extremely important in the geologic record)—uranium-series dating affords precision and accuracy that are mar-ginal in comparison with the chronology available through the use of deep-sea cores. Where a sufficient number of analyses are available on individual terraces, the dates may serve as a crude constraint on correlations to deep-sea cores. However, discussion of whether "the 120-kaBP terrace" of one sequence is the same or a different glacio-eustatic event from "the 130-kaBP terrace" in another sequence must rest on geologic arguments, not on ura-nium-series dating.

Further, one simply cannot trust dates on sparse samples from a single or a few terraces. Such reports are nothing more than a potentially interesting lead suggesting that more work might be desirable. Finally, one simply should not trust uranium-series dates on materials that were col-lected from within the modern freshwater phreatic lense or that can be demonstrated to have resided in a paleophreatic lense at some time in the history of the sample.

Magnetic Stratigraphy Identification of the Brunhes/ Matuyama magnetic boundary in deep-sea cores is funda-mental for late Quaternary chronostratigraphic studies. The independent chronology provided by identification of the Brunhes/Matuyama boundary and the continuous nature of the deep-sea record are the prime reasons for attempting to correlate terrace and shoreline stratigraphy to the deep-sea record. One will never get the simple stratigraphic continuity of deep-sea cores in the complicated nearshore environment. Thus, it is important to develop a deductive research strategy in which the deep-sea record and chro-nology are the *a priori* model that one takes to the nearshore environment.

The SPECMAP Time Scale The SPECMAP group has generated a continuous deep-sea chronology for the last

780 ka, which claims a precision on the order of 2 ka (95 percent confidence level) for major events in the deep-sea $\delta^{18}O$ ice-volume record. The technique utilizes a formal taxonomy for recognizable events in the $\delta^{18}O$ stratigraphy and utilizes orbital tuning to make local adjustments to sedimentation rate (Imbrie *et al.*, 1984; Prell *et al.*, 1986).

The fact that numerous individual records can be incor-porated into the final grand chronology promises continu-ing improvement of both chronology and the resultant $\delta^{18}O$ composite estimate of the glacio-eustatic signal. Further, application of the technology to the entire Ceno-zoic is technologically feasible. In my opinion, this is the chronostratigraphy that shall prevail.

The $\delta^{18}O$ Ice Volume/Temperature Relationship

Calcium carbonate precipitated from sea water records some combination of information concerning the isotopic composition of the water and the temperature at which the calcium carbonate grew. Sizable variation in the isotopic composition of the sea water results from variation in global ice volume. If one can constrain global ice volume, one can read local temperature. Conversely, if one can con-strain local temperature, one can read global ice volume.

This technology is applicable to planktonic foraminif-ers, benthic foraminifers, corals, and molluscs. In order to obtain useful results, one must further constrain strati-graphic variability by working with single taxa. Time-series data based on "mixed assemblages" of planktonic foraminifers, for example, may reflect variation in propor-tions of deep dwelling and shallow dwelling taxa with time. Similarly, there is considerable variation among benthic foraminifer taxa (Graham *et al.*, 1981). Records based on "mixed benthics" or on several different benthics strung together in time series must be considered suspect. Similarly, some taxa can be shown to have considerable variation in isotopic composition as function of size of the specimen (Curry and Matthews, 1981). Whether this re-flects seasonal variation or variation in habitat among juveniles and adults remains to be unraveled. Finally, variation in degree of dissolution of deep-sea sediments as a function of proximity to the lysocline can vary the iso-topic composition of individual taxa by as much as 0.5 per mil (Peterson and Prell, 1985).

The above-cited difficulties notwithstanding, many deep-sea cores yield a remarkably similar global $\delta^{18}O$ signal. This signal, taken in conjunction with magnetic stratigra-phy, provides the data base for stacking and spectral tun-ing, which produces a remarkably precise *a priori* model of the global $\delta^{18}O$ signal (Imbrie *et al.*, 1984; Prell *et al.*, 1986). The isotope-stage terminology of Imbrie *et al.* (1984) is adopted here. Whether this signal should be interpreted as a global ice-volume signal or a temperature signal remains a main point for discussion below.

$\delta^{18}O$ and $\delta^{13}C$ Relationships to Subaerial Exposure Surfaces

Stable isotope data also provide incontrovertible identification of subaerial exposure surfaces within stratigraphic sequences of bank margin carbonates (Allan and Matthews, 1982). Recrystallization of unstable aragonite and high-magnesium calcite in proximity to soil gas imparts a highly negative $\delta^{13}C$ signal to sediments immediately below subaerial exposure surfaces. Similarly, recrystallization of various unconformity-bounded packages of carbonate sediment generally occurs under slightly different conditions of global ice volume and/or climate. Thus, unconformity-bounded packages of carbonate sediment commonly display slightly different $\delta^{18}O$ values from those sediments above and below.

A detailed stratigraphy of sedimentation at or near sea level, followed by subaerial exposure, makes the bank margin carbonate environment a veritable gold mine of information concerning glacio-eustacy (e.g., Major and Matthews, 1983).

THE 18 kaBP TO PRESENT TIME INTERVAL

This time interval encompasses the transition from the last glacial maximum to the present and is within the range of convenient ^{14}C dating.

Latest Pleistocene Maximum Continental Ice Volume

The 18-kaBP perimeter of continental ice sheets is reasonably well known. Calculations based on this perimeter and assumptions concerning ice mechanics of equilibrium glaciers allow estimation of the volume of ice tied up on land during times of maximum glaciation (Denton and Hughes, 1981). Estimates for ice-volume sea-level equivalent tied up in equilibrium ice sheets range from as high as 163 m to as low as 102 m. Importantly, all of these calculations presume the ice sheets were at equilibrium. It is possible for ice sheets to exist for long periods of time at less than these theoretical equilibrium ice volumes. Thus, the various estimates constitute formal statements of the maximum ice volume given different assumptions. It is important that confirmation be sought elsewhere for the amount of ice tied up in continental glaciers.

Lowest, Low Sea-Level Estimates for the 18-kaBP Shoreline

When we seek to confirm maximum sea-level lowering caused by the maximum extent of the 18-kaBP continental ice sheets, we find few high-quality data points below 40 m. Two data points from the Texas Gulf Coast at –57 m and 12,900 ± 400 yrBP for paired valves of the brackish water bay clam, *Rangia cuneata,* tell a very clear story (Curray, 1960). Similarly, a freshwater peat off southern New England at –56.5 m and 12,320 ± 350 yrBP appears trustworthy (Oldale and O'Hara, 1980).

Thus, we are left with a high-quality number for maximum sea-level lowering at 18 kaBP that is both small and young in comparison with the numbers suggested by continental ice-volume calculations cited above. There is a lot of bathymetry suggestive of a sea stand at around –120 m (e.g., Curray, 1960), but it is poorly dated. These features certainly look like good candidates for 18-kaBP shorelines, but this common assertion has not been satisfactorily proven.

We definitely want to know maximum sea-level lowering during the time of maximum ice volume on the continents. This question will be examined below with regard to the transition from isotope stage 6 to isotope stage 5.5 (5e).

Submergence Curves for 12 kaBP to Present

High-quality data are more readily available for the shallow end of the sea-level rise curve. Key papers (Gould and McFarland, 1959; Curray, 1960; Coleman and Smith, 1964; Scholl, 1964; Redfield, 1967; Scholl and Stuiver, 1967; Schnable and Goodell, 1968; Baltzer, 1970; Nelson and Bray, 1970; Thom and Chappell, 1975; Hopley et al., 1978; Lighty et al., 1978; Field et al., 1979) have been reviewed with regard to criteria described above under stratigraphic context of sedimentary materials. With regard to Bermuda, 12 data points meet these criteria and range back to –21 m at 9100 yrBP; Florida, 22 data points ranging back to –27 m at 9400 yrBP; Texas/Louisiana Gulf coast, 15 data points ranging back to –58 m at 13 kaBP; and South Pacific (Australia and New Caledonia), 11 data points ranging back to –37 m at 9700 yrBP. All of these curves show sea level to be within less than 5 m of present sea level by approximately 5000 yrBP. Differences among these curves in this time range are taken to reflect continuing isostatic adjustment to unloading of late Pleistocene glacial ice (e.g., Peltier, Chapter 4, this volume).

Calibration of Deep-Sea $\delta^{18}O$ Signal as Continental Ice Volume

The deep-sea $\delta^{18}O$ record from 18,000 yr to the present is well known and well studied (see Mix and Ruddiman, 1985, for examples). The amplitude of the planktonic signal in high-sedimentation-rate cores averages 1.7 ± 0.1 per mil. Inasmuch as there is no well-dated confirmation of the 18-kaBP shoreline (discussed above), estimates of calibration between the deep-sea $\delta^{18}O$ record and the shoreline submergence record must be based upon ^{14}C correlation of the few good shoreline data (cited above) with the high-quality deep-sea isotopic stacked record of Mix and Ruddiman (1985, Table 4). The four independent estimates thus resulting yield a value of 0.025 per mil per meter for the calibration between deep-sea $\delta^{18}O$ fluctuation and shoreline fluctuation during the Holocene sea-level rise. If this change in $\delta^{18}O$ value as a function of changing sea level were entirely an ice-volume effect, it would require that glacial meltwater be returned to a well-mixed ocean with an isotopic composition on the order of −95 per mil. This is a seemingly outrageously negative number; glaciers today range from −20 to −50 per mil. To escape this dilemma, we might also pursue the possibility that (1) there is systematic offset in the two age models; (2) the shape of the geoid has changed, thus affecting the observed depth difference between paleoshorelines and present sea level; (3) sea-surface warming explains part of the isotopic curve from 13 kaBP to the present; or (4) that surface water of the world ocean (or specifically, the tropical Atlantic study area of Mix and Ruddiman) was contaminated by unusually negative precipitation at the time of glacial melting.

THE 180- TO 18-kaBP TIME INTERVAL

Interest in this time interval stems from two facts; one technological and one scientific. This is the time interval for application of the uranium-series dating techniques. This time interval also encompasses the second to last major glaciation (isotope stage 6) and the last interglacial (isotope stage 5.5, formerly 5e). In examining this time interval, we will pick up further information concerning maximum sea-level fluctuation from glacial to interglacial conditions and we will pick up important calibration concerning the relative importance of ice-volume variation and temperature variation in the observed $\delta^{18}O$ curve from deep-sea cores.

Coral-Reef Terrace Sequences

Absolutely essential to the calibration of the deep-sea $\delta^{18}O$ record as an ice-volume signal is the stratigraphy and geochronology of coral-reef terraces on tectonically emerging islands. On the basis of the record within the range of ^{14}C dating, it appears that only the $\delta^{18}O$ signal in deep-sea cores covaries with the sea-level signal from submerged shorelines. Within the time interval from 180 to 70 kaBP, a number of coral-reef terraces are observed that are in substantial agreement with a well-defined $\delta^{18}O$ signal in deep-sea cores.

Well-Dated Morphostratigraphic Units At least three morphostratigraphic sequences have been relatively well dated using uranium-series methods. These are the uplifted coral-reef terraces of Barbados (Mesolella *et al.*, 1969; Bender *et al.*, 1979), New Guinea (Bloom *et al.*, 1974; Aharon, 1983), and Haiti (Dodge *et al.*, 1983). In each of these areas, distinct morphostratigraphic units can be recognized, which correspond to the Barbados terraces—Rendezvous Hill, Ventnor, and Worthing (approximately 125, 105, and 82 kaBP, respectively). In each of these study areas, the morphostratigraphic units stand independent of radiometric dating. Additionally, the New Guinea terrace sequence contains several younger terraces that are not represented on Barbados or Haiti because the uplift rate of Barbados and Haiti is considerably less than the uplift rate of the New Guinea terrace sequence.

The Constant Uplift Rate Hypothesis Historically, a great deal of literature concerning relative sea level within the time interval 140 to 70 kaBP hinges on application of the "constant uplift rate hypothesis" (Broecker *et al.*, 1968) to Barbados, New Guinea, Haiti, and elsewhere. The "125-kaBP terrace" has been considered to represent a +6-m high stand of the sea. With this terrace as the "known" datum, uplift rates for the various regions were calculated. Most such calculations indicate that sea level at 105 and 82 kaBP was considerably below present sea level. Matthews (1973) showed that relative sea-level estimates can be made independent of the 125-kaBP sea-level assumption where terrace sequences with differing (but constant) uplift rates can be combined into a calculation.

Unfortunately, the constant uplift rate hypothesis is substantially ad hoc. With the work of Stockmal (1983), this assumption seems reasonable for a relatively simple tectonic setting such as Barbados, but this assumption is not easily justified for such complex areas as New Guinea or Haiti. Here, scientific strategy bifurcates.

Historically, I have argued to simply consider tectonic uplift a convenient mechanism for delivering coral-reef terraces into the subaerial environment where they may be easily sampled for geochronologic and isotopic studies. If we had to rely on submarine observations for our knowledge concerning sea-level stands at 105 and 82 kaBP, we would still be groping around in the dark ages of Pleistocene geochronology. We should be thankful for the uplifting of these terraces; but should not rely on the "constant uplift hypothesis" to tell us anything about relative sea level.

Alternatively, some scientists argue that sea-level estimations based on the constant uplift hypothesis constitute an independent check on the isotopic estimates. In recent years, this approach has taken on still greater appeal because of possible uncertainty in isotopic sea-level estimates introduced by the possible existence of thickened

Arctic ice pack (e.g., Broecker, 1975; Williams *et al.*, 1981). In deference to this latter uncertainty, I here give serious consideration to sea-level estimates based on the constant uplift hypothesis. Note, however, that this hypothesis remains substantially ad hoc and can be considered valid at this time only on an empirical, statistical basis.

Dodge *et al.* (1983) provided a detailed comparison between sea-level estimates derived from terrace elevation data in tectonically active areas and the isotopic record from deep-sea benthic foraminifers. They concluded that the deep-sea benthic $\delta^{18}O$ record is two-thirds a bottom-water temperature signal and only one-third a $\delta^{18}O$ ice-volume signal.

There are two things wrong with this conclusion. First, it is misleading and erroneous to compare sea-level information to the benthic $\delta^{18}O$ record (e.g., Prentice and Matthews, 1988). Second, the isotope data of Fairbanks and Matthews (1978) for Barbados terraces (Worthing, Ventnor, and Rendezvous Hill) are slightly inconsistent with data on new samples collected specifically to test the result concerning paleosea-level elevation indicated for Barbados terraces Worthing and Ventnor. These two problems are examined in more detail below.

The $\delta^{18}O$ Record from Coral-Reef Terraces and Deep-Sea Cores

Numerous deep-sea cores reveal a remarkably consistent $\delta^{18}O$ time series from planktonic and benthic foraminifers within this time interval. With the approximate chronology of the deep-sea record established from constant sedimentation rate assumptions and the Brunhes/Matuyama boundary, and with approximate chronology of coral-reef terraces established from uranium-series dating, it is possible to undertake comparison of the $\delta^{18}O$ record from deep-sea cores with the $\delta^{18}O$ record from coral-reef terraces.

Such an effort has been quite successful on Barbados (Fairbanks and Matthews, 1978) and on New Guinea (Aharon, 1983). Both attempts yield similar results; the Barbados record is stressed here because it contains more information concerning maximum glacio-eustatic sea-level fluctuation in the time interval of 170 to 130 kaBP. After discussion of the coral-reef terrace $\delta^{18}O$ record, I will return to integration of these data with the deep-sea $\delta^{18}O$ record and with sea-level estimates based on the coral-reef terrace constant-uplift hypothesis discussed above.

The Barbados Coral-Reef Terrace $\delta^{18}O$ Record Figure 5.1 presents a cross section of the Christ Church region of Barbados indicating the location of numerous cored boreholes and the position of a subaerial exposure surface encountered in these boreholes. Oxygen isotope data have been obtained on numerous samples of reef-crest coral (*A. palmata*) from terraces and from borehole materials above and below the prominent subaerial exposure surface indicated in Figure 5.1. Figure 5.2 presents a plot of $\delta^{18}O$ data versus actual present-day elevation of the *A. palmata* samples.

We presume that uplift of the island of Barbados is relatively small in comparison to sea-level fluctuations that have occurred within the time interval under consideration. Thus, uplift rate, if present, may be considered as "noise" with reversed sign for rising versus falling sea level. Note that the slope of the $\delta^{18}O$ versus elevation relationship for Kendal Hill to borehole 20 (regressive) is quite similar to the slope for the borehole 20-Rendezvous Hill transgressive relationship. The data of Wagner (1983) confirm this relationship of $\delta^{18}O$ versus elevation for recrystallization products of phreatic lenses related to various late Pleistocene stands of sea level.

Figure 5.3 plots $\delta^{18}O$ variation versus age for Barbados *A. palmata* data and for data on planktonic foraminifers from deep-sea cores. The general agreement between these two records is unquestionable; there remains only some small points concerning detail. These are dealt with below.

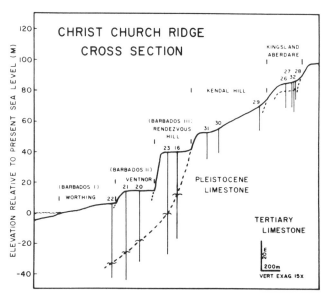

FIGURE 5.1 Cross section of the southern flank of Christ Church Ridge, Barbados, indicating coral-reef terrace relationships, location of cored boreholes, and the existence of a buried caliche profile (dashed line) that records subaerial exposure during deep-sea isotope stage 6. Samples of the surf zone coral (*A. palmata*) were taken from immediately beneath the subaerial exposure surface in RKM 20, 22, and 23 and from immediately above the subaerial exposure surface in RKM 20. Isotopic data for these materials are plotted against elevation in Figure 5.2. (From Fairbanks and Matthews, 1978, with modification.)

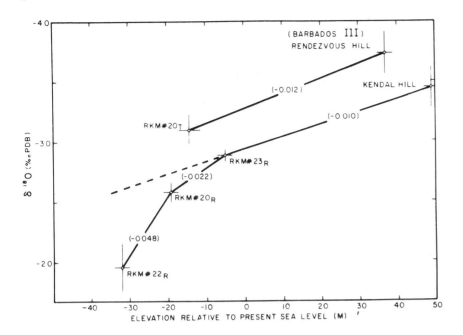

FIGURE 5.2 Plot of mean isotopic composition of Barbados *A. palmata* samples against elevation relative to present sea level. Assuming the rate of tectonic uplift to be small in relation to the rate of glacio-eustatic change, these data constitute a calibration of $\delta^{18}O$ as a function of glacio-eustatic sea-level fluctuation. A relationship of 0.011 per mil per meter is taken as the glacio-eustatic effect. More positive $\delta^{18}O$ values at RK 20 and 22 are satisfied by CLIMAP estimates that this region was approximately 2°C cooler than modern during full interglacial conditions. (After Fairbanks and Matthews, 1978.)

Frozen Bottom-Water Hypothesis The use of the benthic $\delta^{18}O$ record as a possible indicator of ice-volume fluctuations is rooted in the clever argument of Shackleton (1967). At a time when Emiliani (1966a, for example) was talking about a 4° to 6°C colder tropical sea-surface temperature (SST) during glacial times, and small ice-volume component to the $\delta^{18}O$ signal, Shackleton took the trouble to analyze relatively scarce benthic foraminifers. The result was astounding. The benthic $\delta^{18}O$ curve was similar in shape and amplitude to the planktonic $\delta^{18}O$ curve of Emiliani. Bottom water could not possibly have gotten 4° to 6°C colder during glacial conditions because this would put the water below the freezing point. The logical conclusion was that Emiliani had greatly overestimated the temperature effect and greatly underestimated the ice-volume effect in both of these signals. Shackleton proposed that the benthic $\delta^{18}O$ signal was more like two-thirds ice-volume signal and one-third temperature signal, although the precise relation could not be determined from the data at hand.

Importantly, there is no *a priori* reason to assume constant bottom water temperature during the glacial/interglacial cycle. *A priori*, the bottom water shall not freeze, but that is about it. Beyond this, there are no "magic numbers" for bottom-water production. Bottom water is simply that combination of temperature and salinity that is the most dense water. With glacial/interglacial changes in geometry of high-latitude marginal seas, all bets are off concerning precisely what temperature/salinity combination will end up being the glacial world's bottom water. Thus, the initial success of Shackleton (1967) must be viewed as substantially empirical and semiquantitative. If

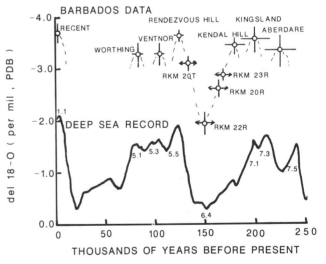

FIGURE 5.3 Comparison of Barbados coral isotope record with the stacked deep-sea planktonic isotope curve of Prell *et al.* (1986). Error bars on isotopic data indicate statistical certainty about the mean at the 95 percent confidence level. Coral isotope data for Worthing, Ventnor, and Rendezvous Hill are from Table 5.2; the remainder are from Fairbanks and Matthews (1978). Note peak-for-peak correlation; Aberdare equals 7.5, and so forward to Worthing equals 5.1. (Similar to Fairbanks and Matthews, 1978, Figure 7; modified to include new data and a more up-to-date deep-sea planktonic $\delta^{18}O$ record.)

one wishes to constrain temperature to be constant, one must hook up to a $\delta^{18}O$ recording system other than benthic foraminifers. There are *a priori* arguments to suggest that the low-latitude sea surface may provide such a recording system.

Low-Latitude Constant Sea-Surface Temperature Hypothesis Sea-surface temperature estimates based on total faunal census data and core top regression equations (CLIMAP, 1976, 1984) provided the (seemingly) final death blow to the Emiliani concept of a substantially cooler global SST during glacial times. Application of this technology to the 18-kaBP maximum glacial reconstruction indicates that some regions of the low-latitude ocean were slightly warmer at 18-kaBP maximum glacial time than they are today. Other large areas were only slightly cooler. Only where major shifts in upwelling occurred are low-latitude SST greatly cooler at 18 kaBP than today. Prell (1985) applied alternative methods and basically confirmed the CLIMAP (1984) results, suggesting, if anything, still *less* tropical cooling than CLIMAP.

Further, there are physical climatologic/oceanographic reasons to predict relatively constant SST of low-latitude ocean surface. The balance between radiation and latent-heat flux dictates a tropical SST of about 28°C (Donn and Shaw, 1977; Newell *et al.*, 1978; among others). Although mechanisms can be envisioned that might cause this number to increase slightly (increased atmospheric CO_2, for example), such mechanisms generate negative feedback (increased frequency of hurricanes, for example) that is poorly understood. At the present state of the art, calculations of atmospheric circulation arrive at 28°C tropical SST; this temperature is in generally good agreement with the CLIMAP empirical estimates for 18 kaBP. In light of existing data and experience, it would be difficult to justify an hypothesis of runaway warmer tropical SST.

The possibility of cooler tropical SST is a different matter. Variation in planetary albedo, atmospheric CO_2, or perhaps even solar insolation could result in a generally cooler planet and with it generally cooler tropical SST. Indeed, Manabe and Broccoli (1985) simulated a glacial world with higher albedo and lower atmospheric CO_2 and thereby calculated a tropical SST on the order of 1° to 2°C cooler than modern. Importantly, this model result is opposed by the empirical temperature estimates of CLIMAP (1984) and Prell (1985) discussed above.

Regardless of the ultimate resolution of this discrepancy between model and empirical results, all global atmospheric circulation model experiments to date suggest that a colder SST in the overall tropical regions would be accompanied by still yet colder polar regions. While this relationship is generally proposed by most working on late Pleistocene materials, the relationship is grossly violated by many working on materials within Tertiary time intervals.

The validity of the constant tropical SST hypothesis is an ongoing research problem. There is room for change. However, tropical SST is a profoundly important number with truly global ramifications. Change should not be proposed lightly. If at some time in the future modeling results move forward to still higher SST under certain conditions, these deviations from constant temperature hypothesis will be based on *a priori* models. Likewise, if the disagreement between model and empirical data concerning 18-kaBP SST shall be resolved in favor of cooler temperatures, these estimates of cooler temperatures will likewise be based on *a priori* models and thoroughly well-reasoned discussion. These new numbers would then become the basis for ice-volume calculations, and so on toward a perfect understanding of Earth history.

With regard to isotopic estimation of sea-level history, the point here is that the low-latitude tropical sea-surface environment offers the best opportunity of constraining temperature variation and thereby reading out the glacio-eustatic ice-volume signal. This water mass is sampled by reef-crest corals and by surface-dwelling planktonic foraminifers. Indeed, if Dodge *et al.* (1983) had chosen to compare their terrace elevation data to the planktonic $\delta^{18}O$ record instead of the benthic $\delta^{18}O$ record, they would have arrived at a much more favorable comparison (a point to which I shall return below).

Maximum Glacio-Eustatic Sea-Level Lowering at Isotope Stage 6 The data presented in Figures 5.2 and 5.3 afford an estimation of the amplitude of the glacio-eustatic signal in late Pleistocene time. The deep-sea $\delta^{18}O$ record clearly demonstrates that isotope stage 2 and isotope stage 6 are comparable examples of maximum glaciation during late Pleistocene time. The Barbados borehole data capture 1.7 per mil $\delta^{18}O$ variation within an elevation variation of approximately 80 m. Allowing for the fact that the Kendal Hill $\delta^{18}O$ value is somewhat heavier than Recent and allowing for local faulting (Wagner, 1983), the amplitude of sea-level variation from full interglacial to data point "RKM 22 (regressive)" could be estimated at 108 m. Making further allowance for the fact that local $\delta^{18}O$ amplitude in nearby deep-sea cores is approximately 2.0 per mil, one has an additional 0.3 per mil to play with as either local temperature signal or global ice-volume sea-level signal.

Note in Figure 5.2 that the isotope versus elevation variation increases dramatically below the data point "RKM 20 (regressive)." This reflects a still more negative $\delta^{18}O$ value for late-stage continental ice, or a larger proportion of floating ice, or a local temperature decline during extreme glaciated conditions. Assuming the effect from data points "RKM 20 (regressive)" to "RKM 22 (regressive)" to be solely an isotopic effect, the full 2.0 per mil amplitude reflects 114-m sea-level lowering relative to present. Alternatively, CLIMAP-18 kaBP ΔT estimates for this area are approximately nil for summer and approximately –2°C for winter. Assumption of a –2°C temperature difference from interglacial to glacial conditions and a 120-m sea-level difference from present to maximum glacial conditions satisfies the isotopic maximum amplitude data in this area. Assuming the 108-m amplitude figure cited above for actual Barbados data with some straightforward corrections, a temperature difference of –3.3°C from interglacial to glacial conditions is required. All three of these estimates are probably within the uncertainty of the data.

Comparison of Isotope Stage 5 Data for Terraces and Deep-Sea Cores As noted above, Dodge *et al.* (1983) erroneously compared terrace elevation data to the benthic rather than the planktonic δ18O record. A second source of error in the Dodge *et al.* (1983) calculation concerns the isotopic data reported by Fairbanks and Matthews (1978) for Barbados terraces Worthing, Ventnor, and Rendezvous Hill. The isotopic data contained in Fairbanks and Matthews (1978) for Barbados surface samples were taken on scraps of sample left over from radiometric dating and petrographic studies. In many cases, these were relatively small samples. Relatively small samples of *A. palmata* present a special problem in that there is strong seasonal banding within this coral. Given a small sample and nondescript geometry of banding, it would be easy to take a nonrepresentative sample for oxygen isotope analysis.

To evaluate this problem, an entirely new set of diamond-drilled, 2-inch-diameter core samples were collected expressly for isotopic study. The middle columns of Table 5.1 present Fairbanks and Matthews data and new data (from Table 5.2) concerning the isotopic difference among Barbados terraces Rendezvous Hill (isotope stage 5.5) as compared to Worthing (5.1) and Ventnor (5.3). The left-hand columns present similar information from Dodge *et al.* (1983) estimates of elevation differences converted to δ18O values by the calibration of Fairbanks and Matthews (1978). The right-hand column presents observed isotopic differences among planktonic foraminifers based on the average of low-latitude deep-sea cores.

With regard to Barbados isotope data, note that the new data indicate slightly less difference between the terraces. At a glance, the difference between the Fairbanks and Matthews (1978) data and the new data is not that exciting. Both data sets put the younger terraces low relative to Rendezvous Hill; the differences among means are within the range of overlapping confidence intervals; in 1978, the situation looked well under control. However, over the years there has been harping criticism concerning the Fairbanks and Matthews estimate that sea level represented by the younger Barbados terraces might have been as much as 45 m below present sea level. To me, the situation was "close enough; move on to other things." To those inclined to be more fastidious, the new data will be somewhat reassuring.

When one compares the coral-reef terrace elevation data of Dodge (left-hand column of Table 5.1) with the Barbados terrace isotope data (middle columns) and with the deep-sea record (right-hand column), one notes remarkable agreement between the new Barbados isotope data and the deep-sea cores. Estimates based on the ad hoc

TABLE 5.1 Comparison of Terrace Elevation Data to Isotope Data from Corals and Low-Latitude Planktonic Foraminifers

Isotope Stage Comparison	Terrace Elevation Data, δ18O equivalent,[a] δ18O (per mil)	Barbados Corals Fairbanks and Matthews (1978), δ18O (per mil)	Low-Latitude New Data,[b] δ18O (per mil)	Planktonic Foraminifer,[c] δ18O (per mil)
5.1-5.5	+0.24 ± 0.1 (3)	+0.54 ± 0.4 (7)	+0.35 ± 0.2 (9)	+0.34 ± 0.2 (5)
5.3-5.5	+0.21 ± 0.1 (3)	+0.52 ± 0.1 (15)	+0.35 ± 0.2 (9)	+0.30 ± 0.1 (5)

[a]Elevation data from Barbados, New Guinea, and Haiti converted to δ18O via Fairbanks and Matthews (1978).
[b]See Table 5.2.
[c]Planktonic δ18O data from the following cores and their respective sources were used: Core 280 (Emiliani, 1958), Core P6304-9 (Emiliani, 1966b), Core P6408-9 (Emiliani, 1978), Core V28-238 (Shackleton and Opdyke, 1973), and Core V22-174 (Thierstein *et al.*, 1977).

NOTE: Data are reported as (mean Δ) ± (confidence about mean, 95% C.L.); number of independent estimates are in parentheses.

TABLE 5.2 Isotope Data on Barbados Corals

Sample Number and Description	δ¹⁸O (per mil)	δ¹³C PDB
Worthing (Barbados I)		
FS-50A	−3.47	+0.19
Duplicate I	−3.31	+0.08
Ave. FS-50A	−3.39	+0.13
FS-51	−3.42	+0.96
Duplicate B	−3.16	+0.92
Ave. FS-51	−3.29	+0.94
FS-52A	−3.61	−1.19
OC-50	−3.01	+0.93
OC-51	−3.22	+0.20
OC-53	−3.20	+0.64
Sandy Lane-1	−2.94	+0.55
Sandy Lane-2	−3.27	−1.14
Sandy Lane-4	−3.64	+0.53
Duplicate H	−3.61	+0.62
Ave. Sandy Lane-4	−3.62	+0.58
Mean	−3.28	+0.17
σ	0.23	0.81
CL₉₅	0.18	0.62
r = +0.45 (not significant)		
n = 9		
Ventnor (Barbados II)		
ANM-20	−3.11	−0.32
ANM-21	−3.03	−0.60
ANM-22	−3.30	−0.37
Duplicate E	−3.30	−0.37
Ave. ANM-22	−3.30	−0.37
FT-50	−3.28	+0.70
Duplicate A	−3.33	+0.44
Duplicate F	−3.38	+0.33
Ave. FT-50	−3.33	+0.49
FT-51	−3.65	−1.30
FT-53	−3.55	−1.88
Duplicate C	−3.25	−1.57
Ave. FT-53	−3.40	−1.73
BAB-10	−3.46	+1.18
BAB-11	−3.00	−0.75
BAB-13	−3.24	−0.18
Mean	−3.28	−0.40
σ	0.21	0.87
CL₉₅	0.16	0.67
r = +0.025 (not significant)		
n = 9		

Sample Number and Description	δ¹⁸O (per mil)	δ¹³C PDB
Maxwell ("New" terrace between Ventnor and Rendezvous Hill; basically part of Rendezvous Hill)		
AEJ-20	−3.65	+0.01
AEJ-21	−3.65	−0.58
Duplicate G	−3.82	−0.52
Ave. AEJ-21	−3.73	−0.55
AEJ-22	−3.82	+1.28
AGP-10	−3.11	−0.28
AGP-11	−3.50	+0.49
Duplicate D	−3.46	+0.56
Duplicate D	−3.61	+0.38
Duplicate D	−3.39	+0.55
Ave. AGP-11	−3.49	+0.50
AGP-12	−3.63	+0.15
Mean	−3.57	+0.19
σ	0.25	0.65
CL₉₅	0.22	0.57
r = −0.40 (not significant)		
n = 6		
Rendezvous Hill (Barbados III)		
AFM-20A	−3.74	+0.38
AFM-22A	−3.75	+0.20
AFM-23	−3.54	0.00
R-50	−3.56	+0.29
Duplicate J	−3.56	+0.46
Ave. R-50	−3.56	+0.38
R-51	−3.44	+0.32
R-52	−3.72	+0.43
AFS-10	−3.60	+0.77
AFS-11	−3.60	−1.43
AFS-12	−3.76	+0.10
Mean	−3.63	+0.11
σ	0.11	0.62
CL₉₅	0.08	0.48
r = −0.04 (not significant)		
n = 9		

NOTE: During the summer of 1977 A. palmata was recollected on Barbados Terraces I, II, and III expressly to confirm the results reported by Fairbanks and Matthews (1978). Whereas their subsurface data were taken on high-quality core materials, the terrace materials were leftover scraps from other projects. As such, samples were rather small and cut in undetermined relation to growth lines. For this reason, it was highly desirable to resample.

Each of the three terraces was sampled three times at each of the three localities. Samples were taken with a 2-inch-diameter diamond core drill. Each core was approximately one foot in length. Samples for isotopic study were taken perpendicular to growth bands to represent the average of the whole coral.

Isotopic data were collected on the Benedum Stable Isotopes Laboratory VG 602D over four days from November 20, 1978, to February 20, 1979. Precision (1σ) on standards (6) was ±0.09 for δ¹⁸O and ±0.18 for δ¹³C. Precision (half range) on random duplicates (9) was ±0.06 for both δ¹⁸O and δ¹³C. Averages were used in calculation of terrace statistics.

constant uplift hypothesis for terrace elevation (Dodge *et al.*, 1983) are only slightly smaller than the actual isotope data for Barbados terraces and the deep-sea cores. The difference is well within the confidence interval for the various data subsets. To bring the means into literal agreement requires tropical sea-surface cooling on the order of only 0.5°C.

Thus within the uncertainty of the data, the low-latitude planktonic $\delta^{18}O$ record is precisely a measure of global ice-volume fluctuation within the sea-level fluctuation range represented among the three younger Barbados terraces. On the basis of isotope data from boreholes, this precise relationship of $\delta^{18}O$ variation to change in elevation of sea level may be extended to the depth of 50 to 70 m below present sea level. Below this depth, the scant data that are available (Figure 5.2) suggest that a local temperature effect may become important in the depths of the glacial portion of cycles.

The Deep-Sea $\delta^{18}O$ Record as an Ice-Volume Signal

In summary then, terrace elevation data, Barbados isotope data, and deep-sea isotope data are in close agreement within the time interval of Barbados terraces Worthing, Ventnor, and Rendezvous Hill (82 to 125 kaBP). Given the uncertainty within each data set, no temperature effect is needed to bring these data into agreement; likewise, no hidden floating ice is required. The calibration of Fairbanks and Matthews (1978) is sufficient to the task in and of itself. Further, the data on which that calibration is based (namely, data points "RKM 23 regressive" and "RKM 20 transgressive" of Figure 5.2) indicate that ice volume alone is a sufficient explanation of the deep-sea $\delta^{18}O$ variation to a depth of sea-level lowering of approximately 50 to 70 m below present sea level, accounting for somewhere between 0.5 and 0.8 per mil of global ocean $\delta^{18}O$ enrichment from interglacial conditions toward glacial conditions.

In the region of Barbados, the lower half of the $\delta^{18}O$ record in coral materials and in nearby deep-sea cores is satisfied by a combination of continued ice-volume buildup and a local temperature lowering of approximately 2°C (a $\delta^{18}O$ effect of 0.5 per mil). Thus, although this combination of numbers arrives at the classical Shackleton (1967) estimate of one-third ΔT effect and two-thirds Δice-volume effect, it is important to note that the temperature effect (if valid at all) is concentrated in the glacial half of the signal. Importantly, to the extent that this temperature effect may represent global cooling, it appears to be decidedly the result of ice-volume buildup (presumably an albedo feedback mechanism); strikingly, this is the reverse of the common wisdom scenario that global cooling initiates continental glaciation.

THE PAST AS THE KEY TO THE FUTURE

The discussion above has dealt largely with geologic information concerning "the numbers" regarding late Quaternary sea-level history. What was the level of the sea at what date in the past; how good are the data; how shall we resolve apparent conflicts? Geologists also have insight and opinion concerning the workings of the Earth as a dynamical system. Two possible scenarios deserve consideration here with regard to the possible ice-volume effects of future climatic warming.

West Antarctic Ice Surge Accompanying Warm Interglacial High Stand

Among the world's major ice sheets, the West Antarctic ice sheet seems to be especially precariously situated today. It is substantially a marine-based ice sheet, and the present-day topography over vast areas lies below the theoretical equilibrium profile. It can be argued that this ice sheet is capable of disintegrating over the next few hundred years. If such a glacial surge should occur, world sea level would rise by approximately 6 m. The geologic question is whether or not such a surge event has occurred in the past.

There is no doubt that sea level at isotope stage 5.5 stood at around present sea level or slightly higher. The question with regard to possible West Antarctic ice surge is whether or not we can be confident that it stood 6 m above present sea level. Moore (1982) provided a convenient summary of ^{230}Th dating from both tectonically active and tectonically stable regions. Seemingly tectonically stable regions giving seemingly reliable estimates for isotope stage 5.5 sea level above present sea level include the Bahamas, Bermuda, western Australia, Aldabra Atoll, and the Yucatan Pennisula, Mexico. Estimates of sea level relative to present range from +2 to +6 m.

Another approach to the relative position of sea level at isotope stage 5.5 would be to compare $\delta^{18}O$ data to Holocene values. However, the differences that we seek to quantify are relatively small in comparison to analytical and geologic precision of the isotopic data, the isotopic difference between modern and +6 m sea-level high stand being only approximately 0.07 per mil. With regard to deep-sea cores, the problem is compounded by nagging problems concerning the reliability of core-top materials. CLIMAP (1984, p. 206) notes the lack of consistency in their data and refers the question to dated coral reefs. With regard to Barbados coral data and New Guinea mollusc data, modern and isotope stage 5.5 equivalent materials yield comparable values within the uncertainty of the estimates. Encouragingly, the New Guinea data do indicate brief number VIIa (isotope stage 5.5 equivalent) to be 0.03 per mil lighter than modern; this is in the right direction, but a difference of only about half the uncertainty of the two estimates.

The isotopic data concerning New Guinea VIIb (Aharon, 1983) are especially troublesome with regard to the surge hypothesis. On the basis of geologic correlation and on the basis of reef-crest $\delta^{18}O$ values, New Guinea reef VIIa is surely isotope stage 5e equivalent. (The fact that its date is slightly older than isotope stage 5.5 must be considered subordinate to these other arguments; see discussion above concerning problems with dating technology.) Aharon (1983) continued to regard New Guinea reef VIIb as a surge event that follows the deposition of reef VIIa. The interesting problem is that reef-crest $\delta^{18}O$ values for VIIb are 0.5 per mil heavier than the values for VIIa. Aharon explained these highly anomolous $\delta^{18}O$ values by calling on dramatic global cooling accompanying a Wilson hypothesis surge event (see Flohn, 1979, for a convenient review of this and similar topics). It is important to note that materials similar to New Guinea reef VIIb have not been observed in any other terrace sequence. Therefore, it is possible that these unusual data have a local explanation and do not require integration into a global scenario. Nevertheless, a scenario such as set forward by Aharon (1983) could be consistent with the deep-sea $\delta^{18}O$ record. Inasmuch as rapid cooling accompanies the surge of isotopically light water into the world ocean, the proposed high stand of the sea is lost somewhere in the transition from isotope stage 5.5 to 5.4.

Taking all of these data literally, one is left with the unsatisfactory proposition that isotope stage 5.5 (equivalent to New Guinea reef VIIa) achieved a level somewhere between +2 and +6 m relative to present sea level by removal of an ice sheet in a manner inconsistent with the Wilson hypothesis, whereas isotope stage 5.5 was followed closely by a glacial surge (and resultant New Guinea reef VIIb) that was consistent with the Wilson hypothesis. At this point, belief or disbelief becomes largely a matter of taste. I would prefer to see the New Guinea reef VII data set enlarged and the result reproduced within some other terrace sequence before I proceed to make choices. At this stage in the science, the simplest choice remains that isotope stage 5.5 sea level is the same or slightly higher than modern for reasons that are not fully understood. While the West Antarctic surge hypothesis remains exciting, the observation could equally be explained by slight deviations of East Antarctic ice sheet from equilibrium conditions by either more rapid flow or insufficient precipitation.

Ice Growth Adjacent to a Warm Ocean

An equally credible scenario for the end of an interglacial can be written around rapid growth of continental ice fed from a nearby warm ocean moisture supply. This scenario is best documented with regard to growth of North American ice sheets at the end of the last interglacial (isotope stage 5.5/5.4 transition) (Ruddiman et al., 1980). By generating both planktonic faunal data and planktonic isotopic data at close spaced sample intervals, it is clearly demonstrated that the isotopic values become relatively heavy while the fauna continues to indicate relatively warm water. The simplest explanation for these observations is that North American ice sheets grew to considerable size (hence, relatively positive oceanic $\delta^{18}O$ values) while the nearby North Atlantic remained relatively warm.

A similar scenario can be envisioned for the Southern Ocean at times throughout the Tertiary. In this case, relatively warm water would be provided to the surface ocean by upwelling to the south of the polar front. Such a mechanism might be similar to the occasional occurrence of ice-free conditions among winter pack ice around Antarctica today (the Weddell polynya; see e.g., Gordon, 1982; Gordon and Huber, 1984). The fact that the polynya remains ice free throughout winter is ample demonstration that upwelling water is transferring large amounts of latent heat to the high-latitude southern atmosphere.

At times in the past, this effect has almost certainly been much larger. For example, it is likely that the deep ocean was approximately 8°C in Oligocene time. Upwelling of such warm water around Antarctica may have provided the local moisture supply for substantial growth of the Antarctic ice sheet at approximately the Oligocene-Eocene boundary (Matthews and Poore, 1980).

Looking to the future, if man's warming of the planet tends to warm intermediate and deep water (Roemmich, Chapter 13, this volume), then one might script an Antarctic ice-growth scenario such as outlined above. However, modern Antarctica has the further complication of large ice shelves that serve to buttress the glacial ice (Lingle, 1984, for example). If these ice shelves were removed by upwelling of warmer water, the initial effect would surely be to increase the flow rate of ice streams. The new equilibrium between these two competing effects (increased precipitation and increased flow rates) is conjectural. The fact that both effects are plausible emphasizes the need to observe Antarctic surface-elevation changes from polar-orbiting satellites.

GLACIO-EUSTACY IN STRATIGRAPHIC PREDICTION

Given the high integrity of the deep-sea, low-latitude planktonic $\delta^{18}O$ record as a glacio-eustatic signal, there exists a significant opportunity for stratigraphy to move beyond a descriptive mode to a predictive mode. The

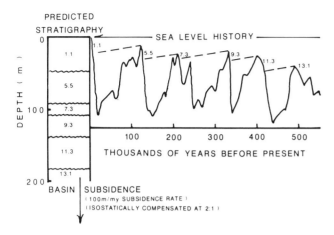

FIGURE 5.4 Stratigraphy predicted for bank-margin carbonates where a well-known (or presumed) sea-level history is superimposed on basin subsidence. Sea-level rise deposits new bank-margin carbonate sediment. Sea-level fall produces subaerial exposure surfaces. Diagonal dashed lines from sea-level high stands indicate subsidence that occurs before the next high stand of the sea. This sort of forward modeling holds great promise for bringing a new level of understanding to stratigraphy and thereby petroleum exploration.

opportunity is before us to model an empirically derived glacio-eustatic signal with mathematical formulations of basin subsidence and isostasy, for example.

By way of example, Figure 5.4 presents a proposed stratigraphy of subaerial exposure surfaces within bank-margin carbonate rocks. The model is generated by interaction of the late Pleistocene average $\delta^{18}O$ curve (Prell *et al.*, 1986) with a regional driving subsidence of 100 m per million years. In such a model, deposition occurs only when rising sea level crosses the subaerial exposure surface of previously existing bank-margin carbonate sediments. As the sea-level curve tops out and heads once again toward glacial low-stand, yet another subaerial exposure surface is formed. Such a diagram is, then, a prediction of the stratigraphy that one would find if one drilled a borehole through a subsiding passive margin (such as the modern reef tract of Belize, Central America) that was subsiding at approximately this rate. While such a prediction in the late Pleistocene is academic, such predictions in older rocks could have extreme importance in petroleum exploration. Indeed, I join Christie-Blick *et al.* (Chapter 7, this volume) in criticizing the alleged seismic stratigraphy record of sea-level changes. This famous curve is nothing more than an ad hoc explanation of the data growing out of enumerable inductive mode studies of seismic profiles. However, whereas Christie-Blick *et al.* propose still further inductive mode investigations to reevaluate the ages of sequence boundaries, the approach

outlined in this chapter would deductively compare independent models for eustatic history (also see Harrison, Chapter 8, this volume) with models for basin subsidence.

Initial attempts at applying a similar research strategy to Tertiary materials have been at least partially successful. Major and Matthews (1983) suggested that the shape of the Vail *et al.* (1977) sea-level curve for middle Miocene is approximately correct, but that the amplitude of their proposed sea-level fluctuations is probably exaggerated by a factor of three.

CONCLUSIONS

1. The low-latitude planktonic $\delta^{18}O$ record surely offers the most promise for delivering a continuous record of glacio-eustatic sea-level fluctuations back through time. On the basis of existing data, it appears that ice-volume fluctuations are sufficient to account for the structure in approximately the more interglacial half of the deep-sea $\delta^{18}O$ signal. The claim that this relationship holds true to a sea-level lowering of approximately 20 or 30 m relative to present is confirmed by isotopic data on surficial samples from coral-reef terraces of Barbados, New Guinea, and elsewhere. The claim that this relationship extends to a sea-level lowering of 50 to 70 m relative to present rests on two data points obtained by Barbados core drilling. These are very important numbers; similar drilling needs to be carried out in other parts of the world to confirm or reject the Barbados data.

2. With regard to the more glacial half of the deep-sea $\delta^{18}O$ signal, we face still further problems with scarcity of high-quality information. Estimates concerning the glacio-eustatic amplitude of isotope stage 6 rest solely on a single cored interval in Barbados RKM 22. Even here, there is no claim that this data point represents maximum low stand of the sea. The amplitude of the glacio-eustatic signal is surely a fundamental property of the dynamic system. We really should know that number quite well; if it turns out to be 80 m as opposed to 130 m, the result would significantly narrow uncertainties concerning continental ice sheets. Once again, Barbados style core drilling in other parts of the world would seem to be indicated. Similarly, renewed attempts to sample and date stratigraphically significant material representing isotope stage 2 from submerged continental margins would appear to be worth unusual effort.

3. With regard to future glacio-eustatic sea-level fluctuations, it is noteworthy that the entirety of the civilization of man has occurred within a single eustatic high stand of the sea that is indeed quite anomalous to the average late Pleistocene condition of the Earth. Without human intervention, surely sea level shall once again fall to intermediate levels. With inadvertent human interven-

tion, the system might go either way; sea-level rise in accordance with the West Antarctic ice surge hypothesis; or sea-level lowering in accordance with the warm ocean ice growth hypothesis. The precise nature of the geologic record with regard to these two possibilities is substantially irrelevent to planning concerning possible sea-level consequences of future climatic warming. Both scenarios are conceivable; we really do not know the time scales involved; best we simply plan ahead to monitor the Earth's ice budget henceforth and forthwith.

4. The advent of predictive stratigraphy would appear to be at hand. It is technologically feasible to construct a detailed glacio-eustatic sea-level curve throughout the Cenozoic. Other long-term eustatic effects can easily be added to this curve. The interaction of a detailed eustatic sea-level curve with basin subsidence models awaits primarily the awakening of the stratigrapher and explorationist to the opportunity at hand.

REFERENCES

Aharon, P. (1983). 140,000-yr isotope climatic record from raised coral reefs in New Guinea, *Nature 304*, 720.

Allan, J. R., and R. K. Matthews (1982). Isotopic signatures associated with early meteoric diagenesis, *Sedimentology 29*, 797–817.

Baltzer, F. (1970). Datation absolue de la transgression Holocene sur la cote ouest de Nouvelle-Caledonie sur des enchantillons de tourbes a paletouiers; interpretation neotectonic, *Acad. Sci. C. R. Ser. D271(25)*, 2251–2254.

Bender, M. L., R. G. Fairbanks, F. W. Taylor, R. K. Matthews, J. G. Goddard, and W. S. Broecker (1979). Uranium-series dating of the Pleistocene reef tracts of Barbados, West Indies, *Geol. Soc. Am. Bull. 90*, 577–594.

Bloom, A. L., W. S. Broecker, M. A. Chappell, R. K. Matthews, and K. J. Mesolella (1974). Quaternary sea level fluctuations on a tectonic coast: New ^{230}Th/^{234}U dates from the Huon Peninsula, New Guinea, *Quat. Res. 4*, 185–205.

Broecker, W. S., D. L. Thurber, J. Goddard, Teh-lung Ku, R. K. Matthews, and K. J. Mesolella (1968). Milankovitch hypothesis supported by precise dating of coral reefs and deep-sea sediments, *Science 159*, 297–300.

CLIMAP Project Members (1976). The surface of the ice-age Earth, *Science 191*, 1131–1137.

CLIMAP Project Members (1984). The last interglacial ocean, *Quat. Res. 21*, 123–224.

Coleman, J. M., and W. G. Smith (1964). Late Recent rise of sea level, *Geol. Soc. Am. Bull. 75*, 833–840.

Cronin, T. M. (1983). Rapid sea level and climate change: Evidence from continental and island margins, *Quat. Sci. Rev. 1*, 177–214.

Curray, J. R. (1960). Sediments and history of Holocene transgression, continental shelf, Northwest Coast Gulf of Mexico, in *Recent Sediments Northwest Gulf of Mexico*, F. Shepard, ed., American Association of Petroleum Geologists, Tulsa, Okla., pp. 221–266.

Curry, W. B., and R. K. Matthews (1981). Equilibrium oxygen-18 fractionation in small size fraction planktic foraminifera: Evidence from Recent Indian Ocean sediments, *Mar. Micropaleont. 6*, 327–337.

Denton, G. H., and T. J. Hughes, eds. (1981). *The Last Great Ice Sheets*, Wiley, New York, 484 pp.

Dodge, R. E., R. G. Fairbanks, and F. Maurrasse (1983). Pleistocene sea levels from raised coral reefs of Haiti, *Science 219*, 1423–1425.

Donn, W. L., and D. M. Shaw (1977). Model of climate evolution based on continental drift and polar wandering, *Geol. Soc. Am. Bull. 88*, 390–396.

Emiliani, C. (1958). Paleotemperature analysis of core 280 and Pleistocene correlations, *J. Geol. 66*, 264–275.

Emiliani, C. (1966a). Isotopic paleotemperatures, *Science 154*, 851–857.

Emiliani, C. (1966b). Paleotemperature analysis of Caribbean cores P6304-8 and P6304-9 and a generalized temperature curve for the past 425,000 years, *J. Geol. 74*, 109–126.

Emiliani, C. (1978). The cause of the ice ages, *Earth Planet. Sci. Lett. 37*, 349–352.

Fairbanks, R. G., and R. K. Matthews (1978). The marine oxygen isotope record in Pleistocene coral, Barbados, West Indies, *Quat. Res. 10*, 181–196.

Field, M. E., E. P. Meisburger, E. A. Stanley, and S. J. Williams (1979). Upper Quaternary peat deposits on the Atlantic inner shelf of the United States, *Geol. Soc. Am. Bull. 90*, 618–628.

Flohn, H. (1979). On time scales and causes of abrupt paleoclimatic events, *Quat. Res. 12*, 135–149.

Gordon, A. L. (1982). Weddell deep water variability, *J. Mar. Res. 40*, 199–217.

Gordon, A. L., and B. A. Huber (1984). Thermohaline stratification below the Southern Ocean sea ice, *J. Geophys. Res. 89*, 641–648.

Gould, H. R., and E. McFarland, Jr. (1959). Geologic history of Chenier Plain, South Western Louisiana, *Trans. Gulf Coast Assoc. Geol. Soc. 9*, 261–270.

Graham, D. W., B. H. Corliss, M. L. Bender, and L. D. Keigwin, Jr. (1981). Carbon and oxygen isotopic disequilibria of recent deep-sea benthic foraminifera, *Mar. Micropaleonol. 6*, 483–497.

Harmon, R. S., T. L. Ku, R. K. Matthews, and P. L. Smart (1979). Limits of U-series analysis: Phase I: Results of the uranium-series intercomparison project, *Geology 7*, 405–409.

Hopley, D., R. F. MacLean, J. Marshall, and A. S. Smith (1978). Holocene-Pleistocene boundary in a fringing reef: Hayman Island North Queensland, *Search 9(8–9)*, 323–325.

Imbrie, J., J. D. Hays, D. G. Martinson, A. McIntyre, A. C. Mix, J. J. Moreley, N. G. Pisias, W. L. Prell, and N. J. Shackleton (1984). The orbital theory of Pleistocene climate: Support from a revised chronology of the marine δ^{18}O record, in *Milankovitch and Climate*, Part 1, A. L. Berger *et al.*, eds., D. Reidel, Dordrecht, pp. 269–306.

Land, L. S., and C. H. Moore (1980). Lithification, micritization and syndepositional diagenesis of biolithites of the Jamaican Island slope, *J. Sediment. Petrol. 50*, 357–370.

Lighty, R. G., I. G. MacIntyre, and R. Stuckenrath (1978). Submerged early Holocene barrier reef South-East Florida Shelf, *Nature 275*, 59–60.

Lighty, R. G., I. G. MacIntyre, and R. Stuckenrath (1982). *Acropora palmata* reef framework: A reliable indicator of sea level in the western Atlantic for the past 10,000 years, *Coral Reefs 1*, 125–130.

Lingle, C. S. (1984). A numerical model of interactions between a polar ice stream and the ocean: Application to Ice Stream E, West Antarctica, *J. Geophys. Res. 89*, 3523–3549.

MacIntyre, I. G., O. H. Pilkey, and R. Stuckenrath (1978). Relict oysters on the United States Atlantic continental shelf: A reconsideration of their usefulness in understanding late Quaternary sea-level history, *Geol. Soc. Am. Bull. 89*, 277–282.

Major, R. P., and R. K. Matthews (1983). Isotopic composition of bank margin carbonates on Midway Atoll: Amplitude constraint on post-early Miocene eustasy, *Geology 11*, 335–338.

Manabe, S., and A. J. Broccoli (1985). A comparison of climate model sensitivity with data from the last glacial maximum, *J. Atmos. Sci. 42*, 2643–2651.

Matthews, R. K. (1973). Relative elevation of late Pleistocene high sea level stands: Barbados uplift rates and their implication, *Quat. Res. 3*, 147–153.

Matthews, R. K., and R. Z. Poore (1980). Tertiary $\delta^{18}O$ record and glacio-eustatic sea-level fluctuations, *Geology 8*, 501–504.

Mesolella, K. J., R. K. Matthews, W. S. Broecker, and D. L. Thurber (1969). The astronomical theory of climatic change: Barbados data, *J. Geol. 77*, 250–274.

Mix, A. C., and W. F. Ruddiman (1985). Structure and timing of the last deglaciation: Oxygen isotope evidence, *Quat. Sci. Rev. 4*, 59–108.

Moore, W. S. (1982). Late Pleistocene sea level history, in *Uranium Series Disequilibrium: Applications to Environmental Problems*, M. Ivanovich and and R. S. Harmon, eds., pp. 481–496.

Nelson, H. F., and E. E. Bray (1970). Stratigraphy and history of the Holocene sediments in the Sabine-High Island area, Gulf of Mexico, in *Deltaic Sedimentation, Modern and Ancient*, J. T. Morgan, ed., Spec. Publ. 15, Society of Economic Paleontologists and Mineralogists, pp. 48–77.

Newell, R. E., A. R. Navato, and J. Hsiung (1978). Long-term global sea surface temperature fluctuations and their possible influence on atmospheric CO_2 concentrations, *J. Pure Appl. Geophys. 116*, 351–371.

Oldale, R. N., and C. J. O'Hara (1980). New radiocarbon data from the inner continental shelf off Southeastern Massachusetts and a local sea-level-rise curve for the past 12,000 yr., *Geology 8*, 102–106.

Peterson, L. C., and W. L. Prell (1985). Carbonate dissolution in recent sediments of the eastern equatorial Indian Ocean: Preservation patterns and carbonate loss above the lysocline, *Mar. Geol. 64*, 259–290.

Prell, W. L. (1985). The stability of low-latitude sea-surface temperatures: An evaluation of the CLIMAP reconstruction with emphasis on the positive SST anomalies, prepared for U.S. Department of Energy, U.S. Government Printing Office, Washington, D.C., 60 pp.

Prell, W. L., J. Imbrie, D. G. Martinson, N. G. Pisias, N. J. Shackleton, and H. F. Streeter (1986). Graphic correlation of oxygen isotope stratigraphy application to the late Quaternary, *Paleoceanography 1*, 137–162.

Prentice, M. L., and R. K. Matthews (1988). Cenozoic ice-volume history: Development of a composite oxygen isotope record, *Geology 16*, 963.

Redfield, A. C. (1967). Post glacial change in sea level in the Western North Atlantic Ocean, *Science 157*, 687–691.

Ruddiman, W. F., A. McIntyre, V. Niebler-Hunt, and J. T. Durazzi (1980). Oceanic evidence for the mechanism of rapid Northern Hemisphere glaciation, *Quat. Res. 13*, 33–64.

Schnable, J. E., and H. G. Goodell (1968). Pleistocene-Recent stratigraphy evolution and development of the Apalachicolan Coast Florida, *Geol. Soc. Am. Spec. Paper 112*, 72 pp.

Scholl, D. W. (1964). Recent sedimentary record in mangrove swamps and its rise of sea level over the southwestern coast of Florida: Part 1, *Mar. Geol. 1*, 344–366.

Scholl, D. W., and M. Stuiver (1967). Recent submergence of southern Florida: A comparison with adjacent coasts and other eustatic data, *Geol. Soc. Am. Bull. 78*, 437–454.

Shackleton, N. J. (1967). Oxygen isotope analyses and Pleistocene temperatures reassessed, *Nature 215*, 15–17.

Shackleton, N. J., and N. D. Opdyke (1973). Oxygen isotope and paleomagnetic stratigraphy of equatorial Pacific core V28-238: Oxygen isotope temperatures and ice volumes on a 10^5 and 10^6 year scale, *Quat. Res. 3*, 39–55.

Stockmal, G. S. (1983). Modeling of large-scale accretionary wedge deformation, *J. Geophys. Res. 88*, 8271–8287.

Thierstein, H. R., K. R. Geitzenauer, B. Molfino, and N. J. Shackleton (1977). *Geology 5*, 500.

Thom, B. G., and J. Chappell (1975). Holocene sea levels relative to Australia, *Search 6*, 90–93.

Vail, P. R., R. M. Mitchum, Jr., and S. Thompson, III (1977). Seismic stratigraphy and global changes of sea level, part 4: Global cycles of relative changes of sea level, in *Seismic Stratigraphy—Applications to Hydrocarbon Exploration*, Memoir 26, American Association of Petroleum Geologists, Tulsa, Okla., pp. 83–97.

Wagner, P. D. (1983). Geochemical Characterization of Meteoric Diagenesis in Limestone: Development and Applications, Ph.D. dissertation, Brown University, Providence, R.I., 386 pp.

Williams, D. F., W. S. Moore, and R. H. Fillon (1981). Role of glacial Arctic Ocean ice sheets in Pleistocene oxygen isotope and sea level records, *Earth Planet. Sci. Lett. 56*, 157–166.

Graphic Analysis of Dislocated Quaternary Shorelines

6

ARTHUR L. BLOOM
Cornell University

NOBUYUKI YONEKURA
University of Tokyo

ABSTRACT

If a sequence of dated emerged shorelines is found at different heights on several transects, the heights of the intermediate shorelines can be plotted against the height of the highest one in each sequence. The regressions of the intermediate shoreline positions yield a set of equations:

$$H_{i,t} = a_i H_{m,t} + b_i, \qquad (6.1)$$

where $H_{m,t}$ is the height of the highest shoreline in the sequence on transect t, $H_{i,t}$ is the height of an intermediate shoreline i on the same transect, a_i is the regression coefficient, and b_i is the intercept. By assuming or establishing the position of sea level at the time of formation of the highest shoreline, the initial heights of the intermediate shorelines can be calculated by substitution in Eq. (6.1) without the previously necessary assumption that the dislocation on any transect had been at a constant rate. A similar procedure gives the initial level of any shoreline in a sequence of submerged features, if their depths are plotted against the depth of the deepest shoreline in the sequence. Correlation coefficients for typical calculations are inflated because $H_{m,t}$ always incorporates all subsequent movements on that transect.

Nevertheless, realistic and testable results can be achieved for intervals of the last 8000 yr and for the last 125,000 yr.

INTRODUCTION

This chapter is about an analytical technique for determining the original altitude of shorelines that are now displaced by tectonic movements. Before we can evaluate the causes of sea-level changes, we must be able to measure them with reasonable confidence. The literature of sea-level change is a jungle of generalizations, misconceptions, faulty interpretations, unsubstantiated age estimates, and just poor science. R. Stuckenrath, Jr. (University of Pittsburgh, personal communication), estimated that fully half of the researchers who received radiocarbon dates from the Smithsonian Institution radiocarbon laboratory do not "believe" the results, meaning that if their *a priori* conclusions were not confirmed by the dates, the dates were doubted. A great amount of experience and judgment is required just to know which published data are valid.

104

A basic problem is that sea level is constantly changing on many temporal and spacial scales (Figure 6.1). Waves cause the water surface to rise and fall a meter or more every few seconds, providing a high-frequency background noise to the secular change of millimeters per year that we hope to document. The diurnal or semidiurnal tide is commonly the largest magnitude sea-level change on any time scale short of 10^4 yr, barring tsunamis and hurricanes.

We consider sea-level changes on two time scales. The first is the scale of glacially controlled worldwide ("eustatic") sea-level fluctuations of 10^4 to 10^5 yr, which covers the time scale of complete glacial-interglacial cycles and their major subdivisions. The second is the scale of 10^3 to 10^4 yr, primarily considering the deglacial hemicycle of the last ice age and the most recent 10,000 yr of geologic history called the Holocene Epoch. On this time scale, glacial eustatic sea level has risen some 120 m as the Northern Hemisphere ice sheets disintegrated, and then it has stabilized or fluctuated slightly with the subsequent isostatic adjustments of the solid earth to the postglacial distribution of surface ice and water masses. Doubts, uncertainty, and unresolved problems are noted in each section, for our present inquiry is to determine what we know with confidence and what we still need to learn. The chapter ends with some suggestions for resolving some of our uncertainties.

THE GLACIAL CYCLE: 10^4 TO 10^5 YEARS

Graphs of sea level for the last 140,000 yr or more are drawn primarily from two sources: (1) coral-reef terraces on tropical coasts that have been uplifted by tectonic forces,

and (2) the deviations from a standard of the $^{18}O/^{16}O$ ratios in foraminiferal tests from deep-sea cores. The methods and their premises are very different, and yet the superficial agreement is good, especially concerning the time scale. The amplitude of interstadial-stadial sea-level change inferred from the oxygen-isotope record of the last ice age is significantly greater than that inferred from the coral-reef record, and we need to understand the cause of that difference. Nevertheless we should stress the overall similarities of the graphs more than their differences (Figures 6.2 and 6.3; see also Matthews, Figure 5.3, this volume).

For coral-reef terraces, a reasonable time control (2σ = 5 percent) is available for 200,000 yr by the uranium-series dating method (Harmon et al., 1979). For reefs, the problem is to convert a known age and a present height into both an initial level and an uplift rate. In the equation $H = ax + b$, present height H is the result of uplift rate a times x (time), and b is the desired initial or starting level. Clearly, the equation cannot be solved knowing only the present height and age of a terrace. However, by assuming that some older reef terrace, such as the last interglacial one that is about 125,000 yr old, formed at an assumed level such as +6 m, and assuming that tectonic uplift is reasonably constant on the time scale of 10^5 yr, the initial level b, and the uplift rate a can be inferred. The results from various islands, where uplift rates vary from 0.1 mm/ yr to more than 5 mm/yr, have been reasonably consistent. For example, one can confidently predict that if a 28,000-yr-old reef terrace is found above present sea level, the uplift rate is greater than about 1.5 mm/yr, because sea level at that time of reef formation was probably at −35 to

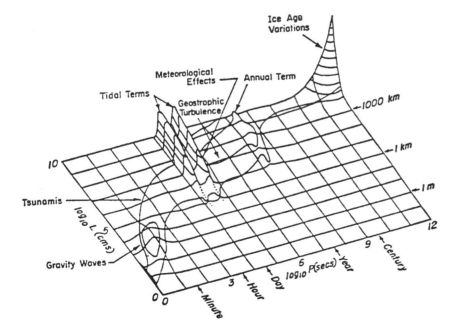

FIGURE 6.1 Schematic diagram of the spectral distribution of sea level (Stommel, 1963, Figure 1).

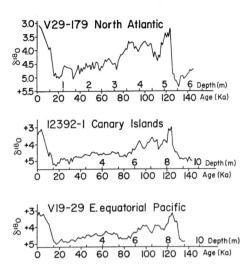

FIGURE 6.2 Three oxygen-isotope curves from high-sedimentation-rate deep-sea cores that show details of the variations of oxygen-isotope ratios for the past 140,000 yr (see text for references).

−40 m and the reef had to rise above present sea level by at least 5000 yr ago to escape the postglacial rise of sea level. Such a prediction can then be tested by using it to predict the height of older and younger terraces in the sequence: for instance, if the uplift rate were 1.5 mm/yr (or the equivalent 1.5 m/ka, or kiloanno), the last interglacial terrace should be nearly 200 m above sea level (125 ka × 1.5 m/ka + 6 m, which is the assumed height of sea level in the last interglacial). Note that the assumed height of 6 m for the last interglacial sea level is only a few percent of the present terrace height on coasts where tectonic movements are rapid, that is, on the order of millimeters per year. As many islands on tropical island arcs have coral-reef terraces measured in hundreds of meters above sea

level, the calculation of the initial position of sea level during the last interglacial is not critical. A range of "a few meters" or "5 to 10 m" above the present gives satisfactory results that lie within the typical survey errors for terrace heights.

Figure 6.4 graphically illustrates the assumptions of the above method: given a sea-level curve and an assumed uplift rate, reefs grow at times of tangency when land and sea level are rising at the same rate, just prior to a sea-level maximum. Sea level then drops while the constructional reef terraces are carried upward to their present position on a hillside. In principle, similar reef-building events should occur just after each sea-level minimum, but so far, only one such reef has been identified. As noted in the

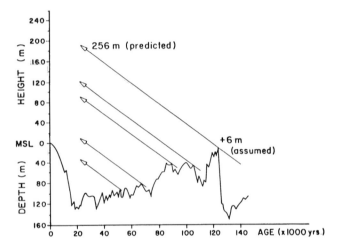

FIGURE 6.4 The hypothetical coral-terrace sequence generated by uniform tectonic uplift of 2 m per 1000 yr superimposed on sea-level oscillations modeled from the oxygen-isotope record of benthic foraminifera in core V29-179 (Streeter and Shackleton, 1979).

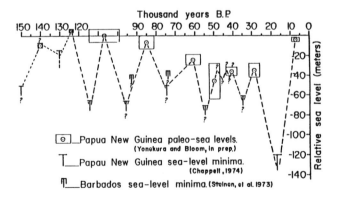

FIGURE 6.3 Revised sea-level curve for the coral-reef terraces on the Huon Peninsula, Papua New Guinea (Bloom and Yonekura, 1985, Figure 6.4).

final section of this chapter, if others could be found by drilling or in natural exposures, stadial and glacial minima as well as interstadial and interglacial maxima sea level could be dated and the amplitudes of sea-level fluctuations could be defined.

In addition to the value of coral-reef models for predicting sea levels of the last 125 ka, an excellent predictive model of the ages of middle Pleistocene and older terraces can be made if one of the lowest terraces in the sequence can be proved to be of last interglacial age, about 125,000 yr old. This situation is very common, because sea level was probably a few meters higher than at present in last interglacial time (+6 m is the widely accepted estimate). On stable coasts or coasts with slow uplift, the last interglacial terrace is always prominent. Downtown Honolulu is built on it. The height of the last interglacial terrace above or below the assumed initial level of +6 m establishes a long-term (10^5 yr) vertical tectonic rate, which can be cautiously extrapolated by an order of magnitude to predict the ages of higher and older terraces dating back to the early Pleistocene. In the few places where a tephrochronology is available, as in New Zealand (Pillans, 1983) or Japan (Machida, 1975); or where amino acid racemization methods can be applied, as in California (Muhs, 1983), or reasonably inferred, as in Baja California, Mexico (Ortlieb, 1980), the predicted ages of older terraces are reasonably well supported. Correlations are often suggested with the odd-numbered oxygen-isotope stages of the deep-sea record, but those correlations should be used with caution, because the earlier oxygen-isotope stages are themselves dated only by interpolation between the time of the last interglacial high sea level and the Brunhes-Matuyama paleomagnetic epoch boundary that is about 730 ka old (Imbrie *et al.*, 1984, p. 282).

TERRACES OF SEA-LEVEL MAXIMA DURING AND SINCE THE LAST INTERGLACIAL

The coral-reef terraces of the Huon Peninsula, Papua New Guinea (Figure 6.3) preserve an exceptionally complete chronology of sea-level fluctuations for at least the last 140,000 yr (Bloom *et al.*, 1974; Chappell, 1974, 1983; Aharon and Chappell, 1986). A great fault splinter on the northern coast of the Huon Peninsula has been rising and tilting in late Quaternary time, broken by numerous minor faults but maintaining its overall morphotectonic integrity. A succession of coral-reef terraces has been built on this block. Where the substrate surface was steep or the terrace-building interval was brief, the reefs were fringing. Where the preexisting slope was gentle or sea level stayed in the same position for a relatively long time (or repeatedly occupied the same level), the reef grew as a barrier seaward of a lagoon that was up to 1 km in width. The internal structure of the reefs shows that they built upward and outward over their own fore-reef taluses (Figure 7 of Chappell, 1974; Chappell and Polach, 1976). Their upper surfaces are usually level and composed of typical Indo-Pacific shallow-water corals and algae. Behind the reef crests, mollusc-rich back-reef and lagoon carbonate sand accumulated. Reef crests as old as 124,000 yr to 140,000 yr have no more than about 1 m of karst relief. Swallow holes and sinking streams mark the lagoon floors. Younger reef terraces show minor gully dissection and dripstone curtains down their fronts. Less than a meter of soil in weathered volcanic tephra has accumulated on the terrace treads.

Previous analyses of the terraces' ages and heights used

an assumed constant uplift history and an assumed sea level of +6 m for the last interglacial stage (124,000 yr) to derive the paleosea-level positions during the multiple interstadials of the latest glaciation. The resulting estimates of interstadial sea levels (Bloom *et al.*, 1974) were consistent with similar estimates for terraces on Barbados, where the average rate of tectonic uplift was only about 10 percent of the rate in Papua New Guinea. The converging estimates were regarded as reasonable for the interstadial sea-level maxima, so that the height of a terrace of similar age elsewhere could be converted into an uplift rate by adding the present terrace height above sea level to its estimated original height (which is for all interstadial terraces at or below present sea level) and dividing the change in height by the age of the terrace.

The weak point of the argument was obviously the assumption that on the time scale of 10^4 to 10^5 yr, tectonic uplift was at a uniform rate. A new graphic approach to this problem has been derived, based on previous Japanese work (e.g., Ota *et al.*, 1968) but similar to the shoreline relation diagnosis of Scandinavian workers (e.g., Donner, 1965). Most of the following paragraphs are directly from Bloom and Yonekura (1985). For illustration, we can use data on the ages of terraces and their heights along six transects on the Huon Peninsula, Papua New Guinea (Tables 6.1 and 6.2). Arbitrary errors of 5 m, 2 m, and 1 m were assigned to the reported terrace heights to permit least-squares error evaluation.

The last interglacial terrace, known as terrace VIIb, has an assumed age of 124,000 yr and was formed when the sea was 6 m higher than at present. For each of the six transects, the heights of all lower and younger terraces are plotted against the height of terrace VIIb (Figure 6.5). The data points on Figure 6.5 are drawn as closed circles large enough to include the probable errors of age and height. If we assume that at a critical location terrace VIIb maintained a height of 6 m for the past 124,000 yr (ignoring erosion), then at this location there would have been no uplift and differences in the elevation of the terraces would reflect changes in sea level only. Thus where the line of best fit for a given terrace intersects the vertical dashed line at $H_{VIIb} = 6$ m, it gives the height of sea level at the time of formation of that terrace.

A regression may be performed to give a better estimate of the relationship. This yields the equation:

$$H_{i,t} = a_i H_{VIIb,t} + b_i, \qquad (6.2)$$

where $H_{VIIb,t}$ is the height of terrace VIIb on transect t; $H_{i,t}$ is the height of an intermediate terrace i on the same intersect; a_i is the regression coefficient; and b_i is the intercept. The intercept gives the elevation of terrace i assuming the elevation of terrace VIIb is zero. However, as we have seen, to calculate sea level at the time each

TABLE 6.1 Measure of Reef-Crest Elevations (in meters) for Six Transects Along the Huon Peninsula, Papua New Guinea

Terrace	Age (ka)	Transects Kanzarua	Blucher	Kwambu	Nama	Sambero	Kambin
VIIb	124	30	80	15	60	50	120
VI	105	50	15	60	15	10	93
V	82	190	55	17	90	80	60
IV	60	125	—	70	48	—	28
IIIa	50–40	90	65	42	—	—	—
IIIb	40	70	41	28	10	10	—
II	28	30	18	7	—	—	—
I	6	15	10	6	5	5	2.5

NOTE: Data include arbitrary errors that were assigned to permit least squares evaluation, as follows: ± 5 m for elevations $H \geq 60$ m; ± 2 m for 60 m $> H \geq 5$ m; and ± 1 m for $H < 5$ m (Bloom and Yonekura, 1985, Table 6.1).

terrace i was formed we must predict its value based on an elevation of 6 m for the terrace VIIb. This is analogous to predicting the heights of the terraces along a transect at the critical location where H_{VIIb} has remained at 6 m and uplift has therefore been zero.

Thus, by substituting the regression-line values of a_i and b_i in the equation with $H_{VIIb} = 6$ m, the paleosea-level estimates for the several Wisconsin-age interstadials can be calculated (Figures 6.3 and 6.5, Table 6.2) without the further assumption of a constant uplift rate. The justification for the method is the very high correlation coefficients for the regression equations (Table 6.2). Only terrace IIa (50,000 to 40,000 yr ago), which has only three measured heights, has a least-squares predicted height error that is significantly greater than the errors arbitrarily assigned to the measured terrace heights on the transects (Table 6.1, Figure 6.3).

The calculated values for terraces I to IV are similar to those listed in Bloom *et al.* (1974, Tables 3 and 4), where they were calculated on the assumption of constant uplift. However, the newly calculated height of sea level during the formation of terrace VI (105,000 yr ago) is 0 m instead of the former value of –15 m, and the calculated height of sea level at the time of terrace V (82,000 yr ago) is –7 m instead of the former value of –13 m. The purpose of this chapter is to review the method, not the results, and so further discussion is deferred. However, the demonstration that a valid mathematical regression technique gives sea-level estimates that are quite similar to those estimates made by assuming constant uplift rate is justification for the assumed constancy of uplift at the time scale of 10^5 yr.

The last interglacial surface is widespread at depths of 6 to 10 m below a Holocene coral veneer on many atolls. If no more than 1 m of limestone has been lost from the reef surface during subaerial exposure, a lowering of the last interglacial reef surface from its assumed original height of +6 m to (for example) a present height of –6 m in about 120,000 yr implies an atoll subsidence rate of about 0.1 m per 1000 yr, a rate appropriate for subsidence of oceanic lithosphere during cooling (Sclater *et al.*, 1971; Bloom, 1980, p. 512).

TECTONIC MOVEMENTS ON THE TIME SCALE OF 20,000 YEARS

The straight-line regression equations with high correlation coefficients demonstrated above justify the assumption of constant uplift rate on the time scale of 10^5 yr, but do not require it on shorter times. Careful analysis of

TABLE 6.2 Regression Equations for Elevations (H_i) of Terrace i against Elevations of Terrace VIIb

H_i	Age (ka)	a_i	b_i	r	SL (m)
H_{VIIb}	124	1.00	0.00	1.000	+6.0 (assumed)
H_{VI}	105	0.77	–4.55	0.999	+0.1
H_V	82	0.60	–10.16	0.999	–6.6
H_{IV}	60	0.46	–26.80	0.999	–24.0
H_{IIIa}	50–40	0.41	–48.26	0.995	–45.8
H_{IIIb}	40	0.32	–40.21	0.981	–38.3
H_{II}	28	0.20	–36.25	0.995	–35.1
H_I	6	0.05	–3.93	0.969	–3.6

NOTE: These equations are used for determination of paleo-sea levels (SL). See Table 6.1 and Figure 6.5 (Bloom and Yonekura, 1985, Table 6.2).

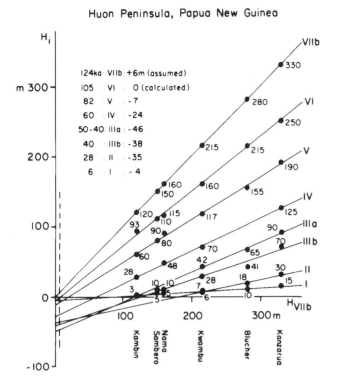

FIGURE 6.5 Regression of height of terrace i (H_i) as a function of height of terrace VIIb (H_{VIIb}) based on six transects on the Huon Peninsula, Papua New Guinea (Bloom and Yonekura, 1985, Figure 6.3).

Table 6.2 shows that there was variation in the Huon Peninsula uplift rate on the time scale of the 20,000-yr sampling interval. For instance, terrace VI is 85 percent the age of terrace VIIb (105 versus 124 ka), but is only 77 percent as high (the value of coefficient a_i, ignoring b_i). Uplift in the interval between 124,000 and 105,000 yr ago was somewhat greater than the long-term average.

Another way of modeling uplift rates on the 20,000-yr time scale is to accept sea-level estimates such as those in Table 6.2 and Figure 6.3 and from them calculate uplift rates for successive increments of dated uplift history. This method is appropriately called bootstrapping in that each increment of uplift is used as the basis for calculating the next older increment. An example (Figure 6.6) is drawn from work by Urmos (1985). The site is Araki Island, a small reef-terraced island 5 km south of the south coast of Santo Island, Vanuatu.

On Araki, a Holocene reef terrace about 5500 yr old is 26 m above present sea level. Assuming, for simplification, that sea level in the region has not changed in the last 5500 yr, the average late Holocene uplift rate is 4.75 m per 1000 yr. Above the large Holocene reef terrace on Araki

is a succession of small stair-step terraces up to the flat reef-capped summit at 237 m. The next dated terrace on the hillside is about 38,000 yr old and is now at a height of about 40 m. Since the rate of uplift for the last 5500 yr is known, and assuming an original paleosea level of −41 m (Figure 6.3), the increment of uplift between 5500 and 38,000 yr is calculated to be about 1.75 m per 1000 yr. The process is repeated for each step back in time, using the previously established estimates of sea level at the time of reef growth. The reef at the top of Araki Island is 105,000 yr old and is now at 237 m. If sea level at the time of origin was at present level (see above), then the average uplift rate for 105,000 yr has been 2.26 m per 1000 yr. However, successive increments of uplift rate range from 1.67 to as high as 4.75 m per 1000 yr (Figure 6.6), a factor of 2.8. In particular, the late Holocene uplift rate has been faster than at any prior time in the last 105,000 yr. We do not know if this is evidence of accelerated Holocene tectonic movement, or an artifact of the short sampling interval. We believe that the tectonic uplift of the region has accelerated in the Holocene, because nowhere in Vanuatu have we found a reef 28,000 to 30,000 yr old, such as has been found on the Huon Peninsula of Papua New Guinea (Chappell and Veeh, 1978). If the rapid uplift of the last 5500 yr had continued for as long as the last 28,000 yr, the interstadial terrace of that age would be far above the Holocene terrace, even though it started at a sea-level position 35 m below sea level (Table 6.2, Figure 6.3). However, the extreme size of the Holocene terrace on Araki Island could be caused by a relatively thin Holocene veneer over an older reef-terrace substrate, as suggested by the dashed trajectory of inferred uplift for a hypothetical 28,000-yr-old terrace on Figure 6.6. The presence of such a substrate under a Holocene veneer has been hypothesized from morphologic evidence in other parts of south Santo Island (Strecker et al., 1984), but will be verified only by future drilling.

The accelerated uplift during the last 5500 yr on Araki Island cannot be directly compared either to the long-term average rate or to older dated increments of uplift. It is possible that during any previous 20,000-yr interval between interstadial high sea levels, much of the total movement was concentrated in brief intervals of 5500 yr or less. There is no way that such jerkiness could be detected with the 20,000-yr sampling interval that is provided by the average spacing of interstadial sea-level oscillations. Therefore, we can only conclude that the last 5500 yr of uplift, during which postglacial sea level has been at its approximate present level, has been unusual by comparison to the 20,000- and 100,000-yr average rates, although we cannot disprove that those average rates consisted of shorter intervals of alternately fast and slow vertical movements.

CONTRADICTORY SEA-LEVEL EVIDENCE FROM OXYGEN ISOTOPES AND CORAL REEFS

Figures 6.2 and 6.3 illustrate the currently unresolved contradiction between sea level interpreted from the oxygen-isotope record and sea level calculated by regression equations on emerged coral-reef terraces. The heights of the several Wisconsin-age interstadial high sea levels in the coral record range from near present sea level to no lower than −46 m, or no more than one-third of full-glacial sea-level lowering. At comparable time intervals, the deep-sea oxygen-isotope record shows enrichment of $\delta^{18}O$ in the range of 50 to 70 percent of the values for full-glacial time. If the $\delta^{18}O$ value is interpreted as primarily controlled by ice volume (Shackleton and Opdyke, 1973; Shackleton, 1977), then the sea-level maxima during the several interstadial intervals would have been twice as low as the values derived from coral-reef studies. Only if cooler temperatures of evaporation caused about 70 percent of the observed $\delta^{18}O$ increase and ice volume caused about 30 percent would the results of the coral-reef regression analysis be similar to the predictions from the oxygen-isotope record.

SEA LEVEL DURING FULL-GLACIAL AND STADIAL INTERVALS

The Pleistocene epoch can be subdivided into glacial and interglacial ages, although the four glacial and three interglacial of the traditional classification is certainly wrong. Within each glacial age, lesser times of ice advances are called stadials and times of retreat are called interstadials (or interstades). It is notable that no stratigrapher has ever offered a subdivision of interglacial ages. It was never found necessary, because interglacial intervals have made up only about 10 percent or less of Pleistocene time and their duration was less that the inherent errors of the methods used to date them. Glacial ages, however, are rich in climate and sea-level detail, whether it is derived from oxygen-isotope or coral-reef studies.

An important question concerns the drop of sea level between the various interstadial sea-level maxima. As should be clear from previous sections of this report, emerged coral-reef terraces record only times just prior to interstadial and interglacial high sea levels, when tectonic uplift was briefly equal to rising sea level. With only two known exceptions, the theoretically possible reefs that should grow during the interval of tangency between tectonic uplift and rising sea level just after each stadial or glacial sea-level minima are unknown. Chappell (1974, 1983) described a cut-and-fill cycle in the uplifted deltaic foreset beds of the Tewai River delta on the Huon Peninsula of Papua New Guinea. Coral reefs that grew on the nearshore topset beds of this gravel delta are now uplifted as much as 400 m. A coral cap on a terrace at 390 m is correlated with terrace VIIb (Figures 6.3 and 6.5, Table 6.1) with an age of about 125 ka. Following this reef-building event at a relatively high sea level, the underlying deltaic foreset beds were eroded down to a present height of 320 m, representing an abrupt drop of sea level of about 70 m. Subsequently, the eroded section was reburied by aggradation up to a present height of about 300 m, and the new deltaic gravels were capped by a reef of series VI, with an age of 105 ka.

In a later paper, Chappell (1983, p. 24) expressed some reservation or ambiguity about the inferred 70-m drop and rise of sea level between 125 ka and 105 ka, but as the record now stands, this single locality may record an abrupt but extreme sea-level drop of 70 m, 50 to 60 percent of a full-glacial cycle, within the 20,000-yr interval between Papua New Guinea reefs VIIb and VI, which correlate very well with oxygen isotope stages 5e and 5c. A similar event in the Barbados record was reported by Steinen et al. (1973), in which sea level would have dropped to −71 ± 11 m in relation to present sea level between Barbados reef stages III (125 ka) and II (105 ka). This was included in a much-cited summary figure of the Papua New Guinea terrace chronology and sea-level record (Bloom et al., 1974; Figure 6.5) but was subsequently shown to be erroneous (Fairbanks and Matthews, 1978, p. 185). The erosional disconformity under Barbados reef II (105 ka) was shown not to separate reef II from underlying reef limestone of stage III (125 ka) but rather from an underlying limestone that was much older. Thus, the inferred sea-level drop between 125 ka and 105 ka could not be demonstrated in Barbados.

The only other place where coral limestone from a low sea level can be shown to have a rational place in a terrace sequence is on Araki Island, near Santo Island in the Republic of Vanuatu. The details of uplift history of this island are not yet published, but Urmos (1985) described a coral sample with an age of 153 ka and a very heavy $\delta^{18}O$ ratio appropriate for full-glacial oxygen-isotope stage 6, in an eroded reef section 180 m above sea level on Araki (Figure 6.6). The height and position of the sample locality would be appropriate for a reef that would have grown during the interstadial sea-level maxima V in the Papua New Guinea sequence, or oxygen-isotope stage 5a (83 ka). However, neither the age nor the isotopic ratio of the sample support this interpretation. As sketched on Figure 6.6, this coral apparently grew during the low sea-level minimum of the penultimate ice age when sea level was estimated at −165 m (oxygen-isotope stage 6; 140 to 160 ka). The island of Araki had then not yet emerged from the sea, but was a rising submarine tectonic block on which reef growth began when the extreme full-glacial low sea level exposed it to the photic zone. The reef was subsequently drowned by the rapid rise of sea level to isotope stage 5e, and was buried by younger reef growth. However, the tectonic uplift of Araki Island caused this ancient, low-sea-level reef to be exposed by erosion during the relatively high sea level about 80,000 yr ago (Figure 6.6).

Scattered in the gray literature of uranium-series dates from various coral regions are other anomalously old dates from relatively high inferred sea-level positions. Figure 6.6 offers a valid theory of how such samples can be explained rather than disregarded as erroneous. An intensive drilling program on selected emerged coral-reef terrace sequence, combined with U-series dating and oxygen-isotope analyses of suitable corals, could conceivably gain some other valid data points for full-glacial and stadial sea-level minima to supplement the current documentation of the intervening interglacial and interstadial maxima. It can be predicted that the amplitude of the oscillations would be greater than the amplitude shown by

the deep-sea $\delta^{18}O$ record, which is inevitably decreased by even the least amount of bioturbation. The climatic implications of such a record would be substantial. For example, the inferred 70 m drop of sea level recorded in the Tewai River delta section of Papua New Guinea suggests a dramatic initial and perhaps brief pulse of ice-sheet growth during oxygen isotope stage 5d that is not seen in the deep-sea record, perhaps because of its brevity. A sharp temperature minimum, corresponding to extreme cold in the areas of snow accumulation, is seen in the Vostok ice core (Lorius *et al.*, 1985). Another hint of the brief, sharp climatic deterioration in stage 5d is recorded in one exceptional palynologic record from France (Woillard and Mook, 1982). If climate can be shown to have changed from a warmer-than-present interglacial 125 ka to an ice volume equivalent to one-half or two-thirds of an ice age within 10,000 yr, the implications for an abrupt onset of the next ice age are serious. Andrews and Mahaffy (1976) showed how difficult it would be to generate large ice sheets and even 5 m of sea-level lowering within a few thousand years as a test of the "instant glacierization" hypothesis, although Mix and Ruddiman (1984) suggested how very rapid ice-sheet growth could occur around the margins of the North Atlantic if a relict mass of warm surface water remained in the region adjacent to abruptly cooled adjacent land masses.

LATE-GLACIAL AND HOLOCENE SEA LEVELS: THE 10^3- TO 10^4-YEAR TIME SCALE

Numerous Holocene sea-level graphs have been published (see Bloom, 1977). Good Holocene sea-level curves are well constrained by radiocarbon dates and have accurate depth or height measurements. However, the search for the elusive postglacial eustatic sea-level curve of global applicability has proved fruitless. Every coastal site seems to have its own unique sea-level history. Regional trends are obvious and predictable (Figure 6.7), but local effects such as the size and harmonic shape of an estuary or bay and their tidal heights change with continued rise of sea level. In the event of a future sea-level rise that averages several meters, similar regional and local deviations can be expected.

The role of isostatic warping of continental margins and ocean basins under the waterload of rising sea level has become a major topic of research (Cathles, 1975; Clark *et al.*, 1978; Peltier, 1982, Chapter 4, this volume). By their nature, coasts are at the margins of the oceans, where differential flexure in late-glacial and postglacial time has been maximal.

The graphic analysis or regression method of determining late Quaternary sea-level maxima (Figure 6.5) can also be applied to late-glacial and Holocene sea-level research

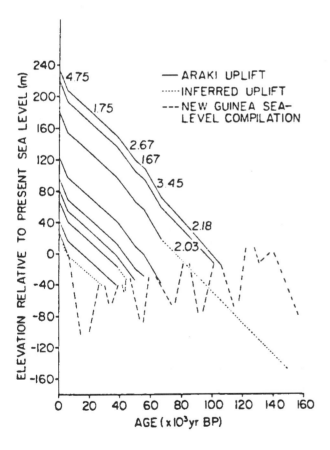

FIGURE 6.6 One model of incremental uplift rates (meters per 1000 yr) for the past 105,000 yr on Araki Island, Vanuatu. Note the unusually rapid late Holocene rate (Urmos, 1985).

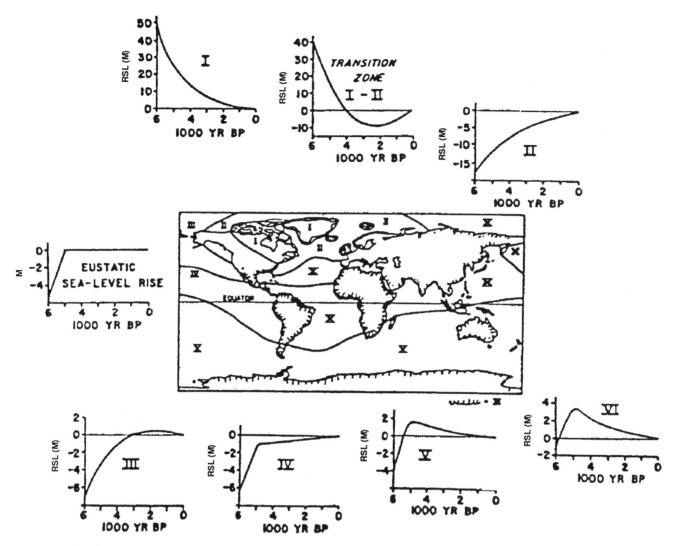

FIGURE 6.7 Distribution of the six predicted sea-level zones resulting from retreat of Northern Hemisphere ice sheets. Within each zone the form of the sea-level response is similar. Typical relative sea-level curves predicted for each zone are included, and they show the wide variety of sea-level expressions possible despite the assumption of no eustatic change since 5000 yrBP (Clark and Lingle, 1979).

(Figure 6.8). Results are tentative and preliminary, but are presented here because of their interesting implications about the relations between sea levels and postglacial isostatic adjustments (Peltier 1982, Chapter 4, this volume). Thirteen well-documented and representative sea-level curves were selected from the published literature to compile Figure 6.8. All have records of 8000 radiocarbon years or more. Some were chosen from Arctic regions where postglacial emergence due to isostatic uplift is the dominant process; others are from the Atlantic coast of the United States and Europe, which are generally areas of postglacial subsidence (zone II of Figure 6.7). Each se-

lected sea-level curve was sampled at 1000-yr intervals, recording the age and vertical height or depth of the sample locality. The height of sea level at each 1000-yr interval was then plotted as a function of the height (or depth) of the 8000-yr-old sea level at that locality (Figure 6.8) and the linear regressions were calculated for each 1000-yr trend relative to the 8000-yr-old trend line (Table 6.3). Unlike Eq. (6.2) and Figure 6.5, in which a value of b, initial sea level, is significant, all of the regression lines in Figure 6.8 pass within a fraction of a meter of the origin. That seems to imply that all 13 samples of each 1000-yr time interval were proportionately spaced; that is, the amount of emergence or submergence recorded for each 1000-yr interval in the last 8000 yr is proportional to the total amount of emergence or submergence at that place. This statement is true even though the total vertical change of relative sea level at the various localities in the past 8000 yr ranges from relative uplift of 70 m at Oslo, Norway, to relative drowning of 20 m on the Delaware coast.

Is it possible that mean sea level has not changed its absolute level (relative to the center of the Earth, for instance) for the last 8000 yr, and that the observed submergence at U.S. and western European ports is due to isostatic adjustments downward that are analogous and proportional to the uplift farther north in more recently deglaciated reasons? The near-zero intercept of all the regression lines in Figure 6.8 would seem to indicate that. If so, the best-documented sea-level histories, those of eastern United States and western Europe, are not directly reporting additional water into the world ocean, but local isostatic re-sponse to water added 8000 yr or more ago. Postglacial isostatic uplift has a half-life of 1000 yr or more, and would give the observed exponentially decreasing submergence observed at all the stations in the lower left side of Figure 6.8, but it is provocative that the submergence should be so rigorously proportional to postglacial uplift elsewhere. Perhaps future sea-level change will be controlled mostly by residual isostatic response to the loading and unloading that was completed as much as 8000 yr ago. Any climatically induced change of real sea level would be added to or subtracted from the trends documented by Figure 6.8.

CONCLUSIONS

1. Paleosea levels for the last 125,000 yr can be determined from tectonically uplifted coral-reef regions, but thus far only for the interglacial and interstadial sea-level

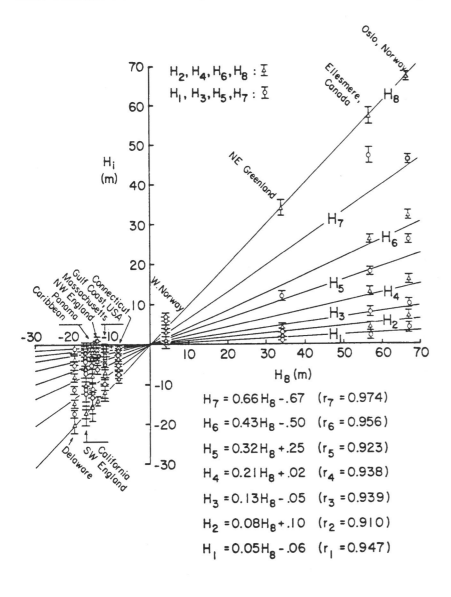

FIGURE 6.8 Regression analysis of sea-level curves from 13 sites (Table 6.1). As in Figure 6.5, height or depth of sea level at each younger 1000-yr interval (H7, H6, etc.) is plotted relative to the height or depth of the 8000-yr-old shoreline (H8) at the site. Odd- and even-numbered increments are plotted by different symbols only to permit easy visual evaluation of the regression lines. Method, equations, and coefficients are similar to those used in constructing Figure 6.5. The important difference here is that all regression lines pass through or very close to the origin, as shown by the small values of coefficient *b* (+0.25 to −0.67 m).

$H_7 = 0.66 H_8 - .67 \quad (r_7 = 0.974)$

$H_6 = 0.43 H_8 - .50 \quad (r_6 = 0.956)$

$H_5 = 0.32 H_8 + .25 \quad (r_5 = 0.923)$

$H_4 = 0.21 H_8 + .02 \quad (r_4 = 0.938)$

$H_3 = 0.13 H_8 - .05 \quad (r_3 = 0.939)$

$H_2 = 0.08 H_8 + .10 \quad (r_2 = 0.910)$

$H_1 = 0.05 H_8 - .06 \quad (r_1 = 0.947)$

TABLE 6.3 Sea-Level History for the Past 8000 Yr, Northern Hemisphere Sites

Location	Relative Sea Level at 1000-yr Intervals since 8000 yrBP							
	1000 yr	2000 yr	3000 yr	4000 yr	5000 yr	6000 yr	7000 yr	8000 yr
Ellesmere, Canada	2 ± 1	4 ± 1	8 ± 1	13 ± 1	18 ± 1	26 ± 1	47 ± 2	57 ± 2
NE Greenland	1 ± 1	2 ± 1	2 ± 1	2 ± 1	3 ± 1	4 ± 1	12 ± 1	34 ± 2
Oslo, Norway	4 ± 1	7 ± 1	10 ± 1	16 ± 1	26 ± 1	32 ± 1	46 ± 1	67 ± 1
Jaeren, Norway	0 ± 1	1 ± 1	1 ± 1	3 ± 1	6 ± 1	3 ± 1	7 ± 1	4 ± 1
Clinton, Conn.	−1.0 ± 0.5	−1.7 ± 1	−2.7 ± 1	−3.6 ± 1	−5.0 ± 1	−6.2 ± 1	−7.3 ± 1	−8.5 ± 1
Plum Island, Mass.	−0.7 ± 0.2	−1.3 ± 1	−2.1 ± 1	−3.2 ± 1	−4.6 ± 1	−7.0 ± 1	−9.5 ± 1	−12.2 ± 1
NW England	1 ± 1	1 ± 1	0 ± 1	−2 ± 1	−2 ± 1	−4 ± 1	−6 ± 1	−14 ± 1
San Francisco	−2 ± 1	−3 ± 1	−4 ± 1	−6 ± 2	−7 ± 2	−10 ± 2	−12 ± 2	−17 ± 3
U.S. Gulf Coast	−0.3 ± 0.5	−0.6 ± 0.5	−1.4 ± 0.5	−2.5 ± 0.5	−3.6 ± 1	−4.8 ± 1	−7.6 ± 2	−12 ± 2
Delaware	−1 ± 1	−3.5 ± 1	−5 ± 1	−8 ± 1	−11 ± 1	−14.5 ± 1	−17 ± 1	−20 ± 2
Carribean Islands	−1 ± 2	−2 ± 2	−2.5 ± 2	−3.5 ± 2	−5.3 ± 2	−7.7 ± 2	−11.5 ± 2	−15.5 ± 2
SW England	0 ± 0.5	−0.8 ± 0.5	−1.3 ± 0.5	−1.9 ± 0.5	−3.4 ± 0.5	−5.6 ± 0.5	−11 ± 1	−17.2 ± 1
Panama Republic	−2 ± 2	−2 ± 2	−5 ± 2	−8 ± 3	−12 ± 3	−12 ± 3	−13 ± 3	−15 ± 4

NOTE: For original sources, see Bloom (1977). Error estimates, added for evaluating the correlation coefficients shown on Figure 6.8, were estimated from statements in the original citations.

maxima that recurred at approximately 20,000-yr intervals. The intervening full-glacial or stadial times of glacier expansion and sea-level minima probably could be determined from coral islands by a future program of drilling and sampling. In addition, it is possible that drowned reefs of former stadial and full glacial times are still exposed on the steep flanks of suitable atolls or shelf reefs at depths on the order of 120 m. These could be studied and sampled from existing submersible research vessels.

2. A few data points now available from coral-reef studies suggest that the amplitude of multiple sea-level oscillations within the last ice age was substantial; perhaps equal to 50 percent or more of the maximum sea-level lowering that occurred about 18,000 yr ago. In particular, a sharp drop of sea level of 70 m may have occurred within about 10,000 yr after the last interglacial (isotope stage 5e) sea-level maximum. To the extent that such a sharp sea-level drop might occur at the beginning of the next ice age, it should be better documented. Only one place on the Huon Peninsula of Papua New Guinea is thought to record the event, but other regions of rapidly uplifted Quaternary coral reefs could be investigated.

3. The deep-sea oxygen-isotope record probably does not display the full amplitude of sea-level or temperature oscillations because of bioturbation. Furthermore, an obvious discrepancy persists between the absolute heights of interstadial sea-level maxima as interpreted from the deep-sea oxygen-isotope record and those calculated from uplifted coral reefs. The oxygen-isotope record of corals agrees with the record obtained from pelagic and benthic foraminifera, and so the problem is not to resolve a contra-diction between the coral-reef and the deep-sea isotopic record, but to determine whether the oxygen isotope ratios document ice volumes or temperature, and whether tectonic uplift models of coral coasts are valid. No new recommendation on this question can be made since the dilemma is well known and vigorously debated. Only the cliche, "More work is needed," can be offered.

4. With the last 10,000 yr, proportional rates of emergence or submergence at a selection of North America and European coastal sites seem to have been remarkably uniform. One interpretation could be that most of the glacier meltwater had returned to the sea by 8000 yr ago, and that the observed sea-level changes at various stations since then do not involve changes in the mass of ocean water, but only continuing isostatic adjustments to displaced ice and water loads (and possibly thermal expansion).

REFERENCES

Aharon, P., and J. Chappell (1986). Oxygen isotopes, sea level changes, and the temperature history of a coral reef environment in New Guinea over the last 10^5 years, *Paleogeogr. Paleoclimat. Paleoecol. 56*, 337–379.

Andrews, J. T., and M. A. W. Mahaffy (1976). Growth rate of the Laurentide ice sheet and sea level lowering (with emphasis on the 115,000 BP sea level low), *Quat. Res. 6*, 167–183.

Bloom, A. L., compiler (1977). *Atlas of Sea-Level Curves*, Int. Geol. Correlation Prog., Project 61, Dept. Geol. Sci., Cornell Univ., Ithaca, N.Y. (litho.), vii + 114 pp.

Bloom, A. L. (1980). Late Quaternary sea level change on South Pacific coasts: A study in tectonic diversity, in *Earth Rheol-*

ogy, Isotasy and Eustasy, N. A. Mörner, ed., Wiley, New York, pp. 505–516.

Bloom, A. L., and N. Yonekura (1985). Coastal terraces generated by sea-level change and tectonic uplift, in *Models in Geomorphology*, M. J. Woldenberg, ed., Allen and Unwin, Winchester, Mass., pp. 139–154.

Bloom, A. L., W. S. Broecker, J. M. A. Chappell, R. K. Matthews, and K. J. Mesolella (1974). Quaternary sea level fluctuations on a tectonic coast: New ^{230}Th/^{234}U dates from the Huon Peninsula, New Guinea, *Quat. Res. 4*, 185–205.

Cathles, L. M., III (1975). *The Viscosity of the Earth's Mantle*, Princeton University Press, Princeton, N.J., 386 pp.

Chappell, J. (1974). Geology of coral terraces, Huon Peninsula, New Guinea: A study of Quaternary tectonic movements and sea-level changes, *Geol. Soc. Am. Bull. 85*, 553–570.

Chappell, J. (1983). A revised sea-level record for the last 300,000 years from Papua New Guinea, *Search 14*, 99–101.

Chappell, J., and H. A. Polach (1976). Relationship between Holocene sea level change and coral reef growth at Huon Peninsula, New Guinea, *Geol. Soc. Am. Bull. 87*, 235–240.

Chappell, J., and H. H. Veeh (1978). ^{230}Th/^{234}U age support of an interstadial sea level of –40 m at 30,000 yrBP, *Nature 276*, 602–603.

Clark, J. A., and C. S. Lingle (1979). Predicted relative sea-level changes (18,000 years B.P. to present) caused by late-glacial retreat of the Antarctic ice sheet, *Quat. Res. 11*, 279–298.

Clark, J. A., W. E. Farrell, and Peltier, W. R. (1978). Global changes in postglacial sea level: A numerical calculation, *Quat. Res. 9*, 265–287.

Donner, J. J. (1965). Shore-line diagrams in Finnish Quaternary research, *Baltica 2*, 11–20.

Fairbanks, R. G., and R. K. Matthews (1978). The marine oxygen isotope record in Pleistocene coral, Barbados, West Indies, *Quat. Res. 10*, 181–196.

Harmon, R. S., T. L. Ku, R. K. Matthews, and P. L. Smart (1979). Limits of U-series analyses: Phase 1 results of the Uranium-Series Intercomparison Project, *Geology 7*, 405–408.

Imbrie, J., J. D. Hays, D. G. Martinson, A. McIntyre, A. C. Mix, J. J. Morley, N. G. Pisias, W. L. Prell, and N. J. Shackleton (1984). The orbital theory of Pleistocene climate: Support from a revised chronology of the marine δ^{18}O record, in *Milankovitch and Climate*, A. L. Berger *et al.*, eds., D. Reidel, Dordrecht, pp. 269–306.

Lorius, C., J. Jouzel, C. Ritz, L. Merlivat, N. Barkov, Y. S. Korotkevich, and V. M. Kotlyakov (1985). A 150,000-year climatic record from Antarctic ice, *Nature 316*, 591–596.

Machida, H. (1975). Pleistocene sea level of south Kanto analysed by tephrochronology, *Roy. Soc. N.Z. Bull. 13*, 215–222.

Mix, A. C., and W. F. Ruddiman (1984). Oxygen-isotope analyses and Pleistocene ice volumes, *Quat. Res. 21*, 1–20.

Muhs, D. R. (1983). Quaternary sea-level events on northern San Clemente Island, California, *Quat. Res. 20*, 322–341.

Ortlieb, L. (1980). Neotectonics from marine terraces along the Gulf of California, in *Earth Rheology, Isostasy and Eustasy*, N. A. Mörner, ed., Wiley, New York, pp. 497–504.

Ota, Y., S. Kaizuka, T. Kikuchi, and H. Naito (1968). Correlation between heights of younger and older shorelines for estimating rates and regional differences of crustal movements, *The Quaternary Research* (Japanese) 7, 171–181.

Peltier, R. (1982). Dynamics of the ice age Earth, *Adv. Geophys. 24*, 1–146.

Pillans, B. (1983). Upper Quaternary marine terrace chronology and deformation, South Taranaki, New Zealand, *Geology 11*, 292–297.

Sclater, J. G., R. N. Anderson, and M. L. Bell (1971). Elevation of ridges and evolution of the central eastern Pacific, *J. Geophys. Res. 76*, 7888–7915.

Shackleton, N. J. (1977). The oxygen isotope stratigraphic record of the late Pleistocene, *Phil. Trans. Roy. Soc. London. ser. B 280*, 169–182.

Shackleton, N. J., and N. D. Opdyke (1973). Oxygen isotope and paleomagnetic stratigraphy of equatorial Pacific core V28-238: Oxygen isotope temperatures and ice volumes on a 10 year and 10^6 year scale, *Quat. Res. 3*, 39–55.

Steinen, R. P., R. S. Harrison, and R. K. Matthews (1973). Eustatic low stand of sea level between 125,000 and 105,000 B.P.: Evidence from the subsurface of Barbados, West Indies, *Geol. Soc. Am. Bull. 84*, 63–70.

Stommel, H. (1963). Varieties of oceanographic experience, *Science 139*, 572–576.

Strecker, M. R., A. L. Bloom, L. M. Gilpin, and F. W. Taylor (1984). Karst morphology of uplifted Quaternary coral limestone terraces: Santo Island, Vanuatu, *Zeits. Geomorphol. 30*, 387–405.

Streeter, S. S., and N. J. Shackleton (1979). Paleocirculation of the deep North Atlantic: 150,000-year record of benthic foraminifera and oxygen-18, *Science 203*, 168–171.

Urmos, J. P. (1985). Oxygen Isotopes, Sea Level and Uplift of Reef Terraces, Araki Island, Vanuatu, M.S. Thesis, Cornell University, Ithaca, N.Y., 123 pp.

Woillard, G. M., and W. G. Mook (1982). Carbon-14 dates at Grande Pile: Correlation of land and sea chronologies, *Science 215*, 159–161.

Seismic Stratigraphic Record of Sea-Level Change

7

NICHOLAS CHRISTIE-BLICK, G. S. MOUNTAIN, and K. G. MILLER
Lamont-Doherty Geological Observatory of Columbia University

ABSTRACT

Seismic stratigraphy is a technique for stratigraphic correlation, which involves the identification of regional unconformities (sequence boundaries) in seismic reflection profiles. These surfaces form during times of relative sea-level fall as a result of abrupt basinward shifts in sites of sediment accumulation, and many have been interpreted to have a eustatic origin. Patterns of progressive onlap punctuated by downward shifts in onlap have been quantified in numerous petroliferous basins and form the basis for a eustatic sea-level curve. The interpretation of sea-level change from seismic stratigraphic data is controversial, however, and the purpose of this chapter is to evaluate critically the method and assumptions by which the sea-level curve has been obtained.

The geometry of stratal surfaces can be determined from seismic sections because these surfaces are among the most abundant reflectors of seismic energy in sedimentary basins. Interference between reflections from different surfaces limits vertical resolution to about one-quarter of the acoustic wavelength (typically on the order of tens of meters), and because the seismic pulse travels as a spherical wave front, there are also limits to horizontal resolution, that is, to the accurate determination of the spatial location and size of features that generate acoustic reflections. Sequence boundaries are recognized by the oblique termination of seismic events (onlap, downlap, toplap, and erosional truncation), and if appropriate velocity and density logs or vertical seismic profiles (VSPs) are available from boreholes, it may be possible to locate a given unconformity in the rock stratigraphy to well within the level of seismic resolution (on the order of meters). Nevertheless, owing to the limits of resolution, many more unconformities are present in a sedimentary basin than can be confidently traced with seismic data. Not all seismic events have primary stratigraphic significance, but nonstratigraphic events generally do not interfere seriously with stratigraphic interpretation. The assumption that unconformities have chronostratigraphic significance is generally a good approximation at an intrabasinal scale for coastal areas and shallow continental shelves. In this chapter, two examples of diachronous unconformities are presented that raise questions about the universal application of seismic sequence analysis to the study of eustatic changes.

Unconformities associated with downward shifts in coastal onlap result primarily from an increase in the rate of sea-level fall or a decrease in the rate of subsidence. A prominent shift in onlap can occur when the rate of sea-level fall is comparable to the rate of tectonic subsidence (<1 cm/1000 yr). Such an onlap shift is not necessarily accompanied by rapid regression of the shoreline. The controversial issue of whether sea level was oscillatory or monotonically decreasing from the Late Cretaceous to the onset of glaciation in the mid-Cenozoic is unresolved by published modeling studies, which did not account for the magnitude of short-term changes in paleobathymetry or sediment supply. The formation of sequence boundaries of probable eustatic origin at times of long-term sea-level rise (such as the Early Cretaceous) suggests that sea-level changes may generally be oscillatory, though of small amplitude. The approximate correspondence of several second-order sequence boundaries with times of major plate-boundary reorganization may reflect regional bias in the data, but in part is probably the result of tectonoeustasy. While the idea of global tectonic events cannot be entirely discounted, no viable mechanism has yet been established for producing correlative unconformities on a global scale. Most tectonic mechanisms for generating sequence boundaries predict dissimilar ages in different basins.

The key to distinguishing boundaries of eustatic and local origin is geochronological resolution. Although there are inherent uncertainties in the time scale, and in the correlation of seismic and rock sections with each other and to the time scale, many second-order boundaries appear to be very persistent from one basin to another. The occurrence of globally synchronous sea-level change is also corroborated by abundant conventional stratigraphic evidence for much of the Phanerozoic. The spacing of third-order boundaries is close to or finer than biostratigraphic resolution, and it may never be possible to resolve the ages of many of these boundaries with sufficient precision for objective correlation between basins.

Amplitudes of eustatic fluctuations cannot be inferred from seismic stratigraphic data alone because coastal aggradation (the vertical component of onlap) is primarily a result of basin subsidence, not sea-level rise; and downward shifts in onlap reflect only the rate of sea-level fall in relation to the rate of basin subsidence. Variations of coastal encroachment (the horizontal component of onlap) are sensitive to lateral gradients in the rate of subsidence, which are in general time-dependent and not necessarily linear. Whichever method is used to gauge changes in coastal onlap, the large component of subsidence cannot be easily or objectively removed to derive the smaller eustatic signal, and similarities in the patterns of coastal onlap for different basins for the most part indicate a similar overall subsidence history. For the purpose of deriving a eustatic sea-level curve, the global onlap chart provides information only about the times of sea-level fall and rise. We conclude that the main limitations of the eustatic curve derived from seismic stratigraphy are (1) the uncritical interpretation of all second- and third-order boundaries as eustatic, (2) the uncertainties about the calibration of many boundaries to the geological time scale, and (3) the largely conjectured inference of amplitudes.

INTRODUCTION

Conventional Stratigraphic Record of Sea-Level Change

Oscillatory changes in sea level relative to the continents, on time scales of <10 m.y., have long been inferred from paleobathymetric variations in facies successions, and from stratigraphic evidence for transgressions and regressions, that is, the alternating landward and seaward migration of the shoreline (e.g., Suess, 1906; Hallam, 1984). If consistent changes in water depth or shoreline location are determined on a regional to intercontinental scale, in a range of siliciclastic to carbonate facies, a eustatic (i.e., global) control is suggested, with the degree of confidence depending on the reliability and resolution of the facies interpretation, the precision of the biostratigraphic correlation, and the extent of such correlation. Examples of this procedure are described by McKerrow (1979; Ordovician and Silurian), Lenz (1982; Ordovician to Devonian), M. E.

Johnson et al. (1985; Silurian), J. G. Johnson et al. (1985; Devonian), Ramsbottom (1979; Carboniferous), Ross and Ross (1985; Carboniferous), Hallam (1981, 1988; Jurassic), and Hancock and Kauffman (1979; Cretaceous). Longer term eustasy and differential vertical movements of the continents (>10 m.y.), as indicated by the degree of continental flooding and the elevation of ancient shorelines, have been discussed by Bond (1978a,b, 1979), Harrison et al. (1981, 1983), and Sahagian (1987) and are summarized in Chapter 8 of this volume by C. G. A. Harrison.

Water depth and shoreline location are sensitive over a wide range of time scales to the rates of tectonic subsidence and clastic sediment supply (or carbonate sediment production) as well as to eustasy. Even if depositional base level is modulated by eustasy, a given eustatic event may not be evident in the facies that are preserved, and times of maximum or minimum water depth or transgression/regression may not be precisely synchronous at different localities (e.g., Parkinson and Summerhayes, 1985).

For example, the time of maximum water depth in a subsiding basin starved of sediment is clearly not the time of maximum transgression because of the time lag between the onset of subsequent regression and the arrival of significant amounts of sediment in the deep basin. Where sedimentation generally keeps pace with subsidence, and sea-level changes are oscillatory, maximum transgression corresponds with times at which the rate of eustatic rise is fastest; where the sediment supply is low, transgression may occur even during a eustatic fall, providing that the rate of net subsidence exceeds the rate of eustatic fall (Pitman and Golovchenko, 1983). Analyzing the effects of sea-level change on shoreline position in differentially subsiding passive continental margins, Pitman (1978) and Pitman and Golovchenko (1983) argued that under conditions of falling sea level and constant shelf gradient, a phase lag exists between the time of an instantaneous change in the rate of sea-level fall and the time at which the shoreline reaches a new equilibrium position. According to their analysis, lags are on the order of millions of years for a typical shelf, with longest phase lags corresponding to broad shelves of steep gradient and to ones characterized by slow subsidence.

In view of these considerations, the degree of synchroneity evident in the conventional stratigraphic record of short-term sea-level change is remarkable. Such synchroneity reflects a bias toward data from shallow-water continental platforms of low gradient. It also suggests that

sea level varies continuously at a broad range of frequencies rather than episodically and that the equilibrium shoreline positions inherent in the model of Pitman (1978) and Pitman and Golovchenko (1983) are rarely attained. Moreover, the assumption that the point of zero sedimentation rate either coincides with the shoreline (Pitman and Golovchenko, 1983) or lies at a fixed distance landward from it (Pitman, 1978) is unrealistic. The estimates of lag times are therefore hard to evaluate.

Seismic Stratigraphic Record of Sea-Level Change

Seismic stratigraphy is an approach to the investigation of sea-level fluctuations that is less sensitive than conventional stratigraphy to variations in sediment supply (Vail et al., 1977, 1980, 1984; Vail and Hardenbol, 1979; Vail and Todd, 1981; Vail, 1987). Developed for interpreting seismic reflection profiles (Figure 7.1), seismic stratigraphy makes use of regional surfaces of erosion or nondeposition known as unconformities or sequence boundaries. These form during times of relative sea-level fall, when patterns of sedimentation shift abruptly, generally toward the basin (near-horizontal bold lines in Figure 7.1). Vail et al. (1977) suggested that many sequence boundaries are of the same age in different parts of the world and are therefore due primarily to a global process, eustasy. They also developed a technique for quantifying the amplitude of relative sea-level change from the sawtooth patterns of

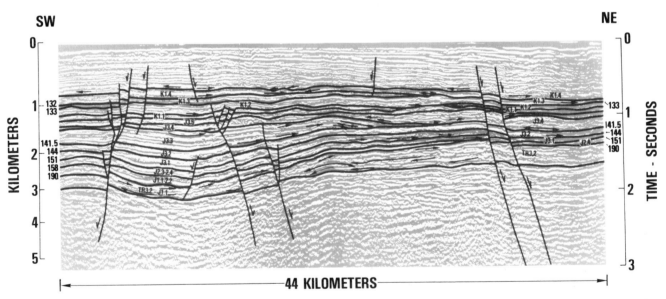

FIGURE 7.1 Seismic section northeast of Beatrice Field, Inner Moray Firth, North Sea, United Kingdom, showing interpretation of seismic sequences defined by the termination of seismic events (full arrows). Numbers with TR, J, and K prefixes identify Triassic, Jurassic, and Cretaceous sequences. Numbers on either side of the section are estimated ages of sequence boundaries in millions of years. Depth is given in kilometers and two-way travel time for seismic waves. Cross-cutting bold lines are inferred faults, with half arrows indicating apparent sense of displacement. From Vail et al. (1984).

onlap observed on seismic sections (see Figure 7.10). By comparing results from different basins, they obtained an estimate of global sea-level changes popularly known as the Vail sea-level curve. This led to considerable discussion as to why sea level should rise slowly and fall quickly. In subsequent publications, the original curve is termed a chart of "relative change of coastal onlap" (e.g., Vail and Todd, 1981; Vail *et al.*, 1984) and is the basis for a smoothly varying "eustatic curve," which takes into account the fact that discontinuous changes in onlap are accompanied by gradual changes in shoreline position. The most recent version, synthesizing for the first time the entire Mesozoic and Cenozoic, and incorporating the results of detailed well-log and outcrop studies, is that of Haq *et al.* (1987).

The so-called sea-level curve has been widely quoted in the geological literature and in recent textbooks (e.g., Kennett, 1982; Miall, 1984; Boggs, 1987), but apart from the obvious criticism that much of the supporting data has not yet been published, the possible limitations of the curve are not generally appreciated. In particular, the original coastal onlap chart is still commonly but incorrectly reproduced as the "sea-level curve." Recent articles by Brown and Fisher (1980), Watts (1982), Hallam (1984, 1988), Steckler (1984), Thorne and Watts (1984), Watts and Thorne (1984), Parkinson and Summerhayes (1985), Miall (1986), Summerhayes (1986), Burton *et al.* (1987), and Hubbard (1988) have raised several important issues for seismic stratigraphic interpretation, such as the origin and chronostratigraphic significance of seismic reflections, the precision with which unconformities can be identified and calibrated, the influence of tectonics in the formation of sequence boundaries, the significance and quantification of onlap (especially to derive amplitudes of eustatic fluctuations), and regional bias in the "global" curve. In the light of these criticisms, the purpose of this chapter is to evaluate the seismic stratigraphic method as a tool for investigating sea-level change. The main conclusion of this chapter is that seismic stratigraphy provides important information about the timing of sea-level fluctuations, on a time scale of millions of years, but little about magnitudes.

SEISMIC IMAGING OF STRATAL GEOMETRY

The gross geometry of unconformities and other stratal surfaces can be determined in seismic sections because these surfaces are commonly an important source of acoustic impedance contrasts. Acoustic impedance, or the product of rock density and the velocity of seismic waves traveling through the rock, determines to what extent seismic energy is reflected from a given surface. The range of velocities in common sedimentary rocks is greater than the range of densities, and velocity is consequently more important in

controlling the strength of reflections. For example, lithified sandstone and shale are characterized by densities of approximately 2.32 and 2.42 g/cm^3, respectively, and by corresponding velocities of 3.6 and 4.2 km/s Nafe and Drake, 1963; Dobrin, 1976). The amplitude of a reflection from a planar contact between these two rock types is about 10 percent of that of the incident wave. Nonlithified deep-sea sediments commonly possess velocity and density contrasts more subtle than these values, and the amplitudes of reflections from the more prominent stratal boundaries in such settings are consequently much weaker, perhaps 1 to 2 percent that of the incident wave. For weak reflections, the detection of stratal boundaries is thus critically dependent on the distinction of reflected seismic energy from background noise. This is a function of geologic setting, the quality of the recording and processing system, and the experience of the interpreter.

Most reflection events seen on a seismic section are composites of reflections from individual interfaces, but a comparison between seismic sections and corresponding well-log cross sections suggests that in many cases the configuration of reflections mimics the configuration of stratal surfaces at the level of resolution permitted by the seismic data (Vail *et al.*, 1977). A common but generally incorrect expectation is that reflections correspond to the boundaries of major lithological units, even where these cut across stratal surfaces (e.g., Hallam, 1984). Most lateral changes in sedimentary facies involve such gradual variations in acoustic impedance that they are not a significant source of reflections.

SEISMIC RESOLUTION

Despite major advances in the recording and processing of seismic reflection profiles over the last two decades, basic laws of physics limit the precision with which acoustic images portray the geometry of subsurface stratal boundaries. To gauge the limits of reflection profiling in practical terms, it is helpful to understand the processes that govern vertical and horizontal resolution. Vertical resolution concerns the ability to distinguish two closely spaced reflecting surfaces, regardless of whether these are boundaries of different beds or the upper and lower surfaces of the same bed. Horizontal resolution concerns the ability to detect narrow features and then to image them in their proper location.

Vertical Resolution

The limits to vertical resolution, thoroughly discussed by Widess (1973), Neidell and Poggiagliolmi (1977), Sheriff (1977, 1985), and Mahradi (1983), can be illustrated by considering reflections from a thin tabular layer of shale enclosed in a thick sandstone layer of lower impedance. Assuming the density and velocity of shale given above, a wave of frequency 20 Hz passing through the shale has a wavelength of 210 m. Experiments and synthetic models have shown that for layers as thin as one-quarter wavelength (about 50 m for the shale), local peaks corresponding to reflections from the upper and lower surfaces are discernible, and the time separation between the peaks is an accurate measure of layer thickness (Widess, 1973). However, for layers thinner than 1 wavelength, addition of the two reflections results in a composite wave whose peak amplitude is sensitive to bed thickness (Neidell and Poggiagliolmi, 1977). As bed thickness is decreased, the amplitude attains a minimum value at a thickness of one-half wavelength, and it increases again to a maximum at one-quarter wavelength. For thicknesses less than one-quarter wavelength, the distinct contributions from the upper and lower surfaces cannot be identified, and the shape of the composite wave continues to change until the layer thickness is one-eighth wavelength. For still thinner layers, the wave shape stabilizes but peak amplitude decreases uniformly with decreasing layer thickness. It does not matter whether a thin layer is a single unit or a composite of numerous thin beds. As long as the aggregate thickness is less than one-eighth wavelength, the reflected energy is characterized by a nonunique composite waveform. For example, if the shale layer were 1 m thick, it would produce a reflection whose amplitude was roughly 0.5 percent that of the incident wave. This bed would produce the same reflected waveform if its thickness were 5 m or 50 cm. Only the amplitude would vary. Unfortunately, to use these amplitude variations as measures of layer thickness, a nearby contact with a layer thicker than 1 wavelength is needed for calibration. This is rarely possible.

The limit to vertical resolution generally increases with acoustic velocity and burial depth and decreases with increasing acoustic frequency. Acoustic velocity tends to increase during burial as a result of progressive compaction and cementation. Seismic energy is also attenuated by reflection from successive interfaces, so that deep reflectors are less easily distinguished from noise than shallow ones. Assuming a monotonic seismic source of frequency 20 Hz, the practical one-quarter wavelength limit to resolution cited above for shale is 50 m. For rocks with high acoustic velocity, such as limestone (about 5.0 km/s), the one-quarter wavelength limit is higher (about 65 m). Fortunately, tuned arrays of marine seismic sources in common use are of relatively broad band and generate components of many tens of hertz. To the extent that sufficient high-frequency signal is generated and recorded, the practical vertical resolution is on the order of 10 to 50 m.

Horizontal Resolution

A convenient way of understanding the concept of horizontal resolution is to think of acoustic pulses traveling in the Earth as spherical wave fronts. Each impedance change encountered by a downgoing pulse acts as a point-source reflector that returns energy as a similar three-dimensional wave front. It is therefore possible to receive an echo off a buried feature at a considerable horizontal distance from the feature. In marine seismology, streamers consisting of hydrophones summed together in a linear array minimize engine noise and reflections returned from features directly ahead of or behind the streamer, but they do not eliminate side echoes.

The problem of the resolution of size and geometry can be explained by reference to Fresnel zones. The pulse from a seismic source is typically several cycles long, and the reflections originating from early and late portions of the pulse therefore interfere with each other. Early arrivals add constructively, but later arrivals are increasingly out of phase and approach total cancellation when the time lag corresponds to one-half wavelength. Arrivals at greater time delays alternately add and subtract, but their contribution is relatively small. Because the seismic pulse is a three-dimensional spherical wave front, the initial part of a pulse that reflects constructively from a planar surface has a measurable cross-sectional area called the first Fresnel zone (Sheriff, 1977).

As a result of this phenomenon, the seismic pulse can be regarded as having a "footprint" whose size depends on (1) the fundamental frequency of the source, which determines the critical one-half wavelength distance, and (2) the depth to the reflector, which determines the radius of the wave front, and hence the radius of the first Fresnel zone. For a seismic source centered at 20 Hz, and a stratigraphic section with an average acoustic velocity of 2.25 km/s, a reflection from a depth of 4 s (two-way travel time) has a diameter of about 500 m (Sheriff, 1985). This simple example shows how objects not directly beneath the seismic streamer can reflect considerable energy. Because the width of the footprint depends on the pulse frequency, sources of low frequency tend to detect reflectors out of the plane of the seismic section more readily than those of high frequency.

The radius of the first Fresnel zone also governs the width of subbottom reflectors that can be accurately detected. Synthetic models have shown that strong returns can be expected from features less than one Fresnel zone wide, but little information is preserved about the size or shape of such bodies (Sheriff, 1985). All objects appreciably narrower than one Fresnel zone have nearly identical reflections.

RECOGNITION OF UNCONFORMITIES IN SEISMIC SECTIONS

Unconformities are recognized in seismic sections by the oblique termination of seismic events (see Figure 7.1). Events that terminate against an underlying boundary indicate stratal onlap or downlap (Figure 7.2). Onlap involves updip terminations, unless the stratigraphic section has been subsequently tilted; downlap is generally downdip. Events terminating against an overlying boundary indicate toplap or erosional truncation. Toplap arises from sediment bypassing across the top of a prograding sedimentary wedge, whereas erosional truncation involves the removal of previously deposited sediment. These stratigraphic relations commonly change laterally along a boundary, and seismic events locally parallel unconformities, especially at correlative conformities, where there is effectively no depositional hiatus. A reflection is generated by the unconformity itself only if a significant impedance contrast is present. Such reflections may be discontinuous, especially where discordance of overlying and/or underlying reflectors leads to lateral phase changes (Vail *et al.*, 1977, 1980; Vail and Todd, 1981).

Unconformities possessing the fundamental geometrical properties of onlap, downlap, toplap, and erosional truncation vary considerably in lateral extent, from hundreds of meters to hundreds or thousands of kilometers, and they separate depositional sequences ranging in thickness from meters to thousands of meters. Observations in outcrop indicate that large-scale sequences commonly contain other sequences of smaller scale or "higher order," although there is probably a continuum of possible scales (Ryer, 1983; Busch and Rollins, 1984). Owing to the restrictions of resolution described above, not all instances of stratal termination against a given unconformity are imaged in seismic data, so that reflections may locally appear concordant even where the corresponding strata are slightly discordant. In addition, apparent seismic toplap and downlap can arise where clinoforms merge into shelf or basin deposits that are too thin to be resolved acoustically (see Figure 8 of Tucholke, 1981). The number of unconformities identified in a seismic section is therefore limited by vertical seismic resolution, and in general many more unconformities are present in a sedimentary basin than can be confidently traced with seismic data. It is usually not possible with seismic data to resolve a given unconformity over the entire area in which it is present.

These limitations are overcome in practice by selection of unconformities on the basis of their lateral persistence and by calibration against the rock record. Although vertical seismic resolution is typically on the order of tens of meters, seismic sections acquired during exploration for petroleum are routinely and reliably tied to borehole or well data by means of synthetic seismograms derived from velocity and density logs, or by using vertical seismic profiles (Sheriff, 1977; Badley, 1985). Where biostratigraphically resolvable or associated with a distinct change in facies or stratal dip, it may be possible to locate a given

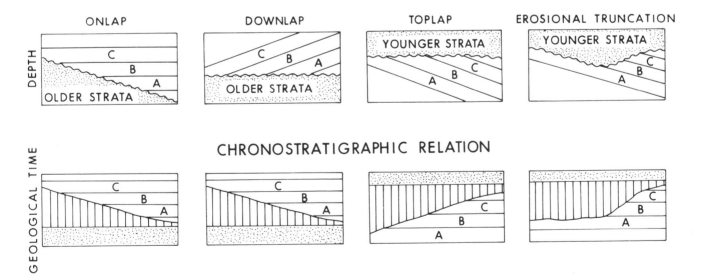

FIGURE 7.2 Stratigraphic and chronostratigraphic relations of onlap, downlap, toplap, and erosional truncation. Vertical ruling indicates the duration of the hiatus represented by the unconformity. After Ramsayer (1979).

unconformity in the rock stratigraphy to at least an order of magnitude better than seismic resolution (see Vail and Todd, 1981). The most common error in seismic stratigraphy arises where such well control is lacking, and an unconformity reflection is traced laterally into a prominent reflection that actually onlaps the unconformity (Vail et al., 1980).

In seismic stratigraphic work, unconformities are delineated primarily by a downward (basinward) shift in the position of coastal onlap. This is not because boundaries with such geometry are necessarily more important than boundaries lacking such a shift in onlap, but for the most part because such boundaries can be traced with the greatest confidence. Indeed, for seismic interpretation, Vail et al. (1984) restricted the term unconformity to those surfaces involving local erosional truncation or subaerial exposure, phenomena commonly associated with a downward shift in onlap. According to this usage, marine surfaces resulting from deep-water sediment starvation and/or dissolution, and representing significant hiatuses but lacking evidence for erosion, are not unconformities. The delineation of unconformities on the basis of onlap and downlap is especially appropriate where "follow cycles" are present. A follow cycle is commonly associated with a strong reflection along an erosional surface, and consists of a second peak on the waveform beneath the principal reflection (Vail and Todd, 1981). Where the follow cycle masks underlying reflections, the unconformity may appear to be stratigraphically lower than its true position.

A downward shift in onlap would seem to be most easily resolved where (1) the gradient of the depositional surface is large (e.g., the continental slope); (2) there is significant differential subsidence; (3) the hiatus represented at a given locality is long; and (4) overall sedimentation rates are high, as in young, rapidly subsiding margins. However, as discussed below, the situation is complicated by the fact that an unconformity produced by a given eustatic event is of greatest regional extent in old, slowly subsiding margins. By considering rates of tectonic subsidence and seismic resolution, Thorne and Watts (1984) concluded that for the passive margin of the eastern United States (an "old" margin), it is unlikely that seismic stratigraphy can resolve unconformities representing a hiatus of less than 4 m.y. This may be overly pessimistic for seismic stratigraphy in general if surfaces as close as one-quarter acoustic wavelength can be resolved, and if unconformities can be traced toward the continent from areas of high subsidence rate, where the reflection terminations are most obvious. Assuming a practical limit to vertical resolution of 50 m and a relatively slow sediment accumulation rate of 2 cm/1000 yr, it should be possible to resolve a hiatus of 2.5 m.y. For broadband seismic sources, the resolution may be considerably better.

SEISMIC REFLECTIONS LACKING PRIMARY STRATIGRAPHIC SIGNIFICANCE

Although seismic sections resemble geologic cross sections, not all seismic events have primary stratigraphic significance. Examples of nonstratigraphic events are those produced by low-angle faults and diagenetic boundaries, together with such features as multiples, coherent noise, diffractions not migrated during data processing to their proper position, and sideswipe (energy returned from outside the plane of the section; see Tucker and Yorston, 1973). Low-angle normal faults and thrust faults in places geometrically resemble some stratigraphic boundaries, but they are not a significant source of confusion in most of the basins used to derive information about sea-level change. Diagenetic boundaries appear to be best developed in fine-grained marl-limestone successions, in which much of the "bedding" may of diagenetic origin, or at least significantly modified by diagenesis (Hallam, 1986; Ricken, 1986). Fortunately, such diagenetic layering mainly parallels depositional layering, and usually does not generate artificial reflection terminations that could be confused with sequence boundaries. Multiples, noise, diffractions, and sideswipe can generally be recognized by an experienced interpreter because they tend to cut across events of stratigraphic origin. Of course, the interpretation of seismic data is not always straightforward, and an example from USGS seismic line 25 of the passive margin of the eastern United States has been discussed by Thorne and Watts (1984).

Line 25 passes through DSDP Site 612 and within 11 km of the COST B-3 well (Figure 7.3). Between these two holes, well-defined reflection terminations on line 25 are present at depths of between 2.3 and 2.0 s (two-way travel time), with at least four instances of apparent onlap against a prominent reflection (event 6 in Figure 7.3) in a lateral distance of only 8 km. In spite of this geometry, no hiatus has yet been detected with biostratigraphic data in COST B-3 at the level of these terminations (Poag, 1980), and only a minor hiatus has been observed at Site 612 (upper lower to lower middle Eocene; Miller and Hart, 1987; Poag and Low, 1987). Moreover, the high amplitude of event 6 at Site 612 can be related to a postdepositional diagenetic front, which is slightly oblique to stratal surfaces, and which separates siliceous nannofossil chalk above from porcellanite-bearing nannofossil chalk below. Two interpretations are possible. (1) The boundary is not a sequence boundary, but a diagenetic front, and the apparent termination of reflections is an artifact of poor seismic resolution (Thorne and Watts, 1984). (2) The boundary is indeed a sequence boundary, but one on which a diagenetic front has been superimposed and for which the biostratigraphic evidence is inconclusive. The existence of a prominent reflection immediately beneath and parallel to event 6 favors the second interpretation. If the boundary were diagenetic, onlapping reflections should continue across it. In addition, most of the biozonation in the COST B-3 well (for which the greatest hiatus should be observed) was based on rotary cuttings rather than cores, and a marked decrease in sediment accumulation rate in the lower and middle Eocene strata encountered in this well is consistent with a hiatus (Poag and Schlee, 1984).

CHRONOSTRATIGRAPHIC SIGNIFICANCE OF UNCONFORMITIES

An unconformity is a buried surface of erosion or nondeposition, whose principal significance for seismic stratigraphy is that it separates younger sediments or sedimentary rocks from older sediments or rocks below (Figure 7.4). The term unconformity thus refers to surfaces not only at a great range of scales but also involving hiatuses of markedly different duration, so that an unconformity at a small scale might be regarded as a conformity at a larger scale. An assumption of the technique of seismic interpretation promoted by Vail *et al.* (1977, 1984) is that a given surface has the same chronostratigraphic significance throughout a given basin; that is, although the duration of the hiatus represented may vary laterally, the sediments above are everywhere younger than the sediments below (Figure 7.4a). The possibility that an unconformity might be diachronous, in the sense that the sedi-

ments above the surface at one locality (e.g., D_1 in Figure 7.4b) might be older than those below it at a different locality (e.g., D_2 in Figure 7.4b), is generally excluded both for observational reasons and because sequence boundaries are regarded by Vail *et al.* as primarily a response to eustatic fluctuations, with rates comparable to or greater than the rate of tectonic subsidence. It is doubtful that high rates of sea-level change are necessary, but the chronostratigraphic assumption is probably a good approximation in coastal areas and on the continental shelves of passive margins. In such settings, the formation of unconformities is largely controlled by variations in the rates of subsidence and sea-level change, and during time intervals equivalent to a depositional sequence (1 to 10 m.y.), subsidence occurs at a relatively uniform rate. The assumption of chronostratigraphic significance may not be valid in the deep oceans, where unconformities are not necessarily related to variations in depositional base level (Tucholke, 1981; Tucholke and Embley, 1984), and in some tectonically active areas, where the rate of tectonic subsidence is not only spatially and temporally variable but changes diachronously within the basin. We do not believe that diachronous unconformities constitute a significant difficulty for seismic stratigraphic interpretation in most of the basins used by Vail *et al.* (1977, 1984) to derive the sea-level curve, but here we briefly discuss two examples of diachronous stratigraphic boundaries because the existence of such boundaries is not recognized by most seismic stratigraphers.

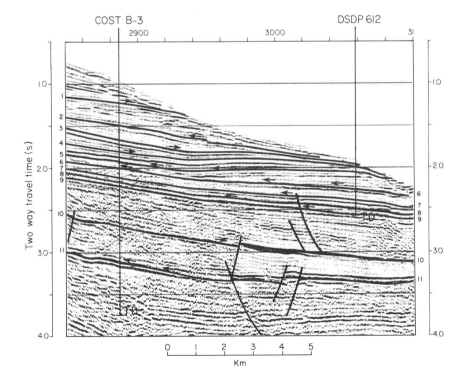

FIGURE 7.3 U.S. Geological Survey seismic line 25 between shot points 2847 and 3100. DSDP Site 612 was drilled at 3558 on line 25. COST B-3 has been projected from 11 km to the northeast to 2885. The numbered bold lines indicate tentative interpretations by A. B. Watts and N. Christie-Blick of sequence boundaries. These are based entirely on reflection terminations (arrows) observed in this single very short segment of the profile, and are subject to revision. After Thorne and Watts (1984).

Diachronous Unconformity on the Blake-Bahama Outer Ridge

Deposition and erosion in the deep ocean are controlled by many processes, including bottom-water current speed, bottom-water chemistry, sediment composition, and sediment cohesion (Tucholke and Embley, 1984). Of these, changes in bottom-water currents appear to be primarily responsible for a diachronous unconformity on the Blake-Bahama outer ridge (BBOR).

Deep-ocean currents are driven by subtle density differences related to temperature and salinity distributions, and they are capable of eroding the sea floor and transporting particles as large as silt and fine sand in suspension (Richardson *et al.*, 1981). A particularly well studied current, the western boundary undercurrent (WBUC), flows south and west along the continental rise of the eastern United States (Heezen *et al.*, 1966). The Coriolis force, which tends to turn the flow to the right in the Northern Hemisphere, is balanced by an opposing pressure gradient, and this results in a quasi-steady geostrophic current that does not necessarily flow in a straight line.

The pressure gradient in a geostrophically balanced system can be supplied by several factors, including seafloor topography. The BBOR on the continental rise off Georgia is a topographic obstruction that exerts a substantial pressure gradient on the WBUC. As the water is diverted around the ridge toward the east (to the left when looking downcurrent), the current speed increases to maintain a constant volume rate of transport, and this leads to an increase in the Coriolis force. The increased Coriolis force in turn increases the tendency for the water mass to turn to the right against the BBOR. This self-regulating system has operated in dynamic balance over the past 25 m.y., and little sedimentation has taken place beneath the core of the flow. In contrast, the region of more gentle seafloor gradient along the crest of the BBOR has not experienced a history of topographically intensified bottom currents. Depositional conditions have been maintained, and so much sediment has fallen out of suspension that the BBOR is now over 2 km high.

The basic elements of the inferred history of the BBOR are shown schematically in Figure 7.5a, which represents a view downcurrent. The BBOR is on the right, and current strength is portrayed by contours of equal speed. Sediment carried in suspension by the WBUC accumulates most rapidly in areas away from the core of the flow, while nondeposition or even erosion takes place beneath the region of greatest current strength. As sediments lap onto the base of the BBOR, the zone of nondeposition migrates upslope (Figure 7.5b). The long-term effect of this process is to produce a diachronous unconformity (Figure 7.5c). Without borehole control, incorrect age relations might be inferred because all of the strata lapping onto the base of the BBOR appear to be younger than any of the strata exposed along the erosional upper flank. In fact, they are of the same age. The error lies in assuming that the erosional surface was created at the same time everywhere.

Past contour-following geostrophic currents are likely to have been important along lower continental slopes and upper rises. In these regions, unconformities may be related both to changes in depositional base level, as in the case of the continental shelves, and to margin-parallel oceanic currents such as those of the BBOR. Some unconformities of the continental slopes may therefore be diachronous, and this possibility needs to be investigated by further seismic stratigraphic studies of sediments deposited in this setting.

Diachronous Unconformity in Alluvial-Fan Sediments Along the San Andreas Fault

An example of a diachronous unconformity in an active tectonic setting has been documented by Weldon (1984) in the vicinity of the strike-slip San Andreas Fault in southern California (Figure 7.6). The Harold Formation, Shoemaker Gravel, and Older Alluvium are Pleistocene alluvial and alluvial-fan sediments, derived from the San Gabriel Mountains southwest of the San Andreas Fault, and deposited northeast of the fault within the Mojave

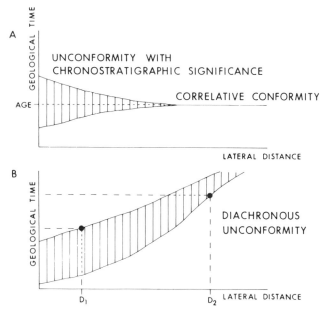

FIGURE 7.4 Chronostratigraphic cross sections of (A), an unconformity with chronostratigraphic significance; and (B), a diachronous unconformity.

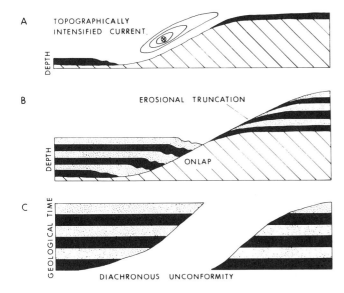

FIGURE 7.5 Development of a diachronous unconformity along the east side of the Blake-Bahama outer ridge (BBOR). (A) View downcurrent, with current strength portrayed by contours of equal speed. (B) Sediment carried in suspension by the western boundary undercurrent (WBUC) accumulates most rapidly in areas away from the core of the flow, while the region of greatest current strength is characterized by nondeposition or erosion. (C) Migration of the zone of nondeposition upslope produces a diachronous unconformity. Conventional seismic sequence analysis would incorrectly predict that the onlapping strata on the left are entirely younger than the truncated strata on the right.

Desert. Sedimentation was accompanied by uplift and tilting, with most of the deformation concentrated in the time interval represented by the angular unconformity between the Shoemaker Gravel and Older Alluvium (Miesling, 1984). Magnetostratigraphic results show that the prominent unconformity between the Shoemaker Gravel and the Older Alluvium is markedly diachronous, occurring within the Matuyama reversed interval in the vicinity of Crowder Canyon, but within the Brunhes normal interval at Phelan Peak and near Puzzle Creek, approximately 23 km northwest of Crowder Canyon. We should expect similar diachronous unconformities on a variety of scales wherever blocks are uplifted or folded, or basins subside in a diachronous fashion, as is common in strike-slip basins such as those of southern California (Christie-Blick and Biddle, 1985), and in basins where sedimentation is accompanied by the propagation of thrust faults (J. Suppe, Princeton University, personal communication, 1986).

ORIGIN OF UNCONFORMITIES

Even the most widespread unconformities are of finite areal extent because they tend to pass laterally into cor-

relative conformities (Figure 7.4a). The formation of an unconformity can therefore be considered in terms of the expansion and subsequent burial of zones of nondeposition or erosion. At the scale of a seismic section, unconformities in shelf and coastal environments are controlled largely by two factors: (1) changes in depositional base level, an imaginary surface asymptotic to sea level, and above which significant sediment accumulation is not possible; and (2) sediment supply, or production in the case of carbonates. On a smaller scale, of course, factors such as the grain size and cohesion of available sediment, the direction and strength of currents, water depth, depth to wave base, and the geometry of the depositional surface influence the development of unconformities. These are subordinate in comparison with depositional base level and sediment supply and are ignored in the following discussion. We also focus primarily on those shelf and coastal environments in which paleoclimatic and paleooceanographic changes are relatively unimportant in the formation of unconformities.

The elevation of base level at a particular locality is a function of the rates of change of tectonic subsidence and sea level. By definition, points at base level are subject to sediment bypassing, and those above base sea level, to erosion. Expansion of the zone of bypassing is therefore promoted by a decrease in the rate of subsidence and by an increase in the rate of sea-level fall. Unconformities tend to become buried when the rate of subsidence increases or the rate of sea-level fall decreases (or sea level is rising). Nondeposition and the development of condensed intervals in deep water tend to occur at times of sea-level rise because at these times available terrigenous sediment is trapped preferentially in nearshore and coastal areas. A decrease in sediment supply or production also promotes the development of unconformities in coastal regions, as in the familiar example of the switching of delta lobes; and a minor downward shift in the position of coastal onlap can be produced by a decrease in regional sediment supply (see below). However, the lowering of depositional base level is a more effective mechanism for the depression of coastal onlap on a regional scale.

A different interpretation of marine unconformities along continental margins has been suggested by Brown and Fisher (1977, 1980). According to them, marine onlap against the continental slope is commonly initiated not by a relative sea-level fall but during times of diminished sediment supply. At these times, sediment is thought to be eroded from the shelf and redeposited in deeper water. This mechanism was invoked to explain instances of marine onlap where no evidence exists for a downward shift in coastal onlap. However, a decrease in sediment supply and corresponding increase in water depth would seem to be unfavorable for the reworking of shelf sediments. This

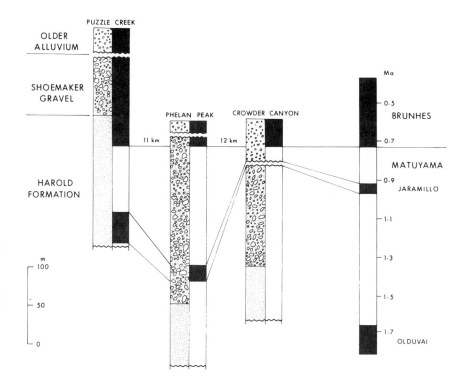

FIGURE 7.6 Magnetic stratigraphy of the Victorville Fan sediments on the northern flank of the San Gabriel Mountains, southern California. Black stripes represent chrons and subchrons of normal polarity; white stripes represent intervals of reversed polarity. All of the units and the angular unconformity between the Shoemaker Gravel and Older Alluvium are younger to the northwest (Puzzle Creek) than to the southeast (Crowder Canyon). After Weldon (1984).

is because storms and tides are less effective in transporting sediment as the shelf becomes deeper. Moreover, as discussed above, not all stratal terminations are acoustically resolved, especially in shelf deposits. Thus the mechanism proposed by Brown and Fisher (1977, 1980) may not be necessary.

Sea-Level Change and Sediment Supply

Conditions for the formation of sequence boundaries in passive continental margin settings have been considered quantitatively by Thorne and Watts (1984). The following is an elaboration of their analysis. To simplify the discussion, we first consider instantaneous changes in the rates of sediment supply and sea-level fall, and time intervals that are sufficiently long for topographic profiles to become dynamically graded. The analysis is then extended to oscillatory sea-level changes and to profiles that are not in dynamic equilibrium.

The rate of tectonic subsidence (Y) of a basin can be written (modified from Steckler and Watts, 1982):

$$\dot{Y} = \Phi\dot{S}^* \left(\frac{\rho_m - \rho_s}{\rho_m - \rho_w}\right) + \dot{W}_d$$
$$+ \dot{\Delta}_{SL} \left(\frac{\rho_m - \rho_w (1 - \Phi)}{\rho_m - \rho_w}\right), \qquad (7.1)$$

where W_d is the rate of increase in water depth; Δ_{SL} is the rate of eustatic sea-level fall (using the sign convention of Thorne and Watts, 1984); S^* is the rate of increase in the sedimentary thickness, corrected for the effects of compaction; ρ_m, ρ_s, and ρ_w are the densities of mantle, sediment, and water; and Φ is the basement response function relating sediment and water loads to tectonic subsidence. Note that ρ_m and ρ_w are constants but that ρ_s actually changes with time as the sediment becomes lithified. If we assume Airy isostasy, $\Phi = 1$, Eq. (7.1) simplifies to

$$\dot{Y} = \dot{S}^* \left(\frac{\rho_m - \rho_s}{\rho_m - \rho_w}\right) + \dot{W}_d + \dot{\Delta}_{SL} \left(\frac{\rho_m}{\rho_m - \rho_w}\right). \quad (7.2)$$

For a point above sea level, Eq. (7.2) must be modified to correct for the sediment load above the datum ($-\dot{h}$):

$$\dot{Y} = \dot{S}^* \left(\frac{\rho_m - \rho_s}{\rho_m - \rho_w}\right) + (\dot{h} + \dot{\Delta}_{SL}) \left(\frac{\rho_m}{\rho_m - \rho_w}\right), \quad (7.3)$$

where $\dot{h}$ is the rate of decrease of elevation with respect to sea level. If the lithosphere has strength, and loads are compensated regionally rather than locally, the basement response function (Φ) in Eq. (7.1) assumes a time-dependent value that varies from less than to greater than unity, but this does not change the overall conclusions of the following discussion.

Equations (7.2) and (7.3) can be rearranged as follows:

$$\dot{\Delta}_{SL} = \dot{Y}\left(\frac{\rho_m - \rho_w}{\rho_m}\right) - \dot{S}^*\left(\frac{\rho_m - \rho_s}{\rho_m}\right)$$

$$- \dot{W}_d\left(\frac{\rho_m - \rho_w}{\rho_m}\right) \qquad (7.4)$$

and

$$\dot{\Delta}_{SL} = \dot{Y}\left(\frac{\rho_m - \rho_w}{\rho_m}\right) - \dot{S}^*\left(\frac{\rho_m - \rho_s}{\rho_m}\right) - \dot{h}. \quad (7.5)$$

The ratio $(\rho_m - \rho_w)/\rho_m$ is approximately equal to 0.7; and $(\rho_m - \rho_s)/\rho_m$ is approximately 0.3.

The expression of a sequence boundary for a ramp setting in a differentially subsiding basin under conditions of falling eustatic sea level ($\dot{\Delta}_{SL} > 0$) is illustrated in Figure 7.7. The point of onlap A (Figure 7.7a) lies on what is here termed a line of critical bypassing, deposition taking place basinward of A and bypassing or erosion occurring on the landward side. At A, $\dot{S}^* = 0$, and

$$\dot{\Delta}_{SL} = \dot{Y}\left(\frac{\rho_m - \rho_w}{\rho_m}\right) - \dot{h}. \qquad (7.6)$$

Stratal surfaces within the accumulating wedge of sediment (shown by dashed lines) are sigmoid and approximately concordant with the upper and lower bounding surfaces (terminology from Vail et al., 1977). The depositional slope changes in the vicinity of the shoreline because sedimentary processes in subaerial and shallow marine environments are different (Swift, 1970, 1976). This slope change corresponds with the depositional coastal break of P. R. Vail (Rice University, personal communication, 1987) and van Wagoner et al. (1987), and is typically located at depths of as much as a few meters below sea level. Here we assume for simplicity that the depositional coastal break coincides with the shoreline, although we recognize the conceptual difference between these two features. In the case of a graded topographic profile, and a stationary shoreline (L_1), $W_d = h = 0$ at all points seaward of A, and

$$\dot{\Delta}_{SL} = \dot{Y}\left(\frac{\rho_m - \rho_w}{\rho_m}\right) - \dot{S}^*\left(\frac{\rho_m - \rho_s}{\rho_m}\right). \quad (7.7)$$

For a profile that is initially graded to a fixed shoreline, a decrease in sediment supply at constant depositional base level leads to transgression of the shoreline [$\dot{W}_d > 0$; Eq. (7.4)]. This in turn results in steepening of the subaerial profile. Dynamic equilibrium is reestablished through enhanced erosion and a shift of the line of critical bypassing (A) toward the basin. A similar argument shows that an increase in sediment supply leads to regression, and a shift of A away from the basin. Thus small shifts in the position of onlap can be produced by changes in sediment input, without changing either $\dot{\Delta}_{SL}$ or $\dot{Y}$. In this simple case, the distance between A and the shoreline is related to the slope of the depositional surface, the lateral gradient of Y, the magnitude of $\dot{\Delta}_{SL}$, the abundance of available sediment, and the time since the system was last perturbed.

A small increase in the rate of sea-level fall results in a downward shift in the position of onlap to B, where the rate of tectonic subsidence ($\dot{Y}$) is greater than at A (Figure 7.7b). The shift is geologically instantaneous because all points landward of B are now subject to erosion ($\dot{S}^* < 0$), or at least to relative uplift ($\dot{h} < 0$), which favors erosion [Eq. (7.5)]. The downward shift in onlap is not a response to a fall in sea level, but to an increase in the rate of fall in sea level. Hence the distance between A and B, however measured, by itself provides no information about the amplitude of sea-level change, contrary to the methodology originally proposed by Vail et al. (1977) for measuring sea-level falls from seismic stratigraphy. It is also clear from Eq. (7.6) that prominent shifts in onlap can result from modest changes in the rate of sea-level fall, comparable to typical rates of tectonic subsidence (<1 cm/1000 yr). The increase in Δ_{SL} and increased sediment supply (from erosion) together cause regression or regression at an increased rate, but slower than the rate of change of the position of onlap. A new equilibrium shoreline position (L_2) is reached when the rate of sedimentation at the shoreline is just sufficient to balance the rate of net subsidence [Eq. (7.7)]. Erosion landward of B does not necessarily produce any marked (acoustically resolvable) discordance, because the surface is initially concordant with the underlying strata, but erosion does result in an apparent shift in the position of A to the erosional edge, A′, reducing the coastal encroachment (the horizontal component of coastal onlap; Vail et al., 1977) apparent in the underlying sequence. The line of critical bypassing (corresponding to points A and B) and the shoreline are thus related but different entities, which move laterally at different rates, and not necessarily in the same direction (see Vail et al., 1977, 1984; Vail and Todd, 1981). We emphasize this important conceptual difference between onlap shifts and transgressions/regressions because it is commonly blurred in the literature (e.g., Pitman and Golovchenko, 1983; Parkinson and Summerhayes, 1985; Summerhayes, 1986).

Figure 7.7c shows the effect of a more rapid increase in the rate of sea-level fall, where the new equilibrium line of bypassing (C) occupies a position that was formerly marine not subaerial (cf. Figure 7.7a). In this case, the line of bypassing shifts instantly to the shoreline (L_1) because all points between A and L_1 are immediately subject to relative uplift and erosion [Eq. (7.5)]. Bypassing then extends rapidly toward C, at a rate that depends on the slope of the depositional surface, the relative magnitudes of Y and Δ_{SL}, and the sediment supply. The point C again corresponds

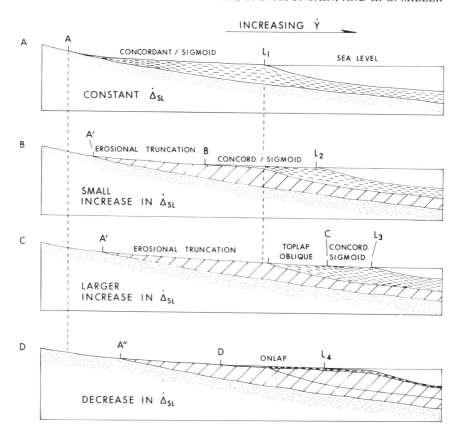

FIGURE 7.7 Formation of an unconformity (sequence boundary) under conditions of falling sea level ($\Delta_{SL} > 0$), for a ramp setting in a differentially subsiding basin, in which the rate of tectonic subsidence (Y) increases to the right. In each panel, the line of critical bypassing (corresponding to points A, B, C, and D) separates areas of deposition from areas of nondeposition and erosion. Points L_1, L_2, L_3, and L_4 indicate the location of the shoreline and approximately the depositional coastal break. Dashed lines within sequences are drawn parallel to stratal surfaces, and diagrammatically indicate concordant and discordant relations with bounding unconformities. When the rate of sea-level fall is constant [see (A)], sediments onlap at the line of critical bypassing; the stratal geometry is sigmoid, and strata are approximately concordant with upper and lower bounding surfaces. An increase in the rate of sea-level fall [see (B) and (C)] produces a downward shift in onlap whose magnitude is determined by the rate of fall and the rate of tectonic subsidence. A small increase in the rate of sea-level fall is shown in (B) (type 2 sequence boundary), and a larger increase, in (C) (type 1 sequence boundary). A decrease in the rate of sea-level fall leads to progressive onlap away from the basin [see (D)]. See text for additional explanation.

to $S^* = 0$ [Eq. (7.6)], and bypassing between L_1 and C leads to the development of toplap and an oblique tangential stratal pattern. Continued regression of the shoreline toward the new equilibrium position L_3 results in erosion landward of C, and deposition of sigmoid strata between C and L_3.

In Figure 7.7b,c, the sequence boundary forms at the time of increased rate of sea-level fall and corresponds to the contact between the units indicated by hatched and dashed-line patterns. The case shown in Figure 7.7b is equivalent to a type 2 sequence boundary, overlain by a shelf-margin systems tract (van Wagoner *et al.*, 1987). The case shown in Figure 7.7c, in which the line of critical bypassing extends beyond and below the initial depositional coastal break (approximately L_1), is equivalent to a type 1 sequence boundary overlain by a lowstand systems tract (van Wagoner *et al.*, 1987).

The depositional coastal break was formerly termed the shelf edge (Vail and Todd, 1981; Vail *et al.*, 1984), and this led to confusion with the shelf break, the familiar physiographic boundary between the shelf and slope of modern passive continental margins (van Wagoner *et al.*, 1987). Seismic stratigraphic interpretations involving

lowering of coastal onlap below the shelf edge during times of minimal continental glaciation were questioned by Pitman and Golovchenko (1983) and by Thorne and Watts (1984) because the rate of tectonic subsidence at the shelf break commonly exceeds the rate of sea-level fall that can be sustained by nonglacial mechanisms for sea-level change. This ceases to be a problem for most ramp settings, such as that illustrated in Figure 7.7, in which the depositional coastal break is at a considerable distance inboard of the shelf break. There may or may not be a problem in the case of continental margins at which the depositional coastal break and shelf break coincide, depending on the rate of tectonic subsidence, and on whether observed onlap below the edge of the shelf is truly coastal or is marine onlap. Pronounced shelf bypassing and submarine-fan sedimentation, explicitly associated with type 1 unconformities, do not require complete exposure of the shelf (May *et al.*, 1983). Significant bypassing probably takes place as soon as the line of critical bypassing intersects the upper parts of submarine canyons, through which sediments can be efficiently transported to the deep marine environment. The shelf may also become subaerially exposed without the line of critical bypassing reaching the shelf break. The current Mississippi delta stands close to the shelf break of the Gulf of Mexico in spite of the Holocene sea-level rise as a result of the extremely high rate of sediment supply. However, in this case progradation was accompanied by significant aggradation, and the line of critical bypassing lies well landward of the shelf break.

Thorne and Watts (1984) concluded that during nonglacial intervals, variations in the rate of sea-level fall are insufficient to produce any unconformities unless the rate of tectonic subsidence is small, as in old passive margins. However, their analysis refers to the outer parts of margins, where the rate of tectonic subsidence is generally large. We have shown above that unless there is a synchronous increase in the rate of tectonic subsidence of appropriate magnitude, an increase in the rate of sea-level fall always produces an unconformity in a passive margin, but the unconformity may not extend very far onto the continental shelf (Figure 7.7b).

After a graded profile has been established (Figure 7.7d), a decrease in the rate of sea-level fall leads to progressive onlap away from the basin (to point D), and to transgression of the shoreline (to L_4). The rates of onlap and transgression may differ because the position of the shoreline is influenced by the rate of sediment supply.

In the discussion above, we have considered only the case of episodic changes in the long-term rate of sea-level fall, and rates of sea-level change no greater than typical rates of tectonic subsidence in passive margins. Quantitative modeling by Pitman (1978), in Figure 4 of Steckler (1984), and by Watts and Thorne (1984) indicates that the gross features of Cenozoic stratigraphy of the Atlantic margin of the United States can be satisfactorily explained by assuming such long-term sea-level changes rather than the higher-amplitude oscillatory rises and falls depicted by Vail *et al.* (1977, 1984) and Haq *et al.* (1987; see Figure 7.8). The efficacy of small sea-level changes in producing sequence boundaries is undeniable. Yet the question of whether Cenozoic sea level was oscillatory or monotonically decreasing prior to the onset of glacioeustasy is not resolved by the modeling studies (e.g., Watts and Thorne, 1984), which did not account for the magnitude of short-term changes in paleobathymetry and sediment supply [Eq. (7.4)]. Sequence boundaries of probable eustatic origin are known not only from times of long-term sea-level fall (such as the Cenozoic) but also from times of long-term rise (such as the Early Cretaceous; Bond, 1979; Watts and Steckler, 1979; Haq *et al.*, 1987). A sequence boundary of early Aptian age is cited by Haq *et al.* (1987) as particularly prominent. Cyclical changes in paleobathymetry and shoreline position can occur during sea-level rise, but prominent downward shifts in coastal onlap require falling sea level ($\dot{\Delta}_{SL} > 0$) or uplift [$\dot{Y} < 0$, Eq.

(7.6)]. To the extent that Early Cretaceous downward shifts in onlap were of eustatic origin, eustatic sea-level fluctuations during that interval must have been oscillatory with the rate of short-term fall at times exceeding the rate of long-term rise. This leads to the conclusion that sea-level changes may generally be oscillatory, although of small amplitude.

The discussion is now extended to consider the effects of oscillatory sea-level change and profiles that do not attain dynamic equilibrium. Under these conditions, sequence boundaries correspond approximately to times of fastest sea-level fall (see Figure 7.8), and the assumption of global synchroneity is still a good approximation (Vail *et al.*, 1984). With reference to Figure 7.7b, expansion of the zone of bypassing begins as soon as the rate of sea-level fall is greater than about 0.7 times the rate of subsidence at A [Eq. (7.6)]. After a finite interval of time, Δ_{SL} attains a maximum value, and the line of critical bypassing reaches B. A subsequent decrease in $\dot{\Delta}_{SL}$ leads to renewed onlap away from the basin. We conclude that type 2 sequence boundaries of eustatic origin should be globally synchronous for all basins connected with the open ocean. For faster rates of sea-level fall, the line of critical bypassing may intersect the depositional coastal break before the rate of sea-level fall reaches its maximum value. We therefore expect type 1 sequence boundaries to be slightly diachronous, although the level of diachroneity (<< one-quater cycle) may be close to or below biostratigraphic resolution even for second-order boundaries. We disagree with the principal conclusion of Summerhayes (1986), reiterated by Hubbard (1988), that most prominent sequence boundaries are likely to have only local significance and not be globally synchronous.

Tectonics

Although this chapter is focused on the seismic stratigraphic record of sea-level change, it is important to consider the influence of tectonics in the development of unconformities, and how the tectonic and eustatic signals may be differentiated. Schwan (1980), Bally (1982), and Watts (1982) noted the approximate correspondence of several of the supercycle (second order) boundaries of Vail *et al.* (1977, 1980, 1984), Vail and Todd (1981), and Haq *et al.* (1987) with times of major plate-boundary reorganization, related to the progressive breakup of the supercontinent Pangea. This correspondence may in part reflect regional bias in the "global" coastal onlap chart, but in part is probably the result of tectonoeustasy associated with changes in the length and rate of crustal accretion at oceanic ridges. Although the full effects of spreading-rate changes are not apparent until more than 70 m.y. after the changes have taken place (Pitman, 1978; Pitman

130

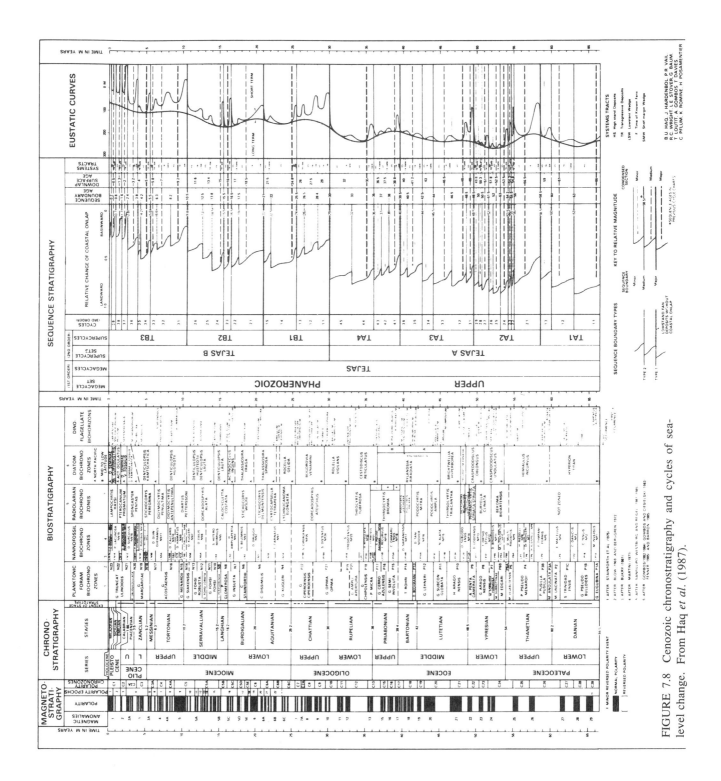

FIGURE 7.8 Cenozoic chronostratigraphy and cycles of sea-level change. From Haq et al. (1987).

and Golovchenko, 1983), the rate of change of sea level is affected immediately (Heller and Angevine, 1985). An alternative interpretation of apparently synchronous sequence boundaries on different continents is that they are a result of global tectonic events (Sloss, 1979; Bally, 1982). While this possibility cannot be entirely discounted, no viable mechanism has yet been established. Even if there is communication of stresses between plates, there seems to be little reason for changes in the rate of basin subsidence to be synchronous on a global scale.

Most "tectonic" mechanisms for generating sequence boundaries predict dissimilar ages from one continent to another, or even on different parts of the same continent. To the extent that major boundaries in passive continental margins are globally synchronous, these mechanisms are of secondary importance, so that the precision with which unconformities can be correlated has become a critical issue. Cloetingh *et al.* (1985), Cloetingh (1986), and Karner (1986) suggested that variations in horizontal stresses in the lithosphere of a few hundred bars can induce vertical motions of tens of meters, at rates comparable to typical rates of tectonic subsidence (1 cm/1000 yr). According to these authors, such motions may be responsible for the third-order boundaries on the coastal onlap chart (those representing base-level changes on a time scale of 1 to 3 m.y.). There appears to be good evidence that appropriate reorganization of in-plane stress takes place, but it has not yet been established that stresses vary frequently enough or that stress changes are of sufficient magnitude to produce most or even many of the observed third-order boundaries. According to the model, compressive stress accelerates basin subsidence and produces uplift of the basin margins; tensile stress retards basin subsidence and enhances subsidence of the margins. The rates of uplift and subsidence are sensitive to the magnitudes of stress variations, which vary from place to place within a given lithospheric plate, to flexural rigidity (a function of age), and to basin geometry. In particular, stratigraphically recorded vertical motions are a composite of motions at each of the density interfaces within the existing sedimentary section and beneath the basin (Karner, 1986). For these reasons, no two basins should behave in the same way, and depending on flexural rigidity, the response of a given basin may vary according to the orientation of its margins in map view. For example, in a large basin an individual unconformity might pass to a correlative unconformity not only in the direction of increasing tectonic subsidence but also parallel to the margin of the basin where its orientation changes through 90°. Indeed, several unconformities in one part of a basin might be exactly half a cycle out of phase with unconformities in an adjacent part of the same basin. Such unusual stratal geometry has

never been described in published seismic stratigraphic or outcrop studies.

Cloetingh (1986) and Karner (1986) both imply that downward shifts in coastal onlap should be associated with compression, and Cloetingh (1986) even attempts to calibrate the coastal onlap chart of Vail *et al.* (1977) in terms of paleostress fluctuation. It is assumed for this purpose that the chart is weighted in favor of the North Atlantic region, which is regarded as being sufficiently small to have experienced the same stress history. Apart from the obvious hazards of such a correlation, compression does not necessarily produce a downward shift in onlap unless eustatic sea level is near stationary or rising. Under compression, the transition from increased basin subsidence to margin uplift occurs near the flexural node. We have shown above that if sea level is falling, the line of critical bypassing assumes a position within the basin. An increase in the rate of tectonic subsidence under this circumstance would lead not to a downward shift in onlap, but to renewed (or continued) onlap toward the basin margin.

Mörner (1976, 1980, 1981) argued that sea-level changes are markedly diachronous in different parts of the Earth and that they result in part from variations in the configuration of the geoid through geological time. The geoid, or equipotential surface of the Earth's gravity field, contains irregularities as great as 180 m, and undoubtedly has varied in the past. We question Mörner's interpretation of available biostratigraphic data, but more important, we think that his reasoning is flawed because he assumes that geoidal changes affect only the oceans, whereas on geological time scales they result in concomitant adjustments of the solid earth (Steckler, 1984). Thus while geoidal changes affect sea level during intervals of thousands of years, they are not relevant to longer-term eustasy.

A fundamental feature of eustatic unconformities is their global persistence in marine basins. In attempting to distinguish eustatic unconformities from those of local significance (related to tectonics, sediment supply, or paleo-oceanographic conditions, for example), the apparent absence of a "global" sequence boundary in a given basin is generally regarded as evidence against a eustatic origin (e.g., Thorne and Watts, 1984; Hubbard, 1988). Such a criterion should be applied with caution. Different seismic stratigraphers may subjectively select different unconformities for correlation in the same basin; many unconformities pass laterally into correlative conformities; and a given unconformity may not be seismically resolved even though known to be present (from biostratigraphic data, for example). On the other hand, unconformities of local origin in different basins may fortuitously be of approximately the same age, and thus incorrectly correlated and assumed to represent a eustatic signal. In order to resolve these problems objectively, it is essential that the stratigraphy of every basin used for comparison is based on an internally consistent detailed interpretation of a grid of seismic sections. Few such interpretations have ever been attempted outside the petroleum industry. It is also imperative to consider the geochronological precision that may be achieved for each sequence boundary.

GEOCHRONOLOGY

Any approach to global seismic stratigraphy requires calibration to geological time through rock stratigraphy, but there are inherent uncertainties in the time scale, and in the correlation of seismic and rock sections with one another and with the time scale. The choice of an appropriate geological time scale is controversial, although every time scale involves similar components. Stratotypes of standard chronostratigraphic units (stages) are correlated with one another in order to establish a chronostratigraphic framework, and these units are then calibrated against a numerical scale. One approach is to use all available radiometric age measurements, including those obtained from "low-temperature" minerals such as glauconite, which commonly differ significantly from ages derived from "high-temperature" minerals (e.g., Odin, 1982; Haq et al., 1987). Another is to correlate stratotypes with changes in geomagnetic polarity, and to calibrate this magnetochronology with the few reliable high-temperature age measurements (e.g., Heirtzler et al., 1968; Berggren et al., 1985; Kent and Gradstein, 1985). Once an age calibration for the chronostratigraphic framework has been chosen, other geological data such as multiple biostratigraphic zonations and geochemical fluctuations can be calibrated to the time scale.

Uncertainties in time scales are caused by errors in isotopic age measurements, and especially by problems in correlation between stratotypes and age measurements. Numerical uncertainties are partly a function of the age of the strata and the techniques employed. For example, the K-Ar technique widely used in Mesozoic and Cenozoic geochronology involves potential errors on the decay constant of less than 2 percent, and a typical accuracy of better than 5 percent (Dalrymple and Lanphere, 1965). Errors in astronomical (Milankovitch) estimates of the ages of boundaries for the latest Quaternary may be less than 5000 years (Imbrie et al., 1984). Comparisons of different time scales suggest that errors are typically about 0.5 to 2 m.y. for the Neogene (cf. Odin, 1982; Berggren et al., 1985), 1 to 7 m.y. for the Paleogene (cf. Odin, 1982; Berggren et al., 1985), 2 to 8 m.y. for the Cretaceous, and as much as 10 m.y. for the Jurassic (Kent and Gradstein, 1985). Numerical calibration of biostratigraphic zones within these intervals is commonly less precise than bio-

stratigraphic correlation. For example, Paleogene foraminiferal zones are typically 1 to 2 m.y. in duration, whereas radiometric precision is about 2 to 3 m.y.

The boundary between the middle and late Miocene provides a good example of correlation and time scale problems. Depending on the author, the age of this boundary varies from 9.5 to 11.5 Ma (Figure 7.9), a range of about 20 percent of the age. Direct calibration of planktonic foraminifera, nannofossils, and magnetostratigraphy has removed some of the ambiguity, and the boundary is now thought to be about 10.4 Ma (Miller et al., 1985a). Differences among earlier estimates were due to miscorrelations. (1) Biostratigraphic correlation of the stratotype Tortonian (basal upper Miocene) was incorrect. Zone NN8 is found in the basal upper Miocene stratotype (Miller et al., 1985a), and this requires Zone NN9 to be well within the upper Miocene (Figure 7.9, column BKV85), and not straddling the boundary as shown in columns VB74, BKD85a, and BKD85b. (2) Correlations of nannofossils and foraminifera with the geomagnetic polarity record were also incorrect. A long normal magnetozone associated with Zone NN9 (Epoch 11, Figure 7.9) was improperly correlated with the Geomagnetic Polarity Time Scale. Instead of being about 11 Ma (columns VB74, BKD85a in Figure 7.9), Zone NN9 must be younger than about 10 Ma (column BKV85).

Virtually all interregional seismic stratigraphic comparisons rely on biostratigraphic correlations. Where synthetic seismograms can be obtained from geophysical logs, or VSPs are available, the major limitations to achievable age resolution have to do with the lack of appropriately positioned boreholes, the use of cuttings rather than cores, and errors or lack of biostratigraphic resolution. Sequence boundaries are most precisely dated at correlative conformities (Figure 7.4), but such conformities may not be penetrated in drilling, or in the absence of sufficient seismic data, may even be misinterpreted as indicating that no unconformity is present. Where two or more unconformities are superimposed, a considerable hiatus may be present, and zonations in such circumstances are usually equivocal. In the case of commercial wells, most biostratigraphic work is based on cuttings, and the stratigraphic significance of such samples is reduced by downhole caving. Problems associated with biostratigraphic errors can be reduced by restricting interregional stratigraphic comparisons to sections that have been studied by the same author, but errors in identification, taxonomy, or calibration of taxa with those studied by others may be significant. Another limitation to biostratigraphic resolution involves the diachrony of taxa (between low and high latitudes, for example). First appearances of taxa are commonly diachronous (Johnson and Nigrini, 1985), and although last appearances are more likely to be syn-

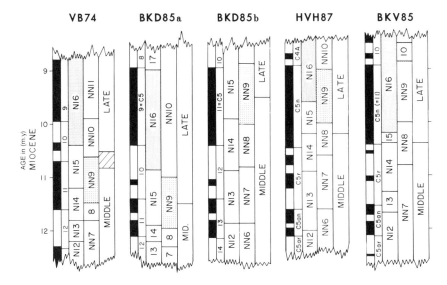

FIGURE 7.9 Comparison of time scales for middle to late Miocene time. Magnetic polarity chronologic units are given in terms of epochs (9 to 14) and chrons (C4 to C5; black, normal polarity; white, reversed polarity). Biostratigraphy: N12 to N16 are planktonic foraminiferal zones; NN7 to NNll are nannofossil zones. Note the variable position of Zone NN9 (coarse stipple) relative to the inferred age, Zone N16 (fine stipple), and magnetic stratigraphy. Symbols for different time scales: VB74, van Couvering and Berggren (1976) based on biochronology; BKD85a, Barron et al. (1985), based on second-order magnetobiostratigraphic correlations and the assumption that Epoch 9 is equivalent to Chron C5n; BKD85b, Barron et al. (1985) assuming correlation of Epoch 11 with Chron C5n; HVH87, Haq et al. (1987) based upon essentially the same geomagnetic time scale as Berggren et al. (1985), and second-order magnetostratigraphic correlations in this time interval; BKV85, Berggren et al. (1985) based on first-order magnetobiostratigraphic correlations of Miller et al. (1985a), calibrated to the geomagnetic polarity time scale.

chronous, even these may be diachronous between sections at different latitudes (Aubry, 1983). In order to use biostratigraphy confidently, ranges of taxa must be calibrated against an independent chronology. Direct calibration to magnetostratigraphy has potential for greatly improved correlations (e.g., Berggren et al., 1985; Miller et al., 1985a). The precision and reproducibility of biostratigraphic picks are not readily quantifiable. Recent Neogene studies have claimed biostratigraphic resolution as good as 100,000 yr (e.g., Keller and Barron, 1983). Such claims are extravagant (see Berggren et al., 1983), but properly calibrated biostratigraphic ranges have potential for Neogene correlations of better than 0.5 m.y. (Berggren et al., 1985).

Our ability to test the scheme of global unconformities proposed by Vail et al. (1977, 1984)—that is, to distinguish between global unconformities and those developed on only a regional or local scale—is limited by our ability to determine the ages of unconformities in continental-margin successions. Biostratigraphic uncertainties of the sort outlined above are often so large that later drilling requires substantial revisions of preliminary findings. A good example is attempts to date the mid-Oligocene unconformity in the continental margin of the eastern United States. Olsson et al. (1980) used well cuttings to suggest that the oldest sediments above a prominent unconformity in that region are about 34 to 31 Ma, and therefore not consistent with a major erosional event predicted by Vail et al. (1977) at about 30 Ma. Subsequent examination of boreholes on the Irish margin, which indicated a hiatus between 34 Ma and 30 Ma, prompted a reevaluation of the

biostratigraphic record from the U.S. margin (Miller et al., 1985b). These workers determined that the hiatus there probably extends to at least 30 Ma, and is thus consistent with the Vail scheme.

In spite of the geochronological problems outlined here and the potential danger for circularity, in which boundaries are assumed incorrectly to be globally synchronous, many second-order sequence boundaries (such as those of the mid-Cenomanian and mid-Oligocene) appear to be very persistent from one basin to another. Although the ages of all boundaries are subject to refinement, the idea that some sequence boundaries record eustatic events remains an appealing working hypothesis. The seismic stratigraphic evidence is supported by a good deal of outcrop evidence, referred to above, for globally synchronous sea-level change through much of the Phanerozoic. It may be difficult to demonstrate a eustatic origin for many third-order sequence boundaries, those derived largely from higher-resolution well-log and outcrop studies (Figure 7.8; Haq et al., 1987). This is because in spite of considerable recent efforts to calibrate the seismic stratigraphic record (Haq et al., 1987), it may never be possible to resolve the ages of many of these boundaries sufficiently well for objective correlation between basins because the spacing of third-order boundaries is close to or finer than biostratigraphic resolution.

INTERPRETATION OF SEA-LEVEL CHANGE

For sequence boundaries and condensed intervals of eustatic origin (perhaps many of the second-order sequences), seismic stratigraphy provides information only about times of rapid sea-level fall and rise, respectively. Seismic stratigraphy provides no direct information about times of eustatic highstands and lowstands; they must be interpolated (Figure 7.8; Vail *et al.*, 1984; Haq *et al.*, 1987). Amplitudes of eustatic fluctuations cannot be inferred from seismic stratigraphic data alone because coastal aggradation (the vertical component of onlap) is primarily a result of basin subsidence, not sea-level rise; and downward shifts in onlap reflect only the rate of sea-level fall relative to the rate of basin subsidence [Eq. (7.6)]. Variations of coastal encroachment (the horizontal component of onlap) are sensitive to lateral gradients in the rate of subsidence, which are in general time dependent and not necessarily linear.

Coastal Aggradation

Basin subsidence is primarily a response to tectonic subsidence, amplified by sediment loading and modified to a limited degree by sea-level change and sediment compaction [Eq. (7.1) integrated with respect to time]. Although Vail *et al.* (1977) used coastal aggradation as a direct measure of eustatic sea-level rise (Figure 7.10), it is not an especially good approximation even of a relative sea-level rise (net subsidence plus eustasy). Four difficulties are as follows:

1. As recognized by Vail *et al.* (1984), the upper part of many sequences consists of alluvial as well as coastal

plain sediments, so that the observed aggradation exceeds the relative sea-level change.

2. For divergent reflection patterns, to be expected in differentially subsiding basins, estimates of the magnitude of aggradation are critically dependent on the path taken across the seismic section, greater values being obtained in basinward locations (Miall, 1986). Incremental measurements of aggradation near the point of onlap (e.g., Vail *et al.*, 1977) lead to minimum estimates of aggradation, but do not eliminate subjectivity from the procedure.

3. Aggradation varies within a basin according to the local rate of subsidence, basin geometry, and the degree of subsequent erosion and compaction. It is not clear how any measurement on a single seismic section can be objectively regarded as the most representative for inclusion in the coastal onlap chart of the basin.

4. The accurate measurement of aggradation within a sequence requires reliable estimates of interval velocities for each of the stratal segments used.

Ideally, coastal aggradation could be corrected for subsidence, compaction, and water-depth changes (e.g., Watts and Steckler, 1979; Hardenbol *et al.*, 1981), but magnitudes of eustatic fluctuations are difficult to estimate even at a single site. This is due to uncertainties in estimating paleobathymetry, particularly where sediments accumulated in water more than 200 m deep; in determining how sediments compacted; in making corrections for sediment loading on lithosphere with finite but poorly known flexural rigidity; and in separating tectonic subsid-

FIGURE 7.10 Procedure for constructing regional chart of relative changes of coastal onlap from estimates of coastal aggradation and downward shifts in coastal onlap—(A) stratigraphic cross section and (B) regional chart of cycles of relative change of coastal onlap. The letters A to E are arbitrary labels for five depositional cycles shown. A supercycle is a group of cycles during which there are only minor downward shifts in onlap. After Vail *et al.* (1977, 1984). See text for an evaluation of this procedure.

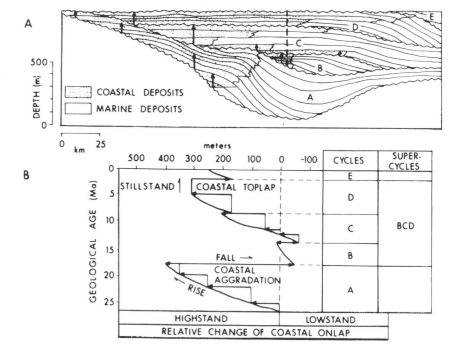

ence from eustasy. Once observed stratigraphic thicknesses have been corrected for water-depth changes, compaction, and sediment loading, the derived subsidence curve can be compared with a best-fit model curve for tectonic subsidence. If high-frequency deviations from smoothly varying long-term tectonic subsidence are largely of eustatic, not local tectonic, origin, the misfit between the two curves is a first approximation to the eustatic signal (e.g., Watts and Steckler, 1979; Hardenbol et al., 1981). However, the estimated amplitudes of eustatic oscillations vary according to assumptions involved in selecting the best-fit model. The model curve may also be biased by long-term eustatic effects, and is not necessarily a true measure of tectonic subsidence. These problems for a single site are compounded if the stratigraphic input is coastal aggradation measured incrementally along a surface, and the errors involved are difficult to assess. For all of the reasons summarized above, the measurement of coastal aggradation is an inappropriate method for estimating eustasy.

Miall (1986) questioned the method for quantifying onlap changes for the additional reason that it appears to ignore the fact that reflections are dipping (Figure 7.10a). Though contrary to geological intuition, for unmigrated seismic data, stratigraphic thickness is most accurately measured on a vertical scale (two-way travel time multiplied by appropriate velocities, or vertical thickness in a depth section). Two-way travel time is the time required for the most direct reflection, which is perpendicular to dipping strata. As a consequence, no correction for stratal dip is required. In migrated data, subsurface points are corrected for spatial mislocation, and aggradation is more properly measured perpendicular to reflections.

Downward Shifts in Onlap

The sawtooth asymmetry of the coastal onlap chart reflects the tendency for intervals of progressive onlap to be punctuated by downward shifts in onlap that appear to be geologically rapid, but the inferred magnitudes of downward shifts have no physical meaning. In Figure 7.10, the coastal aggradation of 400 m measured in sequence A consists largely of differential subsidence during the deposition of A, but the rapid fall of 450 m between cycles A and B includes the differential subsidence during cycles B to D (broken line in Figure 7.10a), which clearly has nothing to do with the formation of the boundary at the top of sequence A. As shown above, a downward shift in onlap is a not a response to a sea-level fall but to an increase in the rate of sea-level fall. Thus even if the downward shift were somehow corrected for the effects of later subsidence and for the various sources of error described in the section above, the shift in onlap would still

provide no direct information about the magnitude of sea-level change.

Coastal Encroachment

Recognizing the problems inherent in measuring coastal aggradation, Vail and colleagues have recently begun to use variations of coastal encroachment (the horizontal component of onlap) to construct coastal onlap charts (P. R. Vail, Rice University, personal communication, 1987). Horizontal distances can be measured accurately on seismic profiles, and for superposed sedimentary cycles the degree of coastal encroachment may allow a qualitative comparison of the eustatic fluctuations associated with each cycle. However, the use of coastal encroachment does not remove the ambiguity associated with the fluvial wedge in the upper part of many sequences, and the significance of coastal encroachment in cycles of markedly different age is uncertain. This is because changes of coastal encroachment are sensitive to lateral gradients in the rate of subsidence, which in general are time dependent and not necessarily linear. Identical eustatic fluctuations might produce very different variations of coastal encroachment at different stratigraphic levels within a basin. Moreover, it is unclear how variations of coastal encroachment (measured laterally) can be quantitatively "corrected" for subsidence, compaction, and water-depth changes (measured vertically), and therefore how resulting coastal onlap charts for different basins can be usefully compared to derive a eustatic signal.

Global Onlap Chart

By qualitatively comparing charts of relative change of coastal onlap for different basins, Vail et al. (1977, 1980, 1984), Vail and Todd (1981), and more recently Haq et al. (1987) derived a global onlap chart for the Mesozoic and Cenozoic (see Figure 7.8). Apart from indicating the timing of global unconformities, assuming that many of the second-order unconformities are global, the significance of such a chart is unclear. It appears to be strongly biased toward North America and Europe (see Figure 11 of Vail et al., 1984), but in our view the most significant problem is that similarities in the patterns of coastal onlap for different basins for the most part indicate similar subsidence history.

Eustatic Curve

In another departure from earlier procedures, attempts have been made recently to estimate the magnitudes of eustatic falls from the degree to which coastal onlap shifts below the depositional coastal break at type 1 unconformities, making corrections for compaction, loading, and water-depth changes (Greenlee *et al.*, 1988). Some of the problems inherent in measuring downward shifts in coastal onlap are thereby avoided. However, it is critically important to demonstrate that the onlapping strata below the depositional coastal break accumulated near sea level rather than in a deeper marine environment. Only rarely can paleobathymetry at the point of onlap be established with confidence owing to the lack of cores and appropriately positioned wells or boreholes. An additional limitation of this approach is that only part of a eustatic fall is sampled because sea level is already falling before the line of critical bypassing reaches the depositional coastal break, and nearshore sediments begin to onlap the sequence boundary before the onset of the next sea-level rise. In spite of these limitations, we nevertheless expect this approach to yield improved estimates of the magnitudes of sea-level falls if procedures are refined and applied to cored boreholes in transects across continental margins.

The technique of Greenlee *et al.* (1988) does not appear to have been used systematically to derive the eustatic curve published by Haq *et al.* (1987). Indeed, their eustatic curve is strikingly similar to a smoothed global onlap curve, in which the ages of inflection points are constrained by the ages of sequence boundaries and condensed intervals (see Figure 7.8 for the Cenozoic segment). In view of the many arguments outlined above for doubting that the onlap curve is a good measure of the amplitudes of eustatic oscillations, we think that the derived eustatic curve contains information for the most part about the timing of eustatic rises and falls. Large-amplitude eustatic oscillations indicated by Haq *et al.* (as much as 100 m or more for some type 1 unconformities) appear to be inferred largely by analogy with Pleistocene sea-level changes (e.g., Vail *et al.*, 1984), and generally not on the basis of firm evidence for every boundary shown. Amplitudes of sea-level falls associated with type 2 sequence boundaries cannot be determined by Greenlee's method, and these must have been inferred from the shape of the global onlap chart.

In summary, the main limitations of the eustatic curve are (1) that all of the observed sequence boundaries (third order as well as second order) are uncritically assumed to be of eustatic origin; (2) that questions persist about the calibration of many boundaries to the geological time scale; and (3) that the inferred amplitudes of sea-level fluctuations are for the most part conjectural.

SUMMARY AND RECOMMENDATIONS

The technique of seismic stratigraphy has led to a fundamental reevaluation of our approach to stratigraphic and sedimentological studies. It provides a new way of interpreting subsurface stratigraphy, and of comparing the stratigraphic record in different basins. The identification of unconformities of apparently global extent has spawned renewed interest in eustasy and vigorous debate about the interpretation of seismic stratigraphic data.

The key to seismic stratigraphy is the identification and calibration of regional unconformities (sequence boundaries), which in most shelf and coastal areas have chronostratigraphic significance. Sequence boundaries associated with a downward shift in coastal onlap develop in response to changes in the rate of subsidence and rate of sea-level change. The distinction of these phenomena hinges largely on the precision with which individual boundaries can be dated and correlated on a regional to global scale. Amplitudes of eustatic fluctuations cannot be inferred from seismic stratigraphic data alone because coastal aggradation is primarily a result of basin subsidence, not sea-level rise; and downward shifts in onlap reflect only the rate of sea-level fall relative to the rate of basin subsidence. Variations of coastal encroachment are sensitive to lateral gradients in the rate of subsidence. Whichever method is used to gauge changes in coastal onlap, the large component of subsidence cannot be easily removed to derive the smaller eustatic signal.

For continued improvement of the seismic stratigraphic record of eustasy, we recommend objective reevaluation of the ages of sequence boundaries in individual basins, with the aim of distinguishing more confidently boundaries of global extent from those of more restricted distribution. Basins should be selected for study on the basis of stratigraphic completeness; simple tectonic history (e.g., passive continental margins lacking diapirism); and the current or future availability of high-resolution seismic sections, fully cored boreholes, and state-of-the-art geophysical logs and VSPs for calibration of the boundaries. The basins should also be of a range of ages and widely separated to avoid regional tectonic bias. Chronostratigraphic control should be obtained by integrating multiple biostratigraphic, isotopic (stable and radiometric), and magnetostratigraphic criteria. For this purpose, basins at mid-latitudes offer the most favorable trade-off between biostratigraphic and magnetostratigraphic techniques. For improved estimates of the magnitudes of sea-level change, a transect of boreholes is required for two-dimensional analysis of tectonic subsidence. The best resolution of paleobathymetry is obtained from shallow-water carbonate platforms (e.g., Kendall and Schlager, 1981).

ACKNOWLEDGMENTS

This chapter was reviewed by W. A. Berggren, A. Hallam, J. Ladd, M. Levy, M. A. Kominz, C. G. St. C. Kendall, W. C. Pitman III, and A. B. Watts. We thank A.

Hallam and B. U. Haq for preprints, and T. S. Loutit, H. W. Posamentier, and P. R. Vail for recent discussions about seismic stratigraphic principles and interpretation. The research was supported by NSF grants OCE 86-00249 (Mountain, Miller, and Christie-Blick), OCE 85-00859 and OCE 87-00005 (Miller), and by the Donors of the Petroleum Research Fund, administered by the American Chemical Society (PRF 16042-G2 to Christie-Blick). Lamont-Doherty Geological Observatory Contribution No. 4271.

REFERENCES

Aubry, M.-P. (1983). Late Eocene to early Oligocene calcareous nannoplanktonic paleobiogeography and the terminal Eocene event, paper presented at the First International Conference on Paleoceanography, Zurich.

Badley, M. E. (1985). *Practical Seismic Interpretation*, International House Resources Development Corporation, Boston, 266 pp.

Bally, A. W. (1982). Musings over sedimentary basin evolution, *Phil. Trans. Roy. Soc. London, Ser. A 305*, 325–338.

Barron, J., G. Keller, and D. Dunn (1985). A multiple microfossil biochronology for the Miocene, in *The Miocene Ocean: Paleoceanography and Biogeography*, J. P. Kennett, ed., Memoir 163, Geological Society of America, Boulder, Colo., pp. 21–36.

Berggren, W. A., M.-P. Aubry, and N. Hamilton (1983). Neogene magnetobiostratigraphy of Deep Sea Drilling Project Site 516 (Rio Grande Rise, South Atlantic), in *Initial Reports of Deep Sea Drilling Project 73*, P. F. Barker, R. L. Carlson, D. A. Johnson et al., eds., U.S. Government Printing Office, Washington, D.C., pp. 675–713.

Berggren, W. A., D. V. Kent, J. J. Flynn, and J. A. van Couvering (1985). Cenozoic geochronology, *Geol. Soc. Am. Bull. 96*, 1407–1418.

Boggs, S., Jr. (1987). *Principles of Sedimentology and Stratigraphy*, Merrill, Columbus, Ohio, 784 pp.

Bond, G. (1978a). Evidence for late Tertiary uplift of Africa relative to North America, South America, Australia and Europe, *J. Geol. 86*, 47–65.

Bond, G. (1978b). Speculations on real sea-level changes and vertical motions of continents at selected times in the Cretaceous and Tertiary Periods, *Geology 6*, 247–250.

Bond, G. (1979). Evidence for some uplifts of large magnitude in continental platforms, *Tectonophysics 61*, 285–305.

Brown, L. F., Jr., and W. L. Fisher (1977). Seismic stratigraphic interpretation of depositional systems: Examples from Brazilian rift and pull-apart basins, in *Seismic Stratigraphy—Applications to Hydrocarbon Exploration*, C. E. Payton, ed., Memoir 26, American Association of Petroleum Geologists, Tulsa, Okla., pp. 213–248.

Brown, L. F., Jr., and W. L. Fisher (1980). Seismic stratigraphic interpretation and petroleum exploration, *Am. Assoc. Petrol. Geol. Cont. Educ. Course Note Ser. No. 16*.

Burton, R., C. G. St. C. Kendall, and I. Lerche (1987). Out of our depth: On the impossibility of fathoming eustasy from the stratigraphic record, *Earth-Sci. Rev. 24*, 237–277.

Busch, R. M., and H. B. Rollins (1984). Correlation of Carboniferous strata using a hierarchy of transgressive-regressive units, *Geology 12*, 471–474.

Christie-Blick, N., and K. T. Biddle (1985). Deformation and basin formation along strike-slip faults, in *Strike-Slip Deformation, Basin Formation, and Sedimentation*, K. T. Biddle and N. Christie-Blick, eds., Spec. Publ. No. 37, Society of Economic Paleontologists and Mineralogists, pp. 1–34.

Cloetingh, S. (1986). Intraplate stresses: A new tectonic mechanism for fluctuations of relative sea level, *Geology 14*, 617–620.

Cloetingh, S., H. McQueen, and K. Lambeck (1985). On a tectonic mechanism for regional sea-level variations, *Earth Planet. Sci. Lett. 75*, 157–166.

Dalrymple, G. B., and M. A. Lanphere (1965). Potassium-argon dating: A summary of principles and techniques, unpublished U.S. Geological Survey paper, 63 pp.

Dobrin, M. B. (1976). *Introduction to Geophysical Prospecting*, 3rd ed., McGraw-Hill, New York, 630 pp.

Greenlee, S. M., F. W. Schroeder, and P. R. Vail (1988). Seismic stratigraphic and geohistory analysis of Tertiary strata from the continental shelf off New Jersey; Calculation of eustatic fluctuations from stratigraphic data, in *The Atlantic Continental Margin, U.S., Decade of North American Geology*, Vol. I-2, R. E. Sheridan and J. A. Grow, eds., Geological Society of America, Boulder, Colo., pp. 437–444.

Hallam, A. (1981). A revised sea-level curve for the early Jurassic, *J. Geol. Soc. London 138*, 735–743.

Hallam, A. (1984). Pre-Quaternary sea-level changes, *Ann. Rev. Earth Planet Sci. 12*, 205–243.

Hallam, A. (1986). Origin of minor limestone-shale cycles: Climatically induced or diagenetic? *Geology 14*, 609–612.

Hallam, A. (1988). A reevaluation of Jurassic eustasy in the light of new data and the revised Exxon curve, in *Sea Level Changes—An Integrated Approach*, C. K. Wilgus, B. S. Hastings, C. G. St.C. Kendall, H. W. Posamentier, J. C. Ross, and J. C. van Wagoner, eds., Spec. Publ., Society of Economic Paleontologists and Mineralogists.

Hancock, J. M., and E. G. Kauffman (1979). The great transgressions of the Late Cretaceous, *J. Geol. Soc. London 136*, 175–186.

Haq, B. U., J. Hardenbol, and P. R. Vail (1987). Chronology of fluctuating sea levels since the Triassic, *Science 235*, 1156–1167.

Hardenbol, J., P. R. Vail, and J. Ferrer (1981). Interpreting paleoenvironments, subsidence history and sea-level changes of passive margins from seismic and biostratigraphy, *Oceanolog. Acta No. SP*, 33–44.

Harrison, C. G. A., G. W. Brass, E. Saltzman, J. Sloan II, J. Southam, and J. M. Whitman (1981). Sea level variations, global sedimentation rates and the hypsographic curve, *Earth Planet. Sci. Lett. 54*, 1–16.

Harrison, C. G. A., K. J. Miskell, G. W. Brass, E. S. Saltzman, and J. L. Sloan II (1983). Continental hypsography, *Tectonics 2*, 357–377.

Heezen, B. C., C. D. Hollister, and W. F. Ruddiman (1966). Shaping of the continental rise by deep geostrophic contour currents, *Science 152*, 502–508.

Heirtzler, J. R., G. O. Dickson, E. M. Herron, W. C. Pitman III, and X. Le Pichon (1968). Marine magnetic anomalies, geomagnetic field reversals, and motions of the ocean floor and continents, *J. Geophys. Res. 73*, 2119–2136.

Heller, P. L., and C. L. Angevine (1985). Sea-level cycles during the growth of Atlantic-type oceans, *Earth Planet. Sci. Lett. 75*, 417–426.

Hubbard, R. J. (1988). Age and significance of sequence boundaries on Jurassic and Early Cretaceous rifted continental margins, *Am. Assoc. Petrol. Geol. Bull. 72*, 49–72.

Imbrie, J., J. D. Hays, D. G. Martinson, A. McIntyre, A. C. Mix, J. J. Morley, N. G. Pisias, W. L. Prell, and N. J. Shackleton (1984). The orbital theory of Pleistocene climate: Support from a revised chronology of the marine $\delta^{18}O$ record, *Milankovitch and Climate*, Part 1, A. Berger *et al.*, eds., D. Reidel, Dordrecht, pp. 269–306.

Johnson, D. A., and C. A. Nigrini (1985). Time-transgressive Late Cenozoic radiolarian events of the equatorial Indo-Pacific, *Science 230*, 538–540.

Johnson, J. G., G. Klapper, and C. A. Sandberg (1985). Devonian eustatic fluctuations in Euramerica, *Geol. Soc. Am. Bull. 96*, 567–587.

Johnson, M. E., Rong Jia-Yu, and Yang Xue-Chang (1985). Intercontinental correlation by sea-level events in the Early Silurian of North America and China (Yangtze Platform), *Geol. Soc. Am. Bull. 96*, 1384–1397.

Karner, G. D. (1986). Effects of lithospheric in-plane stress on sedimentary basin stratigraphy, *Tectonics 5*, 573–588.

Keller, G., and J. A. Barron (1983). Paleoceanographic implications of Miocene deep-sea hiatuses, *Geol. Soc. Am. Bull. 94*, 590–613.

Kendall, C. G. St. C., and W. Schlager (1981). Carbonates and relative changes in sea level, *Mar. Geol. 44*, 181–212.

Kennett, J. P. (1982). *Marine Geology*, Prentice-Hall, Englewood Cliffs, N.J., 813 pp.

Kent, D. V., and F. M. Gradstein (1985). A Cretaceous and Jurassic geochronology, *Geol. Soc. Am. Bull. 96*, 1419–1427.

Lenz, A. C. (1982). Ordovician to Devonian sea-level changes in western and northern Canada, *Can. J. Earth Sci. 19*, 1919–1932.

Mahradi (1983). Physical Modeling Studies of Thin Beds, M.S. thesis, University of Houston, Texas, 100 pp.

May, J. A., J. E. Warme, and R. A. Slater (1983). Role of submarine canyons on shelfbreak erosion and sedimentation: modern and ancient examples, in *The Shelfbreak: Critical Interface on Continental Margins*, D. J. Stanley and G. T. Moore, eds., Spec. Publ. No. 33, Society of Economic Paleontologists and Mineralogists, pp. 315–332.

McKerrow, W. S. (1979). Ordovician and Silurian changes in sea level, *J. Geol. Soc. London 136*, 137–145.

Miall, A. D. (1984). *Principles of Sedimentary Basin Analysis*, Springer-Verlag, New York, 490 pp.

Miall, A. D. (1986). Eustatic sea level changes interpreted from seismic stratigraphy: A critique of the methodology with particular reference to the North Sea Jurassic record, *Am. Assoc. Petrol. Geol. Bull. 70*, 131–137.

Miesling, K. E. (1984). Neotectonics of the North Frontal Fault System of the San Bernardino Mountains: Cajon Pass to Lucerne Valley, California, Ph.D. dissertation, California Institute of Technology, Pasadena.

Miller, K. G., and M. B. Hart (1987). Cenozoic planktonic foraminifers and hiatuses on the New Jersey slope and rise, Deep Sea Drilling Project Leg 95, northwest Atlantic, in *Initial Reports of Deep Sea Drilling Project 95*, C. W. Poag, A. B. Watts *et al.*, eds., U.S. Government Printing Office, Washington, D.C., pp. 253–265.

Miller, K. G., M.-P. Aubry, M. J. Khan, A. J. Melillo, D. V. Kent, and W. A. Berggren (1985a). Oligocene-Miocene biostratigraphy, magnetostratigraphy, and isotopic stratigraphy of the western North Atlantic, *Geology 13*, 257–261.

Miller, K. G., G. S. Mountain, and B. E. Tucholke (1985b). Oligocene glacioeustasy and erosion on the margins of the North Atlantic, *Geology 13*, 10–13.

Mörner, N.-A. (1976). Eustasy and geoid changes, *J. Geol. 84*, 123–151.

Mörner, N.-A. (1980). Relative sea-level, tectono-eustasy, geoidal-eustasy and geodynamics during the Cretaceous, *Cretaceous Res. 1*, 329–340.

Mörner, N.-A. (1981). Revolution in Cretaceous sea-level analysis, *Geology 9*, 344–346.

Nafe, J. E., and C. L. Drake (1963). Physical properties of marine sediments, in *The Sea*, Vol. 3, M. N. Hill, ed., Interscience, New York, pp. 794–815.

Neidell, N. S., and E. Poggiagliolmi (1977). Stratigraphic modeling and interpretation—Geophysical principles and techniques, in *Seismic Stratigraphy—Applications to Hydrocarbon Exploration*, C. E. Payton, ed., Memoir 26, American Association of Petroleum Geologists, Tulsa, Okla., pp. 389–416.

Odin, G. S., ed. (1982). *Numerical Dating in Stratigraphy*, Wiley, New York, 1040 pp.

Olsson, R. K., K. G. Miller, and T. E. Ungrady (1980). Late Oligocene transgression of middle Atlantic coastal plain, *Geology 8*, 549–554.

Parkinson, N., and C. Summerhayes (1985). Synchronous global sequence boundaries, *Am. Assoc. Petrol. Geol. Bull. 69*, 685–687.

Pitman, W. C., III (1978). Relationship between eustacy and stratigraphic sequences of passive margins, *Geol. Soc. Am. Bull. 89*, 1389–1403.

Pitman, W. C., III, and X. Golovchenko (1983). The effect of sea level change on the shelf edge and slope of passive margins, in *The Shelfbreak: Critical Interface on Continental Margins*, D. J. Stanley and G. T. Moore, eds., Spec. Publ. 33, Society of Economic Mineralogists and Paleontologists, pp. 41–58.

Poag, C. W. (1980). Foraminiferal biostratigraphy, paleoenvironments, and depositional cycles in the Baltimore Canyon Trough, in Geologic Studies of the COST No. B-3 Well, United States Mid-Atlantic Continental Slope Area, P. A. Scholle, ed., *U.S. Geol. Surv. Circ. 833*, 44–65.

Poag, C. W., and D. Low (1987). Unconformable sequence boundaries at Deep Sea Drilling Project Site 612, New Jersey transect: Their characteristics and stratigraphic significance, in *Initial Reports of Deep Sea Drilling Project 95*, C. W. Poag, A. B. Watts *et al.*, eds., U.S. Government Printing Office, Washington, D.C., pp. 453–498.

Poag, C. W., and J. Schlee (1984). Depositional sequences and stratigraphic gaps on submerged United States Atlantic margin, in *Interregional Unconformities and Hydrocarbon Accu-*

mulation, J. S. Schlee, ed., Memoir 36, American Assocaition of Petroleum Geologists, Tulsa, Okla., pp. 165–182.

Ramsayer, G. R. (1979). Seismic stratigraphy, a fundamental exploration tool, Offshore Technology Conference Paper OTC 3568, pp. 1859–1862.

Ramsbottom, W. H. C. (1979). Rates of transgression and regression in the Carboniferous of NW Europe, *J. Geol. Soc. London 136*, 147–153.

Richardson, M. J., M. Wimbush, and L. Mayer (1981). Exceptionally strong near-bottom flows on the continental rise of Nova Scotia, *Science 213*, 887–888.

Ricken, W. (1986). *Diagenetic Bedding*, Lecture Notes in Earth Sciences 6, Springer-Verlag, New York, 210 pp.

Ross, C. A., and J. R. P. Ross (1985). Late Paleozoic depositional sequences are synchronous and worldwide, *Geology 13*, 194–197.

Ryer, T. A. (1983). Transgressive-regressive cycles and the occurrence of coal in some Upper Cretaceous strata of Utah, *Geology 11*, 207–211.

Sahagian, D. (1987). Epeirogeny and eustatic sea level changes as inferred from Cretaceous shoreline deposits: Applications to the central and western United States, *J. Geophys. Res. 92*, 4895–4904.

Schwan, W. (1980). Geodynamic peaks in Alpinotype orogenies and changes in ocean-floor spreading during Late Jurassic-late Tertiary time, *Am. Assoc. Petrol. Geol. Bull. 64*, 359–373.

Sheriff, R. E. (1977). Limitations on resolution of seismic reflections and geologic detail derivable from them, in *Seismic Stratigraphy—Applications to Hydrocarbon Exploration*, C. E. Payton, ed., Memoir 26, American Association of Petroleum Geologists, Tulsa, Okla., pp. 3–14.

Sheriff, R. E. (1985). Aspects of seismic resolution, in *Seismic Stratigraphy II: An Integrated Approach to Hydrocarbon Exploration*, O. R. Berg and D. G. Woolverton, eds., Memoir 39, American Association of Petroleum Geologists, Tulsa, Okla., pp. 1–10.

Sloss, L. L. (1979). Global sea level change: A view from the craton, in *Geological and Geophysical Investigations of Continental Margins*, J. S. Watkins, L. Montadert, and P. W. Dickerson, eds., Memoir 29, American Association of Petroleum Geologists, Tulsa, Okla., pp. 461–467.

Steckler, M. (1984). Changes in sea level, in *Patterns of Change in Earth Evolution*, H. D. Holland and A. F. Trendall, eds., Springer-Verlag, New York, pp. 103–121.

Steckler, M. S., and A. B. Watts (1982). Subsidence history and tectonic evolution of Atlantic-type continental margins, in *Dynamics of Passive Margins*, R. A. Scrutton, ed., Geodynamics Series 6, American Geophysical Union, Washington, D.C., pp. 184–196.

Suess, E. (1906). *The Face of the Earth*, Vol. 2, Clarendon, Oxford.

Summerhayes, C. P. (1986). Sea level curves based on seismic stratigraphy: Their chronostratigraphic significance, *Palaeogeogr. Palaeoclimatol. Palaeoecol. 57*, 27–42.

Swift, D. J. P. (1970). Quaternary shelves and the return to grade, *Mar. Geol. 8*, 5–30.

Swift, D. J. P. (1976). Coastal sedimentation, in *Marine Sediment Transport and Environmental Management*, D. J. Stanley and D. J. P. Swift, eds., Wiley, New York, pp. 255–310.

Thorne, J., and A. B. Watts (1984). Seismic reflectors and unconformities at passive continental margins, *Nature 311*, 365–368.

Tucholke, B. E. (1981). Geologic significance of seismic reflectors in the deep western North Atlantic Basin, in *Deep Sea Drilling Project: A Decade of Progress*, J. E. Warme, R. G. Douglas, and E. L. Winterer, eds., Spec. Publ. 32, Society of Economic Mineralogists and Paleontologists, pp. 23–37.

Tucholke, B. E., and R. W. Embley (1984). Cenozoic regional erosion of the abyssal sea floor off South Africa, in *Interregional Unconformities and Hydrocarbon Accumulation*, J. S. Schlee, ed., Memoir 36, American Association of Petroleum Geologists, Tulsa, Okla., pp. 145–164.

Tucker, P. M., and H. J. Yorston (1973). *Pitfalls in Seismic Interpretation*, Monograph No. 2, Society of Exploration Geophysicists, 50 pp.

Vail, P. R. (1987). Seismic stratigraphy interpretation using sequence stratigraphy, Part 1: Seismic stratigraphy interpretation procedure, in *Atlas of Seismic Stratigraphy*, A. W. Bally, ed., Studies in Geology No. 27, American Association of Petroleum Geologists, Tulsa, Okla., pp. 1–10.

Vail, P. R., and J. Hardenbol (1979). Sea-level changes during the Tertiary, *Oceanus 22*, 71–79.

Vail, P. R., and R. G. Todd (1981). Northern North Sea Jurassic unconformities, chronostratigraphy and sea-level changes from seismic stratigraphy, in *Petroleum Geology of the Continental Shelf of North-West Europe*, L. V. Illing and G. D. Hobson, eds., Heyden and Son, London, pp. 216–235.

Vail, P. R., R. M. Mitchum, Jr., R. G. Todd, J. M. Widmier, S. Thompson, III, J. B. Sangree, J. N. Bubb, and W. G. Hatlelid (1977). Seismic stratigraphy and global changes of sea level, in *Seismic Stratigraphy—Applications to Hydrocarbon Exploration*, C. E. Payton, ed., Memoir 26, American Association of Petroleum Geologists, Tulsa, Okla., pp. 49–212.

Vail, P. R., R. M. Mitchum, Jr., T. H. Shipley, and R. T. Buffler (1980). Unconformities of the North Atlantic, *Phil. Trans. Roy. Soc. London, Ser. A 294*, 137–155.

Vail, P. R., J. Hardenbol, and R. G. Todd (1984). Jurassic unconformities, chronostratigraphy, and sea-level changes from seismic stratigraphy and biostratigraphy, in *Interregional Unconformities and Hydrocarbon Accumulation*, J. S. Schlee, ed., Memoir 36, American Association of Petroleum Geologists, Tulsa, Okla., pp. 129–144.

van Couvering, J., and W. A. Berggren (1976). Biostratigraphic basis of the Neogene time scale, in *Concepts and Methods of Biostratigraphy*, J. E. Hazel and E. Kauffman, eds., Dowden, Hutchinson, and Ross, Stroudsberg, Pa., pp. 238–306.

van Wagoner, J. C., R. M. Mitchum, Jr., H. W. Posamentier, and P. R. Vail (1987). Seismic stratigraphy interpretation using sequence stratigraphy, Part 2: Key definitions of sequence stratigraphy, in *Atlas of Seismic Stratigraphy*, A. W. Bally, ed., Studies in Geology No. 27, American Association of Petroleum Geologists, Tulsa, Okla., pp. 11–14.

Watts, A. B. (1982). Tectonic subsidence, flexure, and global changes of sea level, *Nature 297*, 469–474.

Watts, A. B., and M. S. Steckler (1979). Subsidence and eustasy at the continental margin of eastern North America, in *Deep Drilling Results in the Atlantic Ocean Continental Margins and Paleoenvironment*, M. Talwani, W. Hay, and W. B. F.

Ryan, eds., Maurice Ewing Series 3, American Geophysical Union, Washington, D.C., pp. 218–234.

Watts, A. B., and J. Thorne (1984). Tectonics, global changes in sea level and their relationship to stratigraphical sequences at the U.S. Atlantic continental margin, *Mar. Petrol. Geol. 1*, 319–339.

Weldon, R. J. (1984). Implications of the age and distribution of the Late Cenozoic stratigraphy in Cajon Pass, southern California, in *San Andreas Fault—Cajon Pass to Wrightwood*, R. L. Hester and D. E. Hallinger, eds., Pacific Section, volume and guidebook 55, American Association of Petroleum Geologists, pp. 9–15.

Widess, M. B. (1973). How thin is a thin bed? *Geophysics 38*, 1176–1180.

Long-Term Eustasy and Epeirogeny in Continents

8

C. G. A. HARRISON
University of Miami

INTRODUCTION

It is well known that over time intervals of hundreds of millions of years, sea level has fluctuated by several hundreds of meters. The main evidence for this is the changing area of marine sediment deposited on the continents through time, indicating that at certain periods continents have been flooded by sea water much more than they are today, whereas at other periods there appears to have been relatively little flooding. Part of this change is the result of the variable amount of water locked up in continental ice sheets. The waxing and waning of continental ice sheets during the Pleistocene happens on a time scale shorter than those discussed in this chapter, being generally less than 100,000 years. However, there is a long-period time signal in this, in that today, during an interglacial period, there is a significant amount of water locked up in continental ice sheets, and so if we go back to a time when there was no continental glaciation there would be a significant increase in sea level at that time compared with today.

It appears that the most likely cause of large-scale changes in sea level is the variable volume of ridge material, which can produce a signal of several hundred meters. If seafloor spreading increases, then the volume of the ridge crest starts to increase, displacing water and causing additional flooding of the continental areas. The critical

quantity is the area of seafloor produced per unit time. This can be changed either by an increase in spreading rate over a ridge crest of constant length, or by an increase in the length of the ridge crest, or by some combination of the two.

When individual continental flooding curves are studied, it is found that different amounts of sea-level change are required to cause the desired amount of flooding. It is thought that this is because the continents themselves can undergo vertical motion. If this is accompanied by little or no relative horizontal motion or tectonism, the change is termed epeirogeny. Some geologists believe that there is no such thing as epeirogeny, but the evidence, both from the ocean basins (Menard, 1973; Crough, 1979) and from continental areas, is overwhelming. Any area of the Earth can undergo slow vertical motion up to several hundred meters, unaccompanied by any evidence of folding or tectonism.

Many other possibilities exist for changing sea level by tens to hundreds of meters. For instance, a change of the pattern of subduction can cause the ocean basins to become on average younger or older and so produce a sea-level change that is in principle caused by a similar effect to that caused by varying the amount of seafloor produced per unit time interval. If the area of the ocean basin changes because of continental growth or continental destruction, this will also cause a sea-level change. Sedimentation

141

rates have fluctuated through time, and if the ocean basins have more or less sediment in them, this can also cause significant sea-level changes. There are undoubtedly other factors that have not been considered that have the capability of causing significant sea-level changes. In this chapter, I discuss the effects mentioned above, as well as some other effects, in order to arrive at a pattern of sea-level change during the past 200 million years (m.y.).

VERY LONG TIME CHANGES OF FLOODING

Over time scales of hundreds of millions of years, the average amount of continental flooding is expected to remain approximately constant. This is because erosion is effective at reducing continental elevations above sea level, but inefficient at eroding the continental shelves. The rate of chemical erosion is almost independent of continental elevation above sea level and is calculated to be about 8.1 m/m.y. (Holland, 1981). Therefore in several hundred million years even the relatively low lying continental elevations will be eroded to sea level. It is presumed that chemical erosion below sea level is fairly small, except for the minor effect of submarine springs debouching on the continental shelf or slope. Mechanical erosion is more effective at removing the elevations of the high areas of the continents since it is highly correlated with elevation. Mechanical denudation rates given by Holland (1981) lie between 56 and 67 m/m.y. These may be considerably too high for a long-term average due to the influence of man on present-day erosion rates. But even if they are a factor of 2 too high, a time period of only a few hundred million years is necessary to erode much of the high continental areas down to elevations close to sea level, even allowing for isostatic uplift due to the offloading produced by denudation. These time constants appear to be considerably shorter than those derived by Stephenson (1984) for erosion models of the continental lithosphere in which flexure is taken into account. He obtains time constants of 200 m.y. to 400 m.y. for erosion to reduce topographic undulations to $1/e$ of their amplitude. It is more in agreement with the rough estimates of erosion time constants by England and Richardson (1980) of 50 m.y. to 200 m.y.

Aeolian erosion is small in comparison with mechanical erosion or chemical erosion by rivers but could be important as a means of erosion for Africa, which is being eroded by rivers at a much slower rate than the other continents, because of its aridity (Hay and Southam, 1977; Prospero, 1981).

If relative sea level remains perfectly constant for an individual continent, the processes of erosion will serve to plane down areas above sea level, the resulting sediment supply filling up the volume between shelves and the sea surface, such that only a gradual slope between the highest elevations and the edge of the continental shelf is produced. Fluctuations in relative sea level could cause a pumping action whereby some of the sediment deposited on the slope during high stands is eroded away during low stands and deposited on the continental slope or rise, or into the deep ocean basins by the mechanism of turbidity currents.

These arguments suggest that long-term flooding values should not change drastically, even if the area of the continental crust has changed appreciably or if the volume of ocean has grown with time. This conclusion was also reached by Wise (1974), who argued that there were no long-term changes in the amount of continental flooding. Others have argued that there is a secular decrease of flooding during the Phanerozoic that could be explained either by a thickening of the continental crust by about 1 m/m.y. (Hallam, 1971) or, much less likely, by expansion of the Earth (Egyed, 1956)—the expansion occurring in the oceanic regions. Other arguments have been presented (Abbott and Hoffman, 1984) that suggest that both continental area and continental volume have not changed substantially for the past 2 billion years (b.y.). If this is the case, then the larger volume of ridge material that had to exist 2 b.y. ago to allow the Earth to lose the larger amount of radiogenic heat being produced at that time would have caused submergence of much of the continental crust, thus reducing erosion and negating one of the assumptions made in Abbott and Hoffman's model. Since recent data on continental flooding suggest a flooding not much greater than the present day at the beginning of the Phanerozoic, we prefer the model that calls for approximately constant freeboard when averaged over very long time intervals. So the discussion in the rest of this chapter shall be limited to sea-level change over time scales greater than a few hundred thousand years and less than a few hundred million years.

Possible long-term trends in the volume of oceanic water have been summarized by Southam and Hay (1981). Rubey (1951) originally proposed that the volume of oceanic water had grown uniformly with time since the creation of the Earth. Since the average depth of the water in the oceans in meters is approximately the same as the age of the Earth in millions of years, the effect would be to cause a deepening of the oceans at a rate of 1 m/m.y. (0.001 mm/yr). There is also the possibility that the subduction of wet sediments could cause a long-term change in the volume of the ocean water if the rate at which this water is recycled to the oceans is not constant. This is difficult to quantify. Southam and Hay arrived at a figure of about 0.01-mm/yr reduction in ocean depth if none of the water were to be recycled.

METHODS

In this section we derive an expression for the change in freeboard produced by a change in volume of the ridge system. The change may then be used to estimate the change in the area of continent flooded by marine waters if the shape of the hypsographic curve is known.

The first effect that has to be taken into account is that if the ocean floor is loaded by extra water, it will respond isostatically, and so the effect of sea-level change seen by a continent is diminished. This is illustrated in Figure 8.1a. The change in sea level (or more correctly sea depth) is $s = h + d$. The change in freeboard is given by $-h$. For a water density of 1.02 g/cm^3 and an asthenosphere density of 3.4 g/cm^3 the relationship between s and h is given by $h = 0.7s$.

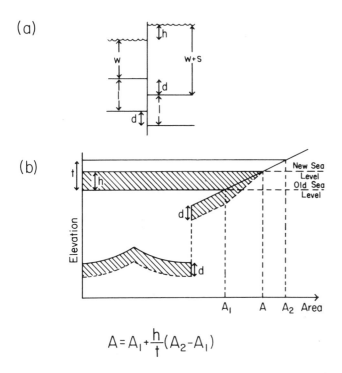

$$A = A_1 + \frac{h}{t}(A_2 - A_1)$$

FIGURE 8.1 (a) Method of calculating the isostatic response of an ocean basin to an increased depth of ocean water, s (equal to $h + d$). (b) Method of relating a volume change to a sea-surface height change. A_1, A_2, and t are taken from the continental hypsographic curve. The change in depth of the oceans is given by $h + d$: d represents the isostatic response of the ocean basins to the extra load of the water; the change in freeboard, h, is less than the change in ocean depth. This model is slightly different from that used by Hays and Pitman (1973), Pitman (1978), and Kominz (1984), who assumed that the continents above present-day sea level did not respond isostatically to the extra load of the water. The volume of extra water or extra ridge crest is shown by the shaded region. The elevation difference between the two cumulative areas A_1 and A_2 is given by t.

Since much of this work involves the calculation of volumes, it is necessary to translate these volumes into sea-level changes. Figure 8.1b shows the principle on which this is done. The shape of the continental area as a function of elevation is given by the continental hypsographic curve (Harrison *et al.*, 1983). A change in the position of the ocean surface of an amount h is caused by a volume change of V, which is shown by the shaded area in Figure 8.1b.

$$V = (h + d)A_1 + \frac{(h + d)}{2}(A_2 - A_1)\frac{h}{t}. \quad (8.1)$$

The modern continental hypsographic curves were used to plot the variation of V with h that is shown in Figure 8.2. A power curve

$$h = 19.694 V^{0.9679} \ (V \geq 10^{16} \ \text{m}^3), \quad (8.2)$$

where h is measured in meters and V in 10^{16} m^3, fits the calculated points with an rms error of 1.6 m. A straight line constrained to pass through the origin, and shown in Figure 8.2, fits the data with an rms error of 3.6 m and has a slope of 17.95×10^{-16} m^{-2}. In converting volume V to freeboard $-h$, we shall use the power curve. If the volume is small or negative, the power curve cannot be used, and as an approximation, the expression $h = 19.1V$ is used to

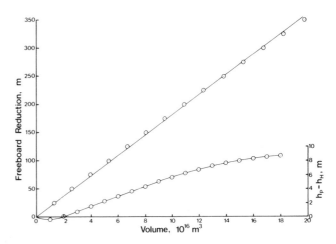

FIGURE 8.2 Relationship between freeboard change (in m) and volume change (in 10^{16} m^3) using modern hypsographic data. The power law curve $h = 19.694 \ V^{0.9679}$ fits the data points to within the size of the small dots. The straight line shown on the graph is the straight line constrained to go through the origin that minimizes the sums of the squared deviations of h. The lower curve and the right-hand ordinate show the deviation between the calculation of Hays and Pitman (1973) (h_p) and the one presented in this paper (h_H). Hays and Pitman (1973) used hypsographic data from Sverdrup *et al.* (1942) and also did not allow for the fact that the newly flooded areas will subside isostatically as sea level increases.

extrapolate to small or negative values of volume. Since negative values of volume are small, this approximation does not introduce any significant inaccuracy.

VOLUME OF MID-OCEAN RIDGES

The most recent estimate of the volume of mid-ocean ridges has been made by Kominz (1984). She analyzed very carefully all the data pertaining to the position of ancient ridge crests and their spreading rates. She investigated the errors produced by inaccurate estimates of spreading rates caused by errors in the time scale of reversals, inaccurate estimates of ridge lengths, the effect of uncertainties in her calculations of variation in areas of oceanic crust, which would be older than 150 m.y. ago (Ma), subducted ridges for which only remanent triple junctions remain, and completely subducted ridge crests.. It should be emphasized that any attempt to calculate ridge volumes for times as long ago as 80 m.y. is fraught with difficulty. Of the oceanic crust that existed then, only 31.2 percent is left, so that large extrapolations need to be made to determine the age-area relationship for crust aged up to 70 Ma. Nevertheless, the work of Kominz is by far the best estimate of mid-ocean ridge volumes.

Her conclusion was that ridge-crest volume changes since 80 Ma have produced an increase in freeboard of 180 m. The analysis of possible errors indicated that the maximum possible sea-level fall could have been 317 m and the smallest could have been –3 m. Table 8.1 presents data on ridge-crest volume calculated from Kominz (1984). The rise in freeboard calculated for the volume of crust 80 Ma is 175.0 m, about 5 m less than Kominz calculated using older hypsographic information.

It has been postulated that the added volume of ridge-crest material produced by increased spreading rates should be counteracted by the effect of increased subduction rates. Hager (1980) suggested that the increased subduction rates should cause the marginal basins in the western Pacific to subside, having an effect opposite to that of the ridge-crest volume. While this may occur to some extent, it is unlikely to cause the ridge-crest volume effect to be entirely obliterated. The area of the marginal basins today is 26.9×10^{12} m² (Sclater et al., 1980). In order to counteract the volume of ridge-crest material, these basins would have to subside on average 3.45 km. In addition, there should today be a strong correlation between marginal basin depth and subduction rate, which does not exist in the magnitude necessary to remove the ridge-crest volume effect. Since

TABLE 8.1 Volume Changes (10^{16} m³)

Time, Ma	Seafloor Spreading	Ice Volume[a]	Pacific Volcanism	CaCO₃ Deposition	Recent Sediments	Sedimentation of Average Crustal Age	Continental Collision	Ocean Cooling	Total	Free-board (m)
0	0	0	0	0	0	0	0	0	0	0
5	–0.713	0.315	0.129	–0.129	–0.381	–0.052	0.133	0.036	–0.662	–12.6
10	–1.331	0.630	0.269	–0.258	–0.381	–0.103	0.265	0.071	–0.838	–16.0
15	–0.816	0.945	0.420	–0.386	–0.381	–0.155	0.398	0.107	0.132	2.8
20	–0.513	1.260	0.585	–0.515	–0.381	–0.206	0.530	0.142	0.902	17.8
25	0.026	1.575	0.763	–0.644	–0.381	–0.258	0.663	0.178	1.922	37.1
30	0.345	1.890	0.955	–0.773	–0.381	–0.309	0.795	0.214	2.736	52.2
35	0.463	2.205	1.164	–0.902	–0.381	–0.361	0.928	0.249	3.365	63.7
40	1.312	2.520	1.390	–1.030	–0.381	–0.413	1.060	0.285	5.103	95.4
45	2.039	2.520	1.634	–1.115	–0.381	–0.464	1.193	0.320	5.746	107.0
50	3.736	2.520	1.899	–1.288	–0.381	–0.516	1.193	0.356	7.519	138.8
55	4.343	2.520	2.186	–1.417	–0.381	–0.567	1.193	0.356	8.233	151.5
60	5.765	2.520	2.497	–1.546	–0.381	–0.619	1.193	0.356	9.785	179.1
65	5.872	2.520	2.833	–1.674	–0.381	–0.671	1.193	0.356	10.048	183.8
70	6.867	2.520	3.197	–1.803	–0.381	–0.722	1.193	0.356	11.227	204.6
75	8.311	2.520	2.697	–1.932	–0.381	–0.774	1.193	0.356	11.990	218.0
80	9.554	2.520	2.197	–2.061	–0.381	–0.825	1.193	0.356	12.553	227.9
100				–2.576						
80[b]	9.554	2.520	2.197	0	–0.381	–0.825	1.193	0.356	14.614	264.1

[a]Corrected for density of continental ice.
[b]80 Ma result omitting the CaCO₃ contribution.

TABLE 8.2 Uplift and Subsidence (Bond, 1979)

Time Interval	Continent	Movement
Miocene-Present	Africa	Uplift, ~90 m
Eocene-Miocene	Africa	Uplift, ~135 m
Campanian/	N. American	Uplift, ~110 m
Maastrichtian-Eocene	Australia	Subsidence, ~110 m
Turonian/Coniacian-Campanian/ Maastrichtian	None	
Albian-Turonian/	Australia	Uplift, ~110 m?
Coniacian	Europe	Uplift, ~110 m?

there is no firm estimate of the magnitude of this effect, it has not been considered further.

ICE-VOLUME EFFECT

A certain amount of water is today locked up in continental ice sheets. In earlier times, when the Earth was warmer than today, the amount so locked up would be minimal, and so an allowance has to be made for this added volume of ocean water in the past. Opinions differ as to the amount of water in today's continental ice sheets. Kennett (1982) estimated a figure of 3×10^{16} m^3, whereas Holmes (1978) estimated a figure of 2.5×10^{16} m^3. Since the density of continental ice is 0.9 g/cm^3, Holmes' figure gives a freeboard change of 43.2 m, whereas Kennett's figure produces a change of 51.5 m.

We use a compromise between these two figures, choosing a volume of 2.8×10^{16} m^3 of continental ice today. The rate of buildup of the ice through time is obviously something that needs to be specified. Kominz (1984) assumed that buildup started 15 Ma. It is generally believed that there was a significant buildup of continental ice starting at the end of the Eocene (Kennett, 1982). We shall therefore assume that the present ice volume commenced accumulating at 40 Ma and has built up since that time uniformly. It should be emphasized that this chapter is not dealing with time scales as short as the Pleistocene glaciation variations. This choice of 40 Ma as the start of significant buildup of continental ice is a compromise. Matthews (1984) suggested a buildup that started 100 Ma and ended 35 Ma. A new curve of freeboard change could be made by reformulating Tables 8.1 and 8.2 to obtain revised volume estimates through time and then using Eq. (8.2) to calculate revised freeboard values. One effect of this would be to make the low stand of sea level calculated to occur 10 Ma even lower (−25 m).

VOLCANIC ACTIVITY

Schlanger et al. (1981) and Watts et al. (1980) have suggested that a large area in the equatorial Pacific Ocean underwent a thermally induced uplift accompanied by large amounts of extrusive volcanic activity. Using a number of lines of different evidence, they came to the conclusion that the thermally induced uplift produced an excess volume of 2×10^{16} m^3 between 110 and 70 Ma. We have assumed a uniform buildup during this 40-m.y. time interval and then an exponential decay with a time constant of 62.8 m.y. with an amplitude of 3200 m (Parsons and Sclater, 1977), which are the same parameters as those for the oceanic crust. Other parameters for the decay of ridge-crest topography have been calculated (e.g., Schroeder, 1984), but we prefer to use the generally accepted values given above.

The depth 70 Ma was 3590 m, and the depth today is 5190 m, giving a total subsidence during this time interval of 1600 m. This allows us to write down two equations for depth at these two time intervals, with two unknowns, being a, the depth to which the Nauru basin will subside after an infinite time, and t, the age from which subsidence is assumed to start in order to obtain the right value for subsidence between 70 Ma and today:

$$3590 = a - 3200 \exp(-t/62.8) \qquad (8.3)$$

and

$$5190 = a - 3200 \exp[-(t + 70)/62.8]. \qquad (8.4)$$

The values of the unknowns are, $a = 5970$ m, and $t = 18.6$ m.y. Since the youngest age that we need to consider is close to 20 m.y., we can use the exponential form for subsidence rather than the form in which subsidence is dependent on the square root of time, which is the correct expression when the age is less than 20 m.y. Since we are interested in volume changes that differ from the present value, we must offset the volumes so calculated by the present volume.

Schlanger et al. (1981) suggested that this large volume of uplift was matched by a similar volume of material on the Farallon plate, all of which material has since been subducted. The islands formed during this period of volcanism on the Farallon plate acted as stepping stones for reefal foraminifera whose distribution without such stepping stones would otherwise be difficult to explain. Schlanger et al. (1981) also added to this volume the effect of other volcanic activity in the ocean basins. Since the additional data that they use are somewhat more speculative as to the size of the effect, they have not been included in our calculation.

This information on volume changes is presented in Table 8.1, along with volume changes produced by spreading activity (Kominz, 1984) and ice volume changes. The total volume change for the effects of seafloor spreading, ice volume, and Pacific volcanism during the past 80 m.y. is 14.271×10^{16} m³, which is equivalent to sea-level fall during this time of 258 m. This is somewhat more than that calculated by Kominz because of the added effect of Pacific volcanism. This is offset somewhat (8 m) by the slightly less steep plot of freeboard versus volume (used in this chapter) than that obtained by the equation used by Kominz (see Figure 8.2).

OCEAN SEDIMENT VOLUMES

Harrison et al. (1981) discussed the possibility that changes in the amount of sediment in the ocean basins could have a significant effect on sea level. Planktonic foraminifera did not evolve into volumetrically important sources of deep-sea sediments until the later Cretaceous. The average thickness of carbonates in the ocean basins today is 300 m (calculated for a carbonate density of 2.7 g/cm³). It has been estimated that 90 percent of this is composed of the tests of pelagic foraminifera. We therefore assume that deep-water carbonate sediments built up uniformly starting 100 Ma such that today there is an additional thickness of 270 m over the ocean basins. The isostatic response to sediments is considerably greater than that for water, since the sediments have a higher density. In order to achieve the same effect on sea level, we can replace the sediments with a water layer, which is $(\rho_a - \rho_s)/(\rho_a - \rho_w)$ times as thick as the sediment, or 79.4 m, where ρ_a is the density of the asthenosphere, ρ_s is the density of the carbonate, and ρ_w is the density of the water. We assume that the sediment layer covers an area equivalent to all oceanic areas below a depth of 1 km, which gives a total volume of 2.576×10^{16} m³. Since this addition of sediment serves to decrease the freeboard as we go back in time, the equivalent volumes are negative. They are tabulated in Table 8.1.

There is a small additional effect due to ocean sediment volumes. It has been shown that sedimentation rates during the past 5 m.y. have been considerably greater than during the earlier Tertiary and Mesozoic (Southam and Hay, 1981). The additional thickness of sediment produced is equivalent to a layer 40 m thick over the ocean floor (Harrison et al., 1981), which gives a volume change of 0.381×10^{16} m³. This has been taken into account in Table 8.1.

A third effect of sediment volume has to do with the change in average age of the ocean basins. The average age has increased by about 20 m.y. during the past 80 m.y. (Harrison et al., 1981). We assume that the noncarbonate

deposition has occurred uniformly through time (except for the recent past, which has already been taken care of). Therefore there is more sediment in the ocean basins today than there was 80 Ma because the ocean basins are on average older today than they were. The average noncarbonate sedimentation rate is 11.65 kg/m² per 1000 yr (Sloan, 1985). For an average age change of 20 m.y. over the whole area of the ocean basins deeper than 1 km, this translates to an equivalent water volume of 0.825×10^{16} m³ (see Table 8.1). This effect is in the opposite direction to that of the ridge-crest volume effect, since the sedimentation on the ocean floor tends to decrease the ridge-crest topographic effect somewhat.

REDUCTION IN CONTINENTAL AREA

Mountain building, by increasing the thickness of the continental crust, serves to decrease its area, and hence increases the area of the ocean basins. This will then cause a lowering of ocean depth and an increase in freeboard. We have attempted to allow for this phenomenon (Harrison et al., 1981, 1985). Most of the effect is caused by the collision of India with the rest of Asia. Additional amounts are contributed by the collision of Arabia with Asia and the collision of the continents on either side of the Alps. The revised reduction in continental area is 3.2×10^{12} m². When the area is multiplied by an average oceanic depth of 3729 m (Menard and Smith, 1966), a volume increase of 1.193×10^{16} m³ is reached. This is apportioned uniformly from 45 Ma to the present and is shown in Table 8.1.

The reduction in continental area was calculated using elevations greater than 500 m in Asia, Europe, India, and Arabia and assuming that these were caused by crustal shortening using an isostatically balanced model to obtain the increase in thickness (Harrison et al., 1985). Allowance was made for elevations above 500 m in these continental areas that were not caused by collision on either side of the Tethys as it closed. The reduction in area is considerably less than that which would be calculated from the timing and speed of collision and the length over which collision occurred (Le Pichon et al., 1986) or from the present rates of consumption, projecting them back into the past (Parsons, 1981). One possibility is that the elevated areas produced by collision have been eroded away, and indeed, the present erosion rates appear to be removing as much continental crust as is being piled up by collision. But a calculation of the sediment deposited in the Bengal fan reveals that this is only one fortieth of that necessary to explain the discrepancy (Harrison et al., 1985). We conclude that the present erosion rates are considerably higher than average erosion rates over the past 40 m.y., possibly because of anthropogenic effects, and that the collision rate of continental crust has been on average much less than that which pertains today, possibly because much oceanic crust existed in the region of collision.

OCEAN COOLING

A small effect should be present because of the cooling of oceanic water since the Cretaceous. Southam and Hay (1981) quote Fairbridge (1961) as giving a 2-m rise of sea level for each 1°C rise in temperature. We shall follow the practice of this paper and calculate the volume change on warming the oceanic waters. One difficulty is that the volume coefficient of expansion of water increases rapidly with rises in temperature and pressure. We have analyzed the effect of a temperature increase, compared with today, of oceanic water going from 2°C to 12°C, this 10° increase being roughly what oxygen isotopic data from Inoceramus indicate (Saltzman and Barron, 1982). Only the water below a depth of 200 m was assumed to increase in temperature. Volume coefficients of expansion were taken from Sverdrup *et al.* (1942). The average coefficient of expansion was determined at a depth by assuming that the coefficient of expansion at any depth may be written:

$$\frac{1}{V}\frac{dV}{dT} = a + bT. \qquad (8.5)$$

It is easy to show that the average coefficient of expansion between temperatures T_1 and T_2 is given by:

$$\frac{1}{V}\frac{dV}{dT} = a + \frac{b(T_1 + T_2)}{2}. \qquad (8.6)$$

The values of a and b are calculated from data presented in Table 9 of Sverdrup *et al.* (1942) for a salinity of 3.5 percent. These average coefficients of expansion were used to determine the depth variation, which was assumed to be of the form:

$$\overline{\frac{1}{V}\frac{dV}{dT}} = p + qx + rx^2, \qquad (8.7)$$

where x is the depth. $(1/V)(dV/dT)$ was evaluated at depths of 0, 2, and 4 km in order to determine the coefficients p, q, and r in Eq. (8.7). The average coefficient of expansion over a depth range from 0.2 km to y km is therefore

$$\sigma_x = \frac{\displaystyle\int_{0.2}^{y}(p + qx + rx^2)\,dx}{y - 0.2} \qquad (8.8)$$

This allows us to use the data from Menard and Smith (1966) on area as a function of depth in the ocean basins to calculate the total volume change on allowing the water

below 200 m to warm up by 10°C. The answer is 0.249×10^{16} m³.

Since thermal expansion of the water increases the water depth without increasing the loading of the deep ocean basins, it is not appropriate to use this volume in the same way as the other volumes are used. The formal calculation of how much freeboard is affected by this effect is complicated. An approximation is to increase the volume by a factor of 1/0.7 and to add up the other volumes given in Table 8.1. The change in freeboard produced by this increase in temperature is then found to be 7.2 m for the ocean surface at its present elevation. This is more than a factor of 2 smaller than the figure given by Fairbridge (1961). In order to obtain such an increase in ocean depth (2 m/K), it would be necessary to have an average coefficient of thermal expansion in a 6-km-deep ocean of 333×10^{-6} K⁻¹. Examination of the table given by Millero (1982) shows that such coefficients of thermal expansion are not achieved for 0°C water until a pressure of over 1 kbar is reached (water depth greater than 10 km). Water at 25°C has such an expansion coefficient at a pressure of 0.5 kbar. It is clear that it is impossible to achieve an average thermal expansion of the amount given by Fairbridge (1961) for a 10°C warming of the present ocean.

The volume change was assumed to begin at 50 Ma, which is about the time that benthic foraminifera oxygen-isotopic data suggest that the bottom water started to cool. Although there have been significant events in this bottom water cooling, it is permissible in this study to apply a uniform rate to the cooling, since the effect is so much smaller than some of the other effects discussed. The effect has been included in Table 8.1.

SUMMARY OF VOLUME EFFECTS

Table 8.1 also shows the total change in freeboard after allowing for all eight factors summarized in Table 8.1. Total change in freeboard during the past 80 m.y. has been 228 m. If the absence of deep-water carbonates during the Cretaceous is made up by the presence of the equivalent volume of shallow-water carbonates, then the change in freeboard will be 264 m. The various volume changes are illustrated in Figure 8.3.

SEA LEVEL MEASURED FROM CONTINENTAL FLOODING

So far, we have used only continental hypsometry to determine the additional area of the oceans as freeboard is decreased, so that we could calculate what the freeboard change is for a change in volume of the ocean basins or the water in them. But a far more significant use can be made of hypsographic curves to determine directly the change in

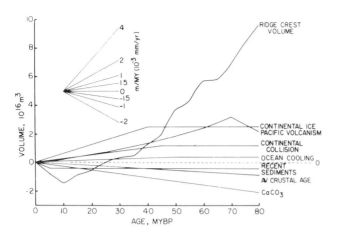

FIGURE 8.3 Volume estimates for six phenomena. A decrease in volume as time progresses produces a sea-level fall (or freeboard increase). Although volume changes are not linearly related to changes in freeboard because of the increase in area of the water as sea-level rises, the straight line in Figure 8.2 can then be used as an approximate linear relationship. Slopes of lines in this figure may then be roughly equated with rates of freeboard change, as shown in the top left-hand corner.

freeboard, using additional data consisting of the amount of continent flooded. This is determined by measuring the area of continent covered by marine sediments through time. This method has formed the basis for a number of measurements of eustatic sea-level change through time (Egyed, 1956; Hallam, 1971). A modern set of paleogeographic maps (Barron *et al.*, 1981) allows us to calculate a

better estimate of the amount of continental flooding and hence eustatic sea-level rise during the Mesozoic and Cenozoic (Harrison *et al.*, 1985). Sea-level rises (i.e., changes in freeboard) calculated in this way are shown on Figure 8.4. Modern hypsographic data (Harrison *et al.*, 1983) for all major continental areas excluding Antarctica and Greenland were used to determine the freeboard reduction necessary to produce the desired amount of flooding. There are several things to note about this curve. First, the amplitude of the change since the Cretaceous (80 Ma) is 153 m. Second, there appears to be a pronounced regression and transgression during the Neogene. A comparison with the data on volume changes (Table 8.1) reveals that the amplitude of the change calculated from flooding is considerably less than that estimated from volume changes, being only 67 percent of the latter. In addition, the volume change data do not show any evidence for the relatively large freeboard at 60 Ma, in comparison with the data on either side.

Wise (1974) showed that if the time slices used to make the paleogeographic maps are lengthened, the area of flooding goes up. This is because all marine deposits occurring within the time slice are included when preparing the paleogeographic map, and small changes in vertical motion between various parts of the continent will be transformed into flooding estimates that are too large for any one instant of time. For this reason, the maps in Barron *et al.* (1981) were made from the closest individual maps in the primary references, with no attempt being made to combine different primary maps if their ages did not agree with the uniformly spaced 20-m.y. ages in the maps in

FIGURE 8.4 Freeboard change measured from amount of continental flooding during the past 180 m.y. The crosses represent mean values from results from six individual continents, and the horizontal bars show the limits of the standard error of this mean value. The open circles show the value calculated from the total amount of flooding and the world hypsographic curve.

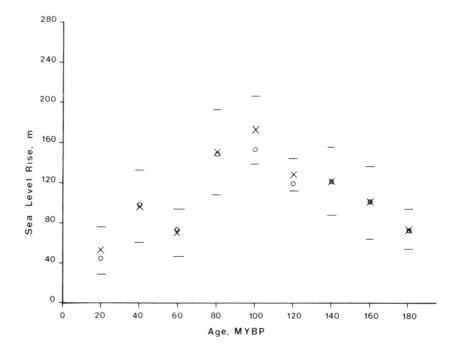

Barron *et al.* (1981). Nevertheless, it is probable that areas of flooding have still been somewhat overestimated, and so the data shown in Figure 8.4 are probably maximum estimates. This means that the discrepancy between Figure 8.4 and Table 8.1 is even larger than it appears at first sight, especially if the absence of deep-water carbonates in the Mesozoic is counteracted by the presence of equivalent volumes of shallow-water carbonates.

The presence of the Paleocene regression presents further problems. Bond (1985) suggested that, in some cases, the evidence for the presence of Paleocene marine sediments has been removed by erosion and that the value measured from paleogeographic maps is probably too low. But we then have to ask why erosion removed the Paleocene sediments in preference to sediments of other ages. It seems likely that erosion could cause removal of young sediments if sea level were suddenly to drop so that these sediments became exposed to aeolian and hydrologic erosion. Thus, even if the record has been obliterated by erosion, the presence of this erosion suggests, in and of itself, that there was some amount of regression during this time. There may, however, be a slight displacement in time of the age of the maximum regression.

It is possible that the Paleocene regression is caused by an amplification of the ridge-crest volume effect, as discussed by Pitman (1978). The position of the shoreline on a subsiding continental margin is highly dependent upon the relative rates of margin subsidence and freeboard increase. Subtle changes in the rate of change of ridge-crest volume can result in large changes in the amount of continental shelf flooded and so possibly cause an effect similar to Paleocene regression. It can be seen in Figure 8.3 that there are some small changes in the rate at which ridge-crest volume is being reduced that might produce the desired effect.

The Paleocene regression has been seen many times before in data of the same sort. Flooding data from Russia, analyzed by Hallam (1977) show a regression of about the same magnitude at the beginning of the Tertiary. However, this pronounced regression is not nearly so obvious on Hallam's (1984) eustatic sea-level diagram.

EPEIROGENY OF THE CONTINENTS

An alternative way of estimating the average change in freeboard is to measure the amount of flooding for individual continents and to use the hypsographic curve for each continent to determine a freeboard change for each continent in turn. Then these estimates may be averaged to produce a mean freeboard curve, which is assumed to be a eustatic curve. This has been done, and the results are shown in Figure 8.4, along with the standard error estimates about each mean value. It can be seen that the result

so estimated is very close to that determined by using the global flooding estimates and the global hypsographic curve. The difference between the two values is never greater than 20 m and is usually much less. The two methods differ because they weight the data from individual continents differently. The global data set weights a continent according to its area, whereas the other method weights each continent equally. But there are obviously major differences between individual continents, as can be seen by the rather large standard error estimates. Bond (1978a,b, 1979) suggested that the differences between individual continents were due to the fact that the average elevations of continents can rise and fall in response to forces beneath them. The deviation of sea level from the mean for each continent is shown in Figure 8.5. This figure illustrates the type of mean vertical motion that each continent must have undergone in order to bring its own "sea-level" curve into agreement with the average. It is not necessary to assume, or likely, that each continent suffers the same vertical motion over its whole area. Rather, the figure shows what the average motion must have been to reach agreement with the mean curve.

Bond (1979) interpreted similar data in a slightly different way. He produced an estimate of the error involved in determining sea-level change for individual continents. The highest sea-level rise was calculated on the basis that the whole of the continental shelves was flooded (the assumption made in this chapter), whereas the lowest sea-level rise was calculated assuming that the present-day shelf only had 50 percent of its area flooded in the past. Bond then made the assumption that the groupings of elevations shown in this figure was not by chance, but reflected real events in sea-level history. Thus, the grouping of elevations between −30 m and 80 m for four of his continents during the Miocene represented the sea level at that time. This enabled him to postulate that since the Miocene, Africa has risen epeirogenically by about 90 m. This correction is then applied to all earlier data from Africa. It is then also observed that the African data, even with this correction, do not agree with the data from the other continents during the Eocene, and so a second correction, of 135 m, is also applied to the African data. Corrections are also applied to Australia, North America, and Europe at various times in order to allow complete agreement between the various curves. A summary of the epeirogenic changes suggested by Bond is shown in Table 8.2. It can be seen that most of the epeirogenic motions suggested by Bond have been uplifts.

Since these motions are corrected before eustatic sea level is calculated, the eustatic sea-level changes suggested by Bond during this time interval will be somewhat less than if the same data had been used with the method of Harrison *et al.* (1983). Bond (1979) attempted a measure-

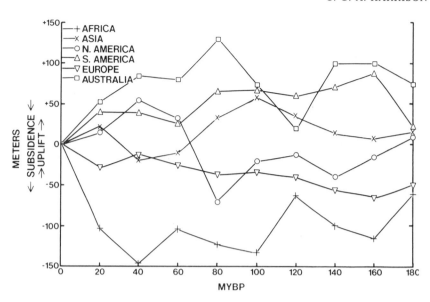

FIGURE 8.5 Sea-level calculations for each of six continents. The mean sea level has been subtracted from each continental value for each time interval, and the results for each continent have been smoothed by taking 0.8 of the central value and 0.1 of the values on either side. Curves that fall going forward in time indicate that the continent, or a large portion of it, has undergone subsidence during the time interval. The North American results were calculated using the North American hypsographic curve excluding Greenland. The curve for Africa has been calculated using the new flooding data for 60 Ma produced by Bond (1985). See Harrison *et al.* (1985).

ment of the error of his sea-level estimations by making two calculations, one in which all of the shelf areas are presumed to be flooded, and another in which only 50 percent of the shelf areas are presumed to be flooded, which will give a lower estimate of sea level. Bond argued that since the real percentage of flooded shelf was likely to be greater than 50 percent, this second figure would give a minimum estimate of sea level. Bond's results show that at 70 Ma, the maximum estimate of sea-level rise is 141 m, and the minimum estimate of sea-level rise is 80 m. These figures bracket an interpolated value for this time interval taken from Figure 8.4 (see also Figure 8.8). The actual values of flooded area measured by Bond must be somewhat larger than those measured by Harrison *et al.*, otherwise there would have been more of a discrepancy between the two results.

We prefer the unbiased calculation that supposes that the modern hypsographic curves are in general like those of previous eras apart from a coherent change discussed in the next section, and that the best estimate of eustatic sea-level change is obtained by the calculation that assumes this, i.e., the one done in Figure 8.4. The epeirogenic movements suggested by Bond (1979) are also seen in the curves (Figure 8.5) except for the European uplift during the Cretaceous.

If the possibility exists that portions of continents can undergo slow epeirogenic movements of amplitude up to several hundred meters, then observations of sea level taken at any one point in a continent are not necessarily of eustatic sea-level change. In order to obtain eustatic sea-level change it is necessary to average over a large number of individual places on a continent or preferably on a series of continents. Measurement of long-term eustatic sea-level changes is thus as fraught with difficulty as the

establishment of the glacio-eustatic signal, where data from different areas often give significantly different results, presumably due to tectonic or epeirogenic changes that have taken place at some or all of the observation points.

Sea-level changes measured in one place have been used to infer eustatic changes. For instance, Sleep (1976) estimated a eustatic sea-level fall of 325 m since the late Turonian-early Coniacian based on the presence of shallow-water marine sediments onlapping onto the continental craton in Minnesota. The present elevation of the craton is 375 m, which was corrected by 50 m to allow for regional uplift due to erosion. Hancock and Kauffman (1979) calculated a sea-level fall of over 600 m since the Campanian. It seems much more likely that this figure is a combination of eustatic effects, plus an epeirogenic elevation of the appropriate land areas since the Campanian.

Other people have attempted to calculate eustatic changes from evidence found in drillings on the continental margins. This is done by finding the present day depth of sediments of a certain age in the continental margins. This has then to be corrected for continental margin subsidence, produced by the slow cooling of the lithosphere following the thermal event that marked the original breakup of the continent. Isostatic adjustments have to be made for the added load of the sediment, and this can be done by using either local isostatic compensation or regional compensation and flexure of the lithosphere. Finally, the depth of the water during the time of deposition has to be estimated from the sedimentology or from benthic fossils found in the sediment. An example of this is shown by Watts and Steckler (1979), who arrived at a freeboard fall of 114 m during the past 75 m.y. This was from a number of wells drilled into the eastern margin of North America, including the Cost B-2 well and four wells off the coast of

Nova Scotia. A more recent analysis gives a fall of 140 m since 83 Ma (Watts and Thorne, 1984). This number is considerably less than that arrived at by a consideration of volumes (Table 8.1). The discrepancy may be removed if it is imagined that this borderland of North America has suffered, in addition to the normal thermal subsidence, an additional long-term epeirogenic subsidence as well. Hallam (1984) pointed out that in some cases evidence for significant differences in the thicknesses of sediments exists from one area of a craton to another. For instance, there is little correlation between the thicknesses of sediments deposited in Dorset and northeast Yorkshire during the Jurassic. This he attributed to taphrogenic influences. The presence of depocenters on continental margins that migrate through time seems to imply a variability in the rate of subsidence, which probably cannot be allowed for in any thermal cooling model, and which therefore may make any estimates of eustatic sea-level change from these margins inaccurate.

Mörner (1976) introduced another complication into our ideas of relative and absolute sea-level changes. He coined the term "geoidal eustasy," by which he means changes of relative sea level produced by changes in the Earth's geoid with time. The present-day geoid departs from the best-fitting oblate spheroid by an amplitude of about ±110 m. The GEM 10B model (Lerch *et al.*, 1979) shows a high of 100 m centered on New Guinea and a low of −120 m in Antarctica. Mörner suggested that the geoid is constantly changing its shape through time. The ocean surface adjusts to the instantaneous position of the geoid, whereas the continental areas take a finite time to respond to the geoid changes. The result is that freeboard changes occur that are not uniform over the surface of the Earth. However, before this mechanism of relative sea-level change can be accepted, it is important to know how fast the geoid can change. Current opinion seems to be that the geoid represents events on the Earth that happened a very long time ago (Crough and Jurdy, 1980; Anderson, 1982; Chase and McNutt, 1982), and so we must expect that changes in the geoid happen very slowly.

COMPARISON OF VOLUME AND FLOODING ESTIMATES

We have seen that the estimates of sea-level change since the Cretaceous calculated from volume estimates of ridge crest, marine sediment, continental ice, thermal bulges, cooling, and ocean-basin area change (242 m) and that from continental flooding (153 m) are different. Although the errors in either of these estimates are large enough to permit this difference, it might be worthwhile to consider what the difference could be caused by, supposing it to be real. The numbers can be made to agree if it is supposed that the continental hypsographic curves were steeper in the past than they are today. In this case, to flood a certain area would require a larger decrease in freeboard than would be calculated by using the present-day hypsographic curve.

Southam and Hay (1981) suggested that the continental hypsographic curve was steeper in the early Mesozoic than it is today because of all the sediment since eroded and deposited onto the passive-margin continental slopes and rises. They produced a hypsographic curve for the early Mesozoic accounting for this effect. A rough estimate of freeboard change from the present day to produce the desired amount of flooding in the Cretaceous is 700 m. If their hypsographic curve is correct, then the amount of erosion that had taken place between the early Mesozoic and the Cretaceous was considerably greater than that which has occurred since. Alternatively, since they did not produce any details of how the hypsographic curve was established except to give one with the required average continental elevation, it is possible that other hypsographic curves could be developed that require a smaller freeboard increase but that still satisfy the requirement of average elevation outlined by Southam and Hay (1981).

An alternative possibility is that the continents have thermally contracted since the splitting up of Pangea. The central regions of Pangea were presumably isolated from major mantle convection currents, allowing the mantle beneath the continental areas to become slightly warmer than the average. When Pangea split apart in response to a new pattern of convection, the mantle would have become ventilated, allowing it to cool and so contract, causing the desired effect (Anderson, 1982). If the contraction of about 100 m occurred in the lithosphere immediately beneath the continents, of thickness about 200 km, then the average temperature drop would be on the order of 15°C to produce the desired effect. A simple model can be used (Carslaw and Jaeger, 1959) in which the base of the lithosphere is suddenly cooled at the time of breakup (180 Ma). In order to obtain a subsidence of 100 m between 100 Ma and today, the cooling at the base of the lithosphere has to be on the order of 130°C to achieve the right magnitude of subsidence.

DISCUSSION OF EPEIROGENIC MOVEMENTS

We have shown that continents have probably moved independently vertically over distances of several hundred meters, with time constants of tens to hundreds of millions of years. It is probable that these movements are caused by convection currents within the mantle or possibly by thermal effects beneath the continental lithosphere. An interesting calculation has been done by Hager *et al.* (1985). They took three-dimensional seismic velocity anomalies within the mantle and converted them to density anomalies. These density anomalies were inverted to obtain the flow field within the mantle. This flow field was in turn used to calculate deviations of the upper and lower surfaces of the mantle. Depending on the viscosity model used in the analysis, the upper surface can be deflected up to several hundred meters with a wavelength of about 10,000 km. This is not a deflection of the geoid, but of the upper surface of the Earth. The deflection of the geoid is considerably less, being in general only a few tens of meters. If this pattern of highs and lows changes as a result of a long-term change in the pattern of convection within the Earth, then we would expect the continents to experience up to several hundred meters of vertical motion, reflected in the amount of flooding that occurs through time. Alternatively, the pattern of convection may remain constant, but the continents may move laterally over the highs and lows, causing the same effect. Interestingly enough, one of the models produced (B. Hager, Massachusetts Institute of Technology, personal communication) shows that Africa is currently situated on a surface high of about 600-m amplitude. This is rather larger than what we would predict from the flooding difference between Africa and the rest of the continents during the past 40 m.y. (Figure 8.5, 150 m) or from the position of the modal elevation in Africa, compared with that from other continents, but it is of the correct order of magnitude. What is necessary to check out this hypothesis is to be able to determine how the convection pattern within the Earth changes with time, and to predict how Africa moved across this convection pattern during the past 100 m.y.

It is interesting to note that Stefanick and Jurdy (1984) calculated that Africa has a much greater density of hot spots than any other portion of the Earth of comparable size. This is true for the 42 hot-spot catalogue of Crough and Jurdy (1980), and for the 117 hot-spot catalogue of Burke and Wilson (1976). This confirms earlier estimates that Africa is a region of intense hot-spot density. For the 117 hot-spot catalogue, most of Africa has a hot-spot density of more than twice the global average. Maybe the relative uplift of Africa with respect to the rest of the continents during the past 40 m.y. is in some way connected with this large density of hot spots.

CHANGING THE AGE DISTRIBUTION OF THE OCEAN BASINS

Changes in the rate of seafloor spreading, as measured by the area of oceanic crust produced per unit time interval, can produce significant changes in the average depth of the basins and thus change sea level, as discussed earlier. The reason is that the age distribution of the oceanic crust is changed by changes in the rate of seafloor spreading. Higher rates of spreading usually mean that there is more young crust present, whereas slow spreading rates usually imply that the crust as a whole is older. Changes in the age distribution can also be affected by the converse process to seafloor spreading, namely subduction. If the subduction process changes the amount of young versus old crust that is subducted per unit time interval, then the pattern of oceanic crustal age will be altered, thus affecting the average depth and the freeboard.

We shall show that this is a potentially powerful method of affecting sea level over time periods of tens of millions of years. The effect of changing the pattern of subduction in the past has been taken into account by the cataloguing of Kominz (1984). What we wish to show is that even with constant spreading rates, the effect of the age of the ocean basins being subducted could have a profound influence on sea level in the future. The pattern of the distribution of the area of oceanic crust as a function of age is one in which there is a monotonically decreasing area as the age is increased (Berger and Winterer, 1974; Harrison, 1980; Sclater et al., 1980). This pattern is most easily produced if there is a constant rate of production of oceanic crust and if subduction occurs uniformly for crust of all ages. The following equations are easy to derive. The area of crust produced per unit time is A km^2/(m.y.); B km^2/(m.y.)^2 is the area of crust subducted per unit time for each unit time age interval; T_{max} is the maximum age of the crust; and T_{av} is the average age of the oceanic crust.

$$A = BT_{max} \qquad (8.9)$$

$$T_{av} = T_{max}/3 \qquad (8.10)$$

$$a = A - Bt, \qquad (8.11)$$

where a is the areal distribution of oceanic crust as a function of age t. The total area of the ocean basins is $AT_{max}/2$. Now, obviously, if B is not a constant, but rather depends on the age of the oceanic crust, then the situation of having a uniformly decreasing area as a function of age will not remain. The oceanic crust will change its age distribution pattern to reflect the rate at which oceanic crust of different age is being subducted.

Parsons (1982) calculated the rate at which oceanic crust is being subducted as a function of the age of the oceanic crust. Details of this are given in Table 8.3. This table presents the same data as are in Figure 3 of Parsons (1982). The numbers in the second column should all be equal, for a steady-state situation to apply, and it is obvious that this is not remotely the case. There are some problems with the interpretation of these figures. They do not produce the same consumption rate (2.506 km^2/yr) as the creation rate (global sum of 3 km^2/yr). Most of the discrepancy is caused by the fact that the oceanic crustal

TABLE 8.3 Subduction Rates

Age Interval (Ma)	Area of Crust Subducted per Unit Age Interval, 10^{-10} km²/yr
0 to 4	40
4 to 9	182
9 to 20	338
20 to 35	291
35 to 52	336
52 to 65	153
Average 0 to 65: 223 ± 49 (standard error)	
65 to 80	45
80 to 95	34
95 to 110	243
110 to 125	25
125 to 140	159
140 to 160	14
160 to 180	22
Average 65 to 180: 77 ± 33 (standard error)	

Note: The first mean is significantly greater than the second mean with a confidence of 97.5 percent.

area is growing at the expense of the continental area, a fact that has been discussed above, and that produces a rate of sea-level fall that can be calculated in the following way. The total generation of seafloor is equal to the total consumption rate of continental and oceanic crust (Parsons, 1981). But part of this consumption, notably that between Arabia and India on the one hand and Asia on the other, causes an increase in the area of oceanic crust, by an amount of 0.229 km²/yr, which will cause a eustatic freeboard increase of 1.8 m/m.y. (0.0018 mm/yr).

In order to calculate the effect of a consumption rate that is not uniform, we have made a calculation in which we start with an oceanic crust that has a uniformly decreasing area versus age plot, and have the total consumption of crust of all ages equal to the production of new crust. The consumption of crust is assumed to vary with age in the same way as the figures in Table 8.3 indicate. The maximum age of crust is assumed to be 180 Ma. The average elevation of the crust is calculated from the normal equations relating elevation to age (Parsons and Sclater, 1977). Then the age distribution after a time lapse of 1 m.y. is calculated by the following procedure: a_i is the area of crust in the ith million year age group, where i runs from 1 to 180; c_i is the rate of consumption per million years for the ith million year age group; b_i is the age distribution after a time lapse of 1 m.y.

$$b_{i+1} = a_i - c_i \qquad (8.12)$$

$$b_i = \sum_{i=1}^{180} c_i . \qquad (8.13)$$

The average elevation for the area depth distribution given by b_i ($i = 1$, 181) may now be calculated. The change in freeboard produced by the c_i's given in Table 8.3 is equal to 2.93 m allowing for a ratio of ocean surface area to oceanic crust area of 0.837 (Menard and Smith, 1966). Thus, if this consumption pattern continues for several tens of millions of years, it could cause a significant change in the freeboard. Over 10 m.y., the system is approximately linear with time. But over longer time periods the effect will average out to less than 2.93 m/m.y. (0.0029 mm/yr).

CRETACEOUS FLUCTUATIONS

The Cretaceous Interior Seaway of North America is a notable feature of the major transgression that occurred worldwide during this time. Evidence exists from sedimentary deposits that there were major fluctuations during the Cretaceous of the depth of the Interior Seaway, which at times stretched from the Gulf of Mexico to the Arctic Ocean. This evidence consists of sets of sedimentary beds that exhibit transgressive-regressive cycles. The transgressive portion of a cycle exhibits upward-fining sequences, from near-shore sands to deeper water pelagic carbonates with relatively little terrigenous sediment. The pattern is repeated in reverse order during the regressive portion of the cycle. It is thought that the maximum depth of water after the major transgressive cycles was several hundred meters. Eicher (1969) calculated a maximum depth of 500 m for the Bridge Creek Limestone member of the Greenhorn Formation (lower Turonian) and the Fairport member of the Carlile Shale (middle Turonian). This was done on the basis of the percent of pelagic foraminifera compared to total foraminifera in these members. An even larger estimate was obtained by estimates of the paleoslope of river drainage basins. Although changes as large as 500 m are probably overestimates of the real situation, it seems unlikely that the sedimentological signal could be produced by a change of less than 100 m. These cycles have been tied into worldwide changes of sea level with a time precision of about 0.3 m.y. (Kauffman, 1977).

What we seek to determine is the range of change in freeboard that might occur if seafloor spreading rates change with a periodicity of about 10 m.y. There are in fact 10 transgressive-regressive cycles recorded during the whole of the Cretaceous, giving a periodicity of 7 m.y., but the

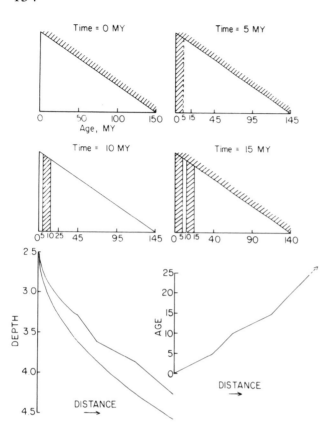

average depth is calculated using the usual age-depth relationships (Parsons and Sclater, 1977). From this, the change in freeboard may be calculated. Spreading is then reduced to its old value for 5 m.y., and the average depth is again calculated. This process is repeated a number of times, in order to determine the magnitude of freeboard change that occurs after each 5-m.y. period of enhanced or reduced spreading rate. This model is illustrated in Figure 8.6.

Calculation of average depth of the oceanic crust allows us to determine the change in depth of the inland sea. Due to the isostatic response of the continental crust by loading of an added water depth, the water depth is increased by a factor of 1/0.7 over the change in freeboard. The result of the calculation is shown in Figure 8.7, where the change in freeboard is shown on the left-hand side, and the change in the depth of water on the continent is shown on the right-hand side. After the first 5 m.y. of rapid spreading, the depth of water has increased by about 110 m. Following the 5-m.y. period of reduced spreading, water depth has fallen by only 22 m. Repetition of the cycles results in ever smaller sea-level rises following the periods of rapid seafloor spreading and ever increasing falls following the periods of reduced seafloor spreading. After a few more cycles, the rises and falls would become equal, and would oscillate with a double amplitude of

FIGURE 8.6 Model of seafloor spreading changes to explain the transgressive-regressive cycles during the Cretaceous. The top four portions show a triangular ocean with a spreading center on the left, and a subduction zone running diagonally from top left to bottom right and hatched. Time is stepped forward at 5-m.y. intervals from (a) to (d). Oceanic crust produced during times of rapid spreading is shaded. Relevant ages are marked along the bottom, and the isochrons are vertical in all cases. Bottom right shows the age-distance plot for the 15-m.y. diagram, in which two phases of rapid spreading are recognized by the large inverse slope of the line. Bottom left shows depth-distance plots for the first diagram (a) and for the last diagram (d). The two bottom diagrams [(e) and (f)] have the same distance scale, which is one-sixth that of the top four diagrams.

major ones, occurring during Aptian to Santonian time, average about 10 m.y. each.

The model used is an ocean that starts off with an area-age distribution that is triangular, representing one in which spreading has been occurring at a constant rate, and in which subduction occurs uniformly for crust of all different ages (see above). The maximum age of the ocean basins is made 150 Ma, representing a slightly faster rate of seafloor spreading than today, appropriate for a time of decreased freeboard. Spreading is then doubled in this triangular ocean for a period of 5 m.y., and the change in

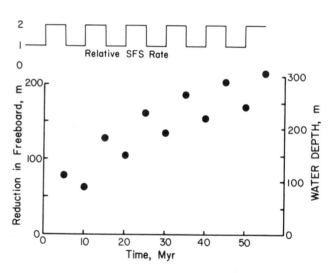

FIGURE 8.7 Changes in sea level produced by fluctuations of seafloor spreading rate. The relative seafloor spreading rate is shown along the top. The decrease in freeboard is shown by the left-hand ordinate. The change in depth of the water flooding the continent is shown by the right-hand ordinate. The figures used to generate this diagram were calculated with an oceanic crust area equivalent to the present area below 2-km depth, and an ocean surface area equal to the present ocean surface plus the area of present land below an altitude of 100 m. The ratio of these two areas is 0.793.

about 60 m around a depth that is on average about 300 m greater than it was when the process started.

Now it appears extremely unlikely that the world's spreading centers could all undergo a factor-of-2 change in spreading rate at 5-m.y. intervals, exactly synchronized with each other, or that an additional portion of ridge crest could turn on and off to produce the extra amount of young crust necessary to give the effect discussed above. Even if this strange phenomenon happened, it would not give the magnitude of effect seen in the paleodepths of the Western Interior Seaway. A more likely possibility is that only one portion of the ridge crest is involved in changing its spreading rate roughly every 5 m.y. If this portion of the world ridge-crest system is, say, 20 percent of the total ridge system, then the eustatic effect would be about 15 m. This is the sort of magnitude of the third-order cycles of Vail *et al.* (1977) with which the transgressive-regressive cycles in the Western Interior Seaway have been correlated, so that a worldwide correlation could be possible. If the changes in spreading occurred in the Pacific, then they would be matched by changes in subduction rates of the Pacific Ocean basin, and in particular in subduction rates at the western margin of North America. It is possible that during increased rates of subduction to the west of the Western Interior Seaway, the shallow basin could be caused to subside tectonically, whereas during times of reduced subduction rate, uplift of the basin could occur. In this way, the tectonic effects could amplify the eustatic sea-level effects of changing ridge-crest volume, to give the large changes in depth inferred from the transgressive-regressive cycles of sedimentation. Unfortunately, since much of the Cretaceous is in the magnetic quiet zone, it would be difficult, if not impossible, to determine whether changes in spreading rate of this sort had occurred. Further information about models of flooding of the Western Interior Seaway can be found in Harrison (1985).

CONCLUSIONS

Various phenomena that can affect sea level over a time scale of tens to hundreds of millions of years have been discussed. During the Neozoic, the most important of these has been the effect of changing ridge-crest volume, which has altered continental freeboard by a volume change of 9.55×10^{16} m^3 over the past 80 m.y. The buildup of continental ice has caused an additional effect of 2.52×10^{16} m^3, but over a shorter time scale of about 40 m.y. An additional effect of 3.20×10^{16} m^3 is provided by the thermal bulge that is thought to have occurred during a time of enhanced volcanism in the Pacific, this being over a time scale of about 70 m.y. Smaller effects are produced by the reduction in continental area during the collision of

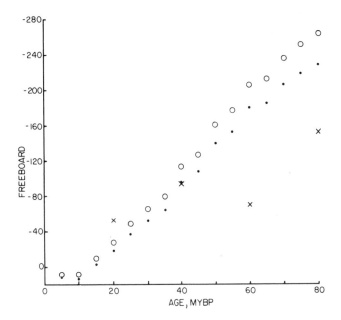

FIGURE 8.8 Changes in freeboard produced by volume effects. The small dots allow for the deep-water carbonate volume. The open dots show freeboard without this effect. The crosses show freeboard changes predicted from amounts of continental flooding, taken from Figure 8.3.

the continents on either side of Tethys (1.19×10^{16} m^3 over a time scale of 45 m.y. and in the same direction as the other three effects) and possibly an effect in the opposite direction of 2.58×10^{16} m^3 produced by deposition of calcium carbonate in the deep ocean starting 100 Ma. The curve of freeboard, calculated from these figures (and other small effects) and using a modern hypsographic curve to obtain the volume of water flooding onto the continents (a small correction) is shown in Figure 8.8. The total freeboard change is 228 m. If the carbonate deposition is not a factor (i.e., if the deep-water carbonates just replaced shallow-water carbonates), then the increase in freeboard is 264 m. These calculations have large errors associated with them, such as those discussed by Kominz (1984).

The area of present land flooded 80 Ma was about 51.5×10^6 km^2. Using the present-day hypsographic curve, a freeboard decrease of 150 m is necessary to flood this area, which is considerably less than that calculated from volume estimates. This suggests that the continental hypsographic curves were somewhat steeper in the past than they are today, requiring larger freeboard decreases to cause the same amount of flooding. It is possible that this decrease in the elevation of the hypsographic curves with time is caused by a cooling of the continents following breakup of Pangea about 200 Ma.

Variations in the amount of freeboard change to flood

different continents by the observed amounts indicates that the continents themselves have suffered vertical motions. These epeirogenic motions have amplitudes of up to hundreds of meters, and occur on time scales of tens of millions of years. Because of these vertical motions, estimates of freeboard change measured at one place may not indicate true eustatic sea-level changes.

Sea-level changes necessary to produce the large effects of water depth seen in the Western Interior Seaway of North America are likely to be caused by an amplification of a eustatic change. Since the transgressive-regressive cycles can be correlated with worldwide sea-level changes, there must be a eustatic signal present. Amplification by tectonic subsidence and uplift, produced by varying speeds of subduction, may be the mechanism whereby tens of meters of eustatic sea-level change can produce hundreds of meters of water depth change within the Western Interior Seaway.

NOTE ADDED IN PROOF

Further discussion about sea-level changes on time scales appropriate to this chapter may be found in Heller and Angevine (1985), Sahagian (1987, 1988), and Harrison (1988).

ACKNOWLEDGMENTS

I have benefitted from discussions with Garry Brass, Bill Hay, Eric Saltzman, and Kim Miskell-Gerhardt. Reviews by Michelle Kominz and Tony Hallam materially improved this paper. Some of this research was supported by grants from the National Science Foundation. Contribution from the University of Miami, Rosenstiel School of Marine and Atmospheric Science.

REFERENCES

Abbott, D. H., and S. E. Hoffman (1984). Archean plate tectonics revisited. 1. Heat flow, spreading rate, and the age of subducting oceanic lithosphere and their effects on the origin and evolution of the continents, *Tectonics 3*, 429–448.

Anderson, D. L. (1982). Hotspots, polar wander, Mesozoic convection and the geoid, *Nature 297*, 391–393.

Barron, E. J., C. G. A. Harrison, J. L. Sloan II, and W. W. Hay (1981). Paleogeography 180 million years ago to the present, *Eclogae Geol. Helv. 74*, 443–470.

Berger, W. H., and E. L. Winterer (1974). Plate stratigraphy and the fluctuating carbonate line, in *Pelagic Sediments on Land and under the Sea*, K. J. Hsü and H. Jenkyns, eds., Spec. Publ. Int. Assoc. Sedimentol. 1, pp. 11–48.

Bond, G. C. (1978a). Evidence for late Tertiary uplift of Africa relative to North America, South America, Australia, and Europe, *J. Geol. 86*, 47–65.

Bond, G. C. (1978b). Speculations on real sea level changes and vertical motions of continents at selected times in the Cretaceous and Tertiary periods, *Geology 6*, 247–250.

Bond, G. C. (1979). Evidence for some uplifts of large magnitude in continental platforms, *Tectonophysics 61*, 285–305.

Bond, G. C. (1985). Comment on "Continental Hypsography" by C. G. A. Harrison *et al.*, *Tectonics 4*, 251–255.

Burke, K., and J. T. Wilson (1976). Hot spots on the Earth's surface, *Sci. Am. 235*, 46–57.

Carslaw, H. S., and J. C. Jaeger (1959). *Conduction of Heat in Solids*, second edition, Clarendon Press, Oxford, 510 pp.

Chase, C. G., and M. K. McNutt (1982). The geoid: Effect of compensated topography and uncompensated oceanic trenches, *Geophys. Res. Lett. 9*, 29–32.

Crough, T. (1979). Hotspot epeirogeny, *Tectonophysics 61*, 321–333.

Crough, S. T., and D. M. Jurdy (1980). Subducted lithosphere, hotspots, and the geoid, *Earth Planet. Sci. Lett. 48*, 15–22.

Egyed, L. (1956). Change of Earth dimension as determined from paleogeographic data, *Geophysica Pura e Applicata 33*, 42–48.

Eicher, D. L. (1969). Paleobathymetry of Cretaceous Greenhorn Sea in eastern Colorado, *Am. Assoc. Petrol. Geol. Bull. 53*, 1075–1090.

England, P. C., and S. W. Richardson (1980). Erosion and the age dependence of continental heat flow, *Geophys. J. R. Astron. Soc. 62*, 421–437.

Fairbridge, R. W. (1961). Eustatic changes in sea level, in *Physics and Chemistry of the Earth*, vol. 4, L. H. Ahrens, F. Press, K. Rankama, and S. K. Runcorn, eds., Pergamon, New York, pp. 99–185.

Hager, B. (1980). Eustatic sea level and spreading rate are not simply related, *EOS 61*, 374.

Hager, B., R. W. Clayton, M. A. Richards, R. P. Comer, and A. M. Dziewonski, (1985). Lower mantle heterogeneity, dynamic topography and the geoid, *Nature 313*, 541–545.

Hallam, A. (1971). Re-evaluation of the paleogeographic argument for an expanding Earth, *Nature 232*, 180–182.

Hallam, A. (1977). Secular change in marine inundation of USSR and North America through the Phanerozoic, *Nature 269*, 769–772.

Hallam, A. (1984). Pre-Quaternary sea-level changes, *Ann. Rev. Earth Planet. Sci. 12*, 205–243.

Hancock, J. M., and E. G. Kauffman (1979). The great transgressions of the late Cretaceous, *J. Geol. Soc. London 136*, 175–186.

Harrison, C. G. A. (1980). Spreading rates and heat flow, *Geophys. Res. Lett. 7*, 1041–1044.

Harrison, C. G. A. (1985). Modelling fluctuations in water depth during the Cretaceous, in *Fine-Grained Deposits and Biofacies of the Cretaceous Western Interior Seaway: Evidence for Cyclic Sedimentary Processes*, E. Kauffman, F. Zelt, and L. Pratt, eds., Soc. Econ. Paleontol. Mineral., Tulsa, Okla., pp. 11–15.

Harrison, C. G. A. (1988). Eustasy and epeirogeny of continents on time scales between about 1 and 100 m.y., *Paleoceanography 3*, 671–684.

Harrison, C. G. A., G. W. Brass, E. Saltzman, J. Sloan II, J. Southam, and J. Whitman (1981). Sea level variations, global sedimentation rates and the hypsographic curve, *Earth Planet. Sci. Lett. 54*, 1–16.

Harrison, C. G. A., G. W. Brass, K. Miskell-Gerhardt, and E. Saltzman (1985). Reply to 'Comment on "Continental Hypsography" by Harrison et al. (1983)' by G.C. Bond, *Tectonics 4*, 257–262.

Hay, W. W., and J. R. Southam (1977). Modulation of marine sedimentation by the continental shelves, in *The Role of Fossil Fuel CO$_2$ in the Oceans*, N. R. Andersen and A. Malahoff, eds., Plenum Press, New York, pp. 569–605.

Hays, J. D., and W. C. Pitman III (1973). Lithospheric plate motion, sea level changes and climatic and ecological consequences, *Nature 246*, 18–22.

Heller, P. L., and C. L. Angevine (1985). Sea-level cycles during growth of Atlantic-type oceans, *Earth Planet. Sci. Lett. 75*, 417–426.

Holland, H. D. (1981). River transport to the oceans, in *The Sea*, Vol. 7, C. Emiliani, ed., Wiley, New York, pp. 763–800.

Holmes, A. (1978). *Principles of Physical Geology*, third edition, Wiley, New York, 730 pp.

Kauffman, E. G. (1977). Geological and biological overview: Western interior Cretaceous basin, in *Cretaceous Facies, Faunas and Paleoenvironments across the Western Interior Basin*, E. G. Kauffman, ed., The Mountain Geologist, vol. 14, pp. 75–99.

Kennett, J. P. (1982). *Marine Geology*, Prentice-Hall, Englewood Cliffs, N.J., 813 pp.

Kominz, M. (1984). Oceanic ridge volumes and sea-level change—an error analysis, in *Interregional Unconformities and Hydrocarbon Accumulation*, J. S. Schlee, ed., Am. Assoc. Pet. Geol. Mem. 36, pp. 109–127.

Le Pichon, X., P. Huchon, and E. Barrier (1986). Pangea, geoid and the evolution of the western margin of the Pacific Ocean, in *Formation of Ocean Margins*, N. Nasu, ed., Terra Scientific Publication Co., Tokyo.

Lerch, F. J., S. M. Klosko, R. E. Laubscher, and C. A. Wagner (1979). Gravity model improvement using GEOS-3 (GEM 9 and 10), *J. Geophys. Res. 84*, 3897–3916.

Matthews, R. K. (1984). Oxygen isotope record of ice volume history: 100 million years of glacio-eustatic sea-level fluctuation, in *Interregional Unconformities and Hydrocarbon Accumulation*, J. S. Schlee, ed., Am. Assoc. Pet. Geol. Mem. 36, pp. 97–107.

Menard, H. W. (1973). Epeirogeny and plate tectonics, *EOS 54*, 1244–1255.

Menard, H. W., and S. M. Smith (1966). Hypsometry of ocean basin provinces, *J. Geophys. Res. 71*, 4305–4325.

Millero, F. J. (1982). The thermodynamics of seawater, Part I. The PVT properties, *Ocean Sci. Eng. 7*, 403–460.

Mörner, N.-A. (1976). Eustasy and geoid changes, *J. Geol. 84*, 123–151.

Parsons, B. (1981). The rates of plate creation and consumption, *Geophys. J. R. Astron. Soc. 67*, 437–448.

Parsons, B. (1982). Causes and consequences of the relation between area and age of the ocean floor, *J. Geophys. Res. 87*, 289–302.

Parsons, B., and J. G. Sclater (1977). An analysis of the variation of ocean floor bathymetry and heat flow with age, *J. Geophys. Res. 82*, 803–827.

Pitman, W. C., III (1978). Relationship between eustasy and stratigraphic sequences of passive margins, *Geol. Soc. Am. Bull. 89*, 1389–1403.

Prospero, J. M. (1981). Eolian transport to the world ocean, in *The Sea*, Vol. 7, C. Emiliani, ed., Wiley, New York, pp. 801–874.

Rubey, W. W. (1951). Geological history of sea water, *Geol. Soc. Am. Bull. 62*, 1111–1147.

Sahagian, D. (1987). Epeirogeny and eustatic sea level changes as inferred from Cretaceous shoreline deposits: Applications to the central and western United States, *J. Geophys. Res. 92*, 4895–4904.

Sahagian, D. (1988). Ocean temperature-induced change in lithospheric thermal structure: A mechanism for long-term eustatic sea level change, *J. Geol. 96*, 254–261.

Saltzman, E. S., and E. J. Barron (1982). Deep circulation in the late Cretaceous: Oxygen isotope paleotemperatures from Inoceramus remains in DSDP cores, *Paleogeogr. Paleoclim. Paleoecol. 40*, 167–181.

Schlanger, S. O., H. C. Jenkyns, and I. Premoli-Silva (1981). Volcanism and vertical tectonics in the Pacific basin related to global Cretaceous transgressions, *Earth Planet. Sci. Lett. 52*, 435–449.

Schroeder, W. (1984). The empirical age-depth relation and depth anomalies in the Pacific ocean basin, *J. Geophys. Res. 89*, 9873–9883.

Sclater, J. G., C. Jaupart, and D. Galson (1980). The heat flow through oceanic and continental crust and the heat loss of the Earth, *Rev. Geophys. Space Phys. 18*, 269–311.

Sleep, N. H. (1976). Platform subsidence mechanisms and "eustatic" sea-level changes, *Tectonophysics 36*, 45–56.

Sloan, J. L., II (1985) Cenozoic Organic Carbon Deposition in the Deep Sea, M.S. thesis, University of Miami, Fla.

Southam, J. R., and W. W. Hay (1981). Global sedimentary mass balance and sea level changes, in *The Sea*, Vol. 7, C. Emiliani, ed., Wiley, New York, pp. 1617–1684.

Stefanick, M., and D. M. Jurdy (1984). The distribution of hot spots, *J. Geophys. Res. 89*, 9919–9925.

Stephenson, R. (1984). Flexural models of continental lithosphere based on the long-term erosional decay of topography, *Geophys. J. R. Astron. Soc. 77*, 385–413.

Sverdrup, H. U., M. W. Johnson, and R. H. Fleming (1942). *The Oceans*, Prentice-Hall, Englewood Cliffs, N.J.

Vail, P. R., R. M. Mitchum, and S. Thompson III (1978). Relative changes of sea level from coastal onlap, *Am. Assoc. Petrol. Geol. Bull. 63*, 1–63.

Watts, A., and M. S. Steckler (1979). Subsidence and eustasy at the continental margin of Eastern North America, in *Deep Drilling Results in the Atlantic Ocean: Continental Margins and Paleoenvironment*, M. Talwani, W. W. Hay, and W. B. F. Ryan, eds., Maurice Ewing Series 3, American Geophysical Union, Washington, D.C., pp. 218–234.

Watts, A. B., and J. Thorne (1984). Tectonics, global changes in sea level and their relationship to stratigraphical sequences at the US Atlantic continental margin, *Mar. Petrol. Geol. 1*, 319–339.

Watts, A. B., J. H. Bodine, and N. M. Ribe (1980). Observations of flexure and the geological evolution of the Pacific ocean basin, *Nature 283*, 532–537.

Wise, D. U. (1974). Continental margins, freeboard and the volumes of continents and oceans through time, in *The Geology of Continental Margins*, C. A. Burk and C. L. Drake, eds., Springer-Verlag, New York, pp. 45–58.

PROCESSES AND FEEDBACKS

Could Possible Changes in Global Groundwater Reservoir Cause Eustatic Sea-Level Fluctuations?

9

WILLIAM W. HAY and MARK A. LESLIE
University of Colorado

ABSTRACT

The total pore space in sediments is about 116×10^6 km^3; this volume constitutes a water reservoir significantly larger than the estimated volume of 24×10^6 km^3 for present-day ice caps and glaciers. Discounting sediments that currently reside 2000 m below sea level, and including only sands, sandstones, and carbonates as part of the aquifer system, there is about 25×10^6 km^3 of pore space in the groundwater system, which might respond to changing inputs and outputs; this corresponds to a change in sea level of over 76 m without or 50 m with isostatic adjustment. The time scale for effecting major changes in the volume of the groundwater reservoir is probably 10^4 to 10^5 yr; possible mechanisms include (1) changes in overall precipitation on the continents, (2) temporal changes in the volume of pore space in sediments on the continents, and (3) changes in the infiltration and discharge rates with time. Because significantly larger volumes of porous sediments may have been present on the continents at several intervals in the geologic past, there may have been a positive feedback between larger porous sediment volumes and greater infiltration rates to amplify the sea-level signal of climatic changes. Although this possible mechanism for sea-level change cannot be demonstrated to have happened, it implies episodic diagenetic changes and periodic high fluid flows, which would affect hydrocarbon migration; studies of these mechanisms might serve as independent tests of the hypothesis.

INTRODUCTION

The problem of explaining geologically rapid eustatic sea-level changes on an Earth without major ice caps is vexing because it has been difficult to understand where the water that left the ocean was stored. One possible storage reservoir that has not been previously investigated is the pore space in sediment on the continental blocks. Although there is general agreement that the volume of water in the world ocean is about 1370×10^6 km^3 (=13,700 $\times 10^{20}$g; Korzun, 1978, gives 1338×10^6 km^3, which may be the most authoritative figure), estimates of the volumes of water in other reservoirs differ widely. Estimates of the volume of ice currently on the surface of the Earth vary from 30×10^6 km^3 (Gavrilenko and Derpgol'ts, 1971) to 200×10^{20}g (Garrels and MacKenzie, 1971), with 24×10^6 km^3 cited by L'vovich (1974) and Korzun (1978). (For H$_2$O, 1 km^3 = 10^{15} cm^3 = 10^{15} g; 10^6 km^3 = 10^{21} g.) Estimates of the amount of water in lakes and rivers are even more variable, from 10×10^6 km^3 (Gavrilenko and Derpgol'ts, 1971), which is clearly too high by almost two orders of magnitude, to 0.03×10^6 km^3 (Garrels and MacKenzie, 1971), with intermediate values being given by L'vovich (1974; 0.29×10^6 km^3). Estimates of volumes of water in the pore space of sediments are generally higher than the volume currently in ice, from 330×10^6 km^3 (Garrels and MacKenzie, 1971) to 190×10^6 km^3 (Gavrilenko and Dergopol'ts, 1971) to 64×10^6 km^3 (L'vovich, 1974) to a minimum of 23.4×10^6 km^3 estimated for groundwater by Korzun (1978). From their calculations of sediment volumes and porosities, Southam and Hay (1981) indicate a total pore volume in sediments of 116×10^6 km^3. Clearly, the pore space in sediments currently contains a large enough volume of water whereby any large-scale changes in the amount of pore space filled with water could affect sea level. The question is whether changes of the magnitude required are possible.

In the equilibrium situation, generally assumed in hydrologic studies, the amount of recharge to the groundwater reservoir is balanced by the discharge of water to the rivers and oceans so that the groundwater reserves remain constant, and there is no net change in the elevation of the groundwater table. However, if either inflow or outflow changes relative to the other, then the volume of water in the groundwater reservoir will change to compensate for the change in supply or release; an increase in groundwater recharge relative to groundwater discharge will result in an increase in the volume of water in the reservoir and a rise of the groundwater table; and a decrease in recharge relative to discharge will result in a decrease of water in the reservoir and a fall of the groundwater table. Because the amount of water residing in the groundwater reservoir is related to the amount of water available to the oceans, large increases or decreases in the groundwater reservoir may raise or lower global sea level. By gradually filling up the reservoir with infiltrated meteoric water or by releasing groundwater to the oceans through runoff, the sediments on the continents may play an important role in controlling the amount of water that is available to the world oceans. This study investigates the potential reservoir capacities of the major sedimentary bodies of the continents now and in the past, and the role they may have played in affecting global sea level.

THE PRESENT DAY HYDROLOGIC CYCLE AND RUNOFF

The hydrologic cycle is the process of global water circulation by which evaporation of water results in precipitation back onto the ocean or onto land areas in the form of rain or snow. The water that falls on land is later

TABLE 9.1 Annual Global Water Balance (modified after data from Budyko, 1974; L'vovich, 1974)

Element of Water Balance	Volume (km^3)	Percent of Total
Precipitation, land	111×10^3	21.3
Precipitation, ocean	410×10^3	78.7
Evaporation, land	67×10^3	12.9
Evaporation, ocean	454×10^3	87.1
Runoff, surface	30×10^3	68.2
Runoff, subsurface	14×10^3	31.8

Note: Total precipitation equals total evaporation. Runoff coefficient (proportion of precipitation that becomes runoff) = 39 percent. Subsurface runoff coefficient (proportion of precipitation that becomes subsurface runoff) = 12 percent.

returned to the oceans through reevaporation, which may be direct or through transpiration, precipitation, or runoff (see Table 9.1). Over 111×10^3 km^3 of water is estimated to fall annually on land in the form of precipitation. About 67×10^3 km^3, or 60 percent of the total precipitation on land, is evaporated, and 44×10^3 km^3 is left on the land surface in the form of potential runoff.

Of this amount, 30×10^3 km^3, or 69 percent, of potential runoff is stored temporarily as snow and ice or is shed directly by the land in the form of rivers or glaciers, and is commonly referred to as surface runoff. The other 14×10^3 km^3, or 31 percent, of the potential runoff, is absorbed into the pore spaces and fissures of the sediments and rocks of the continent and moves downward to the groundwater system. Here, it becomes part of the groundwater reservoir, eventually to be discharged and reappear in marshes, streams, and rivers; i.e., about two-thirds of the flow of rivers measured at the points where they enter the sea is surface runoff and about one-third of the flow is groundwater discharge. The global average residence time for shallow groundwater that is discharged into rivers is not at all well constrained, but has been estimated by L'vovich (1974) to be about 330 yr.

Although there are some notable exceptions, such as the Snake River basalts, we consider the groundwater system to be limited essentially to the pore space in the major sedimentary bodies on the continental blocks. The water-bearing sediments may act both as conduits for the transmission of water through the Earth's crust and as long term storage reservoirs for the groundwater (Todd, 1980). Consequently, the potential variations in the amounts of water that can be stored in, transported through, or released from the groundwater reservoir depends on the physical characteristics of these sedimentary bodies.

SEDIMENTARY RESERVOIRS OF THE CONTINENTAL BLOCKS

The sites of residence of sediments on the Earth's continental blocks can be grouped into three major categories: cratonic, geosynclinal, and the coastal plains and continental shelves (Southam and Hay, 1981). The sedimentary bodies at these sites cover approximately 80 percent of the land surface (Ronov, 1982) and have a total volume of over 790×10^6 km³ (Southam and Hay, 1981; Ronov and Yaroshevsky, 1977). Their dimensions and characteristics are summarized in Table 9.2.

Cratonic Sediments

Cratonic sediments fringe the stable shield areas of the continental interior and the adjacent continental platforms. Because the central shield areas contain little or no sediment cover, the majority of the cratonic sediments reside on the peripheral continental platforms (Southam and Hay, 1981). Using data compiled by Gilluly *et al.* (1970) and Ronov and Yaroshevsky (1977), Southam and Hay (1981) calculated the combined total area of the world's cratonic shields and cratonic platforms to be 96.3×10^6 km², the cratonic shields occupying an area of 29.4×10^6 km² and the platforms having an extent of 66.9×10^6 km². However, because most of the sedimentary cover resides on the peripheral areas of the platforms, they estimated that only 55×10^6 km² of the total cratonic area contains appreciable sediment cover. They assumed the average global thickness of the cratonic sediments to be 3 km so that the total volume is 165×10^6 km³. This accounts for almost 21 percent of the total sedimentary volume on the continental blocks (see Table 9.2).

Ronov (1982) calculated the percentages of the major rock types occurring on the continental blocks based on volume estimates of the different rock types of North America, Europe, and the Soviet Union (see Table 9.2). He estimated that almost half of the cratonic platform sediments are clays and shales with carbonates and sandstones almost 25 percent each. These three rock types make up nearly 93 percent of the total volume of cratonic platform sediments, the remainder being mostly volcanics and evaporites.

Geosynclinal Sediments

Geosynclinal sediments, occupying elongate regions of the Earth's continental crust, usually represent sites of thick sediment accumulation on passive or active continental margins subsequently exposed by uplift and erosion as a result of plate tectonic processes. Geosynclines have been estimated by Ronov and Yaroshevsky (1977) to cover an area of 59×10^6 km² and contain sediments to an average depth of 9 km. They represent over 67 percent of the sediment found on the continental block.

Ronov (1982) compiled percentage estimates of the main rock types found in the geosynclines as shown in Table 9.2. The relative proportions of 40.9 percent clays and shales, 19.2 percent carbonates, and 19.2 percent sands

TABLE 9.2 Average Dimensions, Porosity, Pore Volumes, and Composition of the Major Continental Sedimentary Reservoirs (modified after data from Ronov and Yaroshevsky, 1977; Southam and Hay, 1981; Ronov, 1982)

Sedimentary Reservoir	Area	Thickness	Reservoir Volume[a]	Sediment Volume[b]	Estimated Porosity	Pore Volume	Shale	Sandstone	Carbonate	Volcanic	Other
							Volume and Percentage of Major Rock Types				
Cratonic platforms	55	3	165	157	20%	31.6	76.4 46.3%	36.8 22.3%	40 24.3%	7.26 4.4%	4.4 2.7%
Geosynclines	59	9	531	422	13%	54.9	217 40.9%	102 19.2%	102 19.2%	108.8 20.5%[c]	1.06 0.2%
Passive margin, shelves, and coastal plains	31	3	95	91	20%	18.2	44 46.3%	21 22.3%	23 24.3%	4.2 4.4%	2.5 2.7%
Total	146	5.4	791	670	15.6%						

Note: Thickness is in km, areas are in 10^6 km², and volumes are in 10^6 km³.

[a] Includes volcanic rocks.

[b] Minus volcanic fraction; Ronov (1982) assumed the passive margin shelves and coastal plains to have a composition equivalent to the cratonic sediment.

[c] A compromise between 19.4% (Ronov, 1982) and 21.9% (Ronov and Yaroshevsky, 1977).

and sandstones are very similar to those of cratonic sediments, but together they account for only 79 percent of the sedimentary volume of the geosynclines; the remainder is almost entirely volcanics.

Coastal Plain and Continental Shelf Sediments

Most of the world's coastal plains and continental shelves form the periphery of the continental blocks along the passive trailing margins of the continents formed by the breakup of Pangea. They are underlain by sediment that has been eroded from the continental interior. Southam and Hay (1981) calculated the total area of the coastal plains and continental shelves to be 31.8×10^6 km². The sediments form a wedge that is thin inshore and thickens offshore. Emery and Uchupi's (1972) map of sediment thickness shows an average maximum of 6 km for the shelves. From this measure, Southam and Hay (1981) estimated the average thickness to be 3 km and hence the sediment volume to be 95.4×10^6 km³, representing approximately 12 percent of the total sediment on the continental blocks.

Because the coastal plain and continental shelf sediments were the least well known in terms of composition, Ronov (1982) assumed that the distribution of rock types would closely approximate the general composition of the continental cratons and platforms and hence used identical rock type percentages as shown on Table 9.2.

SEDIMENTS AS HYDROLOGIC RESERVOIRS

Contained within these bodies of sedimentary material is a substantial volume of pore space (Table 9.2). The pore space is not equally distributed between the different rock types but is mostly in the aquifers, i.e., rock bodies that are sufficiently permeable and porous such that they are able to yield large quantities of groundwater (Todd, 1980). According to Muskat (1937), Manger (1963), Morris and Johnson (1967), and others, the most common aquifers are unconsolidated gravels and sands, sandstones, and limestones. Rocks of lesser permeability and porosity may act as barriers to groundwater flow; they are termed "aquifuges" if they neither store nor transmit water, "aquicludes" if they store but do not transmit water, or "aquitards" if they store water but transmit only small amounts of water over long periods of time (Davis and DeWiest, 1966). In this chapter, we consider all clays and shales and all volcanic rocks to be aquifuges and assume that they do not participate in the transmission or storage of underground water.

In the potential aquifers, the sands and carbonates, the permeability and porosity of the rocks decrease steadily with depth of burial and age (Chilingar, 1964; Maxwell, 1964; Choquette and Pray, 1970). This is due chiefly to the increases in lithostatic pressure and temperature with depth, both of which act to reduce pore space and close off the interconnecting pore throats. A number of other fac-

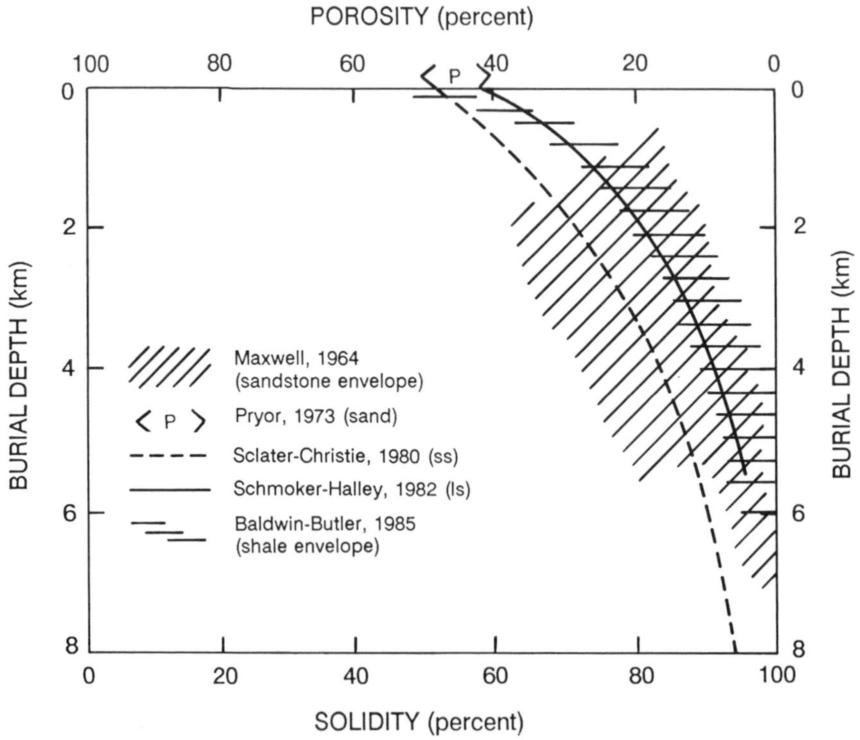

FIGURE 9.1 Porosity versus burial depth for sandstone, limestone, and shale. (After Baldwin and Butler, 1985).

tors also act to reduce porosity and permeability with depth, such as the increase in quartz solubility and clay mineral authigenesis; however, these are all related to and dependent upon increased temperatures and pressures (Blatt *et al.*, 1972).

Using porosity versus depth data from a number of studies (Dickinson, 1953; Maxwell, 1964; Baldwin, 1971; Pryor, 1973; Sclater and Christie, 1980; Schmoker and Halley, 1982), Baldwin and Butler (1985) constructed general sediment compaction curves for the three major rock types (Figure 9.1). Pryor (1973) reported that the porosities of Holocene river point bar and beach and dune sands range from 41 percent to 49 percent, respectively. Sclater and Christie (1980) studied the subsurface sandstones of the North Sea, giving porosity-depth values up to a depth of 10 km. Maxwell (1964) studied the porosity characteristics of Paleozoic and Cenozoic quartzose sandstones; his data cover a wide range of values and vary up to 25 percent at any given depth, producing the "sandstone envelope" values of Baldwin and Butler (1985). The sandstone curve of Sclater and Christie (1980) is almost the midline of the Maxwell sandstone envelope values of Baldwin and Butler (1985). Hence, we consider the Sclater and Christie curve to approximate the average porosity of sandstones for any given depth up to 5 km (see Figure 9.1). The projection of the Sclater-Christie curve to the surface intersects Pryor's range of values for surficial sands at the lower end of the porosity range, at approximately 45 percent.

Sands retain most of their original porosity down to a depth of 1 km. Porosities of approximately 48 percent at the surface show little change for the initial 100 m of burial, and then begin to decrease slightly with depth: to 45 percent at 300 m and 37 percent at 1 km. Porosities decrease rapidly at burial depths greater than 1 km; porosity values are 5 percent or less at depths approaching 10 km.

Schmoker and Halley (1982) studied the porosity of limestones in south Florida and plotted porosity-depth values for depths to 5.5 km. They indicated that limestones have high initial porosities (over 40 percent) that decrease in much the same way as do sandstones, but limestones exhibit 5 to 10 percent lower average porosities than sandstones at any given depth.

POTENTIAL WATER-BEARING CAPACITY OF THE MAJOR SEDIMENT RESERVOIRS

Using the values for porosity cited above, we have calculated the possible ranges of pore-space volume versus depth for the sandstone and carbonate (aquifer) fraction of the major sedimentary bodies to estimate their water-storage capacity, i.e., the amount of water that might be withheld from the oceans as a result of temporary storage. The available pore volume can be divided by the area of the ocean basins (325×10^6 km^2) to give the hypothetical sea-level change that would occur if all of the pore space were empty and then filled with water, or vice versa.

If the sandstone and carbonate fractions of the global cratonic sediments have an average porosity of 20 percent, the available pore space is 15.4×10^6 km^3 and the potential sea-level change is ±47 m. Considering only the aquifer fraction of the geosynclines (the sandstones and carbonates) the total pore volume is 26.5×10^6 km^3, equivalent to a potential sea-level change of ±81.5 m. The sandstone and carbonate fractions of the shelf and coastal plain reservoir have a volume of 44.5×10^6 km^3 and, if assumed to have an average porosity of 20 percent (Atwater and Miller, 1965), have a combined pore space volume of 8.9×10^6 km^3, for a potential sea-level change of ±27.4 m.

WATER-BEARING CAPACITY AND CONTINENTAL ELEVATION

The total water-bearing capacity of the aquifers on the continental blocks is thus 50.8×10^6 km^3, enough to change sea level by ±156 m. However, this situation is hypothetical, and in it, all the sandstones and limestones are ideal aquifers, possessing equal porosities, permeabilities, and other hydrological properties and forming a single homogenous reservoir, capable of responding to the complete filling-up or emptying of their pore spaces with water, thereby raising or lowering global sea level. This scenario might be possible if all of the sediment bodies of the continents were to reside above sea level so their pore spaces were not permanently filled with water. This is obviously not the case; the continental shelf sediments are at present submerged and saturated with water while the adjacent coastal plain sediments are slightly above sea level and have a groundwater content reflecting the local climate. The submerged continental shelf sediments cannot store more water than they do at present, but they could release water if sea level were to fall. Coastal plain sediments could store more water if sea level were to rise and could release water if sea level were to fall. Clearly, sediments in this geologic setting can play a modulating role in sea-level changes by releasing water as sea level falls and filling with water as sea level rises, but the maximum possible effect is equivalent to less than 30 m of sea-level change.

The average elevation of the continents is less than 1 km above sea level, but the average thickness of the major sedimentary bodies that reside on the continents is about 3 km (see Table 9.2). Since most sediment on the continental blocks lies below sea level, except possibly in some interior basins enclosed by high basement, it is permanently saturated with water. Only those sediments that lie above sea level can be potentially filled or emptied of groundwater, and only the aquifers are able to absorb, store, and transmit water through their pore spaces and thus participate in the process. However, the sediments above sea level are younger than average, hence less compacted and more porous. Specifically, the porosity and depth curves for sandstones of Sclater and Christie (1980) and curves for limestone of Schmoker and Halley (1982) suggest that average porosities of 30 to 40 percent are reasonable for such sediments buried to depths of 1 km or less.

A HYPOTHETICAL MODEL OF CONTINENTAL ELEVATION AND SEA-LEVEL CHANGE

The potential water-bearing capacity of sediments on the continental blocks and the possible effect on worldwide sea level can be evaluated as a model taking into account continental elevation and sea level. Because the present-day average elevation of the continents excluding ice-covered Antarctica is approximately 750 m (Southam and Hay, 1981), a lowering of sea level to the current global shelf break would add another 200 m to the continental elevation, so that the average continental elevation would be almost 1000 m above sea level. Assuming the sediments to be randomly distributed, and assuming a higher than average porosity of 40 percent for the near-surface sediments, this average elevation would indicate that over 24.7×10^6 km^3 of pore space would reside above sea level in the major sedimentary aquifers at a sea-level stand 200 m below that of today. If this pore space were initially empty but then filled by infiltration of precipitation, a further global drop in sea level of over 76 m would result, but would be reduced to 50 m as isostatic adjustment occurred.

In reality, it is impossible for the groundwater table to be reduced to sea level even if the hydrologic cycle were to cease; capillary forces alone would cause the water table to be some finite distance above sea level. Because the water in the upper part of the groundwater reservoir, i.e., that participating most actively in aquifer flow, is fresher than sea water, it is lighter, and therefore its surface must be above sea level. A lens of freshwater, being lighter, would of course depress the freshwater-saltwater interface in the groundwater system, and cause discharge of saltwater, which must eventually return to the ocean. Although this complication is important for the freshwater-saltwater mass balance, it is immaterial in the discussion of sea level and will not be discussed further here. Figure 9.2 shows the effects of varying the elevation of the groundwater table to simulate different degrees of satura-

tion and thereby to estimate the different volumes of unsaturated sediment that might respond to the introduction of infiltrated groundwater. For example, assuming an average continental elevation of 1000 m, and an average groundwater table of only 200 m elevation above sea level, there is an average of 800 m of unsaturated sediment above the groundwater table. A 40 percent porosity in these sediments would yield a total aquifer pore volume of 19.9×10^6 km^3 that, if filled with water, could lower sea level initially by more than 61 m, or 40 m after isostatic adjustment.

RESIDENCE TIMES

We can estimate a residence time for water in the aquifer system by assuming that there is no subsurface discharge to the rivers and oceans, only surface infiltration into the empty sedimentary reservoirs until they are filled. As shown in Table 9.1, the present annual volume of water precipitated on land is 111×10^6 km^3/yr, but only 12 percent of this, or 13.5×10^6 km^3/yr, infiltrates into the subsurface reservoir. Figure 9.3 shows residence times in years obtained by dividing the volume of pore space contained within the aquifers by the annual infiltrate volume; the filling times vary from about a 100 to a 1000 yr. This situation is not realistic because there is always subsurface discharge out of the groundwater reservoir to the rivers and oceans to compensate for the incoming infiltrate recharge, and the rate of discharge must increase with increasing hydrostatic head. From this we can guess that the length of time required to fill or empty a groundwater reservoir after a step function change in the global hydrologic cycle would probably be on the order of tens of thousands to hundreds of thousands of years.

MECHANISMS FOR CHANGING THE VOLUME OF GROUNDWATER

It is obvious that groundwater levels depend on climatic conditions because climate determines the amount of precipitation that will fall onto the continents. High rates of precipitation onto the continents will increase the volume of water that infiltrates the groundwater reservoir and subsequently will cause the groundwater table to rise. An increase in the volume of water held by the continents means that there is less water available to the oceans and the sea level drops. Similarly, decreased precipitation onto the continents means decreased volumes of water available to the groundwater reservoir, resulting in lowering of the groundwater table; hence, less water is retained by the continents, more water becomes available to the oceans, and the sea level rises. Consequently, changes in the climate regimes of the Earth over time can have an

167

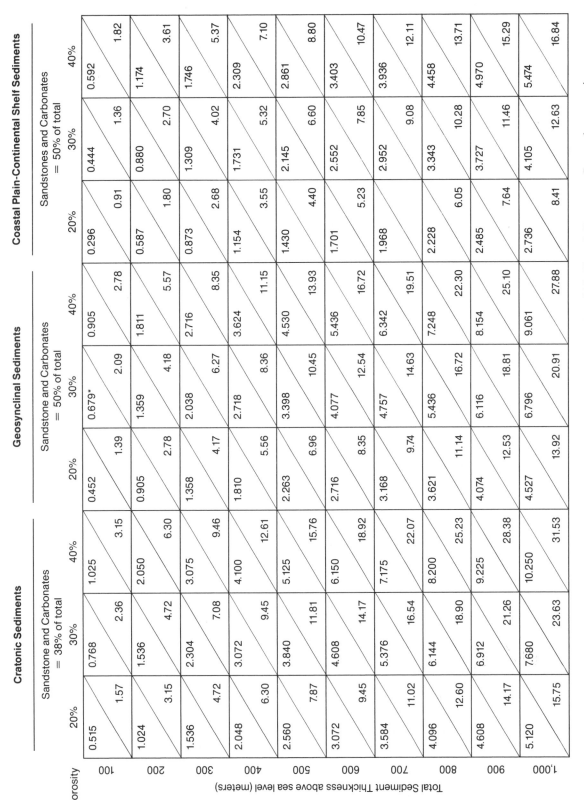

FIGURE 9.2 Pore volumes (10^6 km^3) and equivalent sea-level change (meters) for sandstone and carbonate (aquifer) fractions of varying total thickness of sediment above sea level assuming aquifer porosities of 20, 30, and 40 percent. Pore volumes are in upper left-hand portion of each box and equivalent sea-level changes are in lower right-hand portion of each box.

Sandstone and Carbonate Fraction

Thickness of Variable Groundwater Lens (meter)	Porosity		Range	
	10%	20%	30%	40%
900	5.578 / 413	11.167 / 827	16.755 / 1241	22.349 / 1655
800	4.962 / 367	9.945 / 736	14.923 / 1105	19.906 / 1474
700	4.35 / 322	8.72 / 646	13.085 / 969	17.453 / 1293
600	3.736 / 277	7.489 / 555	11.237 / 832	14.989 / 1110
500	3.12 / 231	6.253 / 463	9.383 / 695	12.516 / 927
400	2.501 / 185	5.012 / 371	7.521 / 557	10.033 / 743
300	1.879 / 139	3.767 / 279	5.651 / 418	7.537 / 558
200	1.255 / 93	2.516 / 186	3.775 / 279	5.035 / 373
100	0.629 / 47	1.26 / 93	1.891 / 140	2.522 / 187

FIGURE 9.3 Pore volume (10^6 km³) of sediments residing above groundwater table and time (years) required to fill at an infiltration rate of 13.5×10^3 km³/yr (assuming no discharge). Pore volumes are in upper left-hand portion of each box, and times are in lower right-hand portion of each box.

effect on sea level by storing water in the subsurface continental reservoir or releasing water to the oceanic reservoir through subsurface runoff.

Presumably, any long-term change in the amount or temporal distribution of precipitation on land will induce some change in the size of the groundwater reservoir. The major question is whether global or extensive changes in the hydrologic cycle do occur in such a way as to significantly alter the amount of precipitation on land or the way in which it is temporally distributed. Experiments by W. W. Hay, E. J. Barron, and S. Thompson using the numeri-

cal Community Climate Global Atmospheric Circulation Model at the National Center for Atmospheric Research for several different idealized paleogeographies have suggested that the global hydrologic cycle may vary considerably depending on the distribution of land, sea, and mountain ranges. In these experiments, runoff varied by an order of magnitude, with the present-day situation being close to the maximum. Clearly, if conditions did change so that runoff became an order of magnitude less than at present and the climate remained stable for many thousands of years, the groundwater reservoir would become depleted and a net transfer of water to the sea would occur. It should be noted that these model experiments assumed an atmosphere with present-day composition (i.e., relatively low CO_2 concentrations). Other compositions might change the hydrologic cycle significantly; specifically, with higher CO_2 content, the amount of precipitation on land might increase significantly. Evaporation rates would also be higher, but it is not at all clear what would happen to infiltration rates under such conditions.

If rainfall on land were concentrated into relatively short periods rather than being more evenly distributed throughout the year, more instantaneous runoff would occur and infiltration of the groundwater reservoir would become less effective. However, it is not clear whether this could produce a change in the volume of the groundwater reservoir large enough to affect sea level.

A second possibility for changing the volume of groundwater lies in the temporal variations in the pore space contained in aquifers above sea level. Ronov's (1982) data have been analyzed by Hay and Wold (1986) who noted significant variations in the abundances of both shallow-water carbonates and nonmarine rocks with time. Such sediments formed in abundance in the mid-Paleozoic, late Paleozoic–early Mesozoic, and mid-Cretaceous. At these times their abundance was more than twice that of such sediments at the present time. The present is a peculiar moment in the history of the Earth because the relatively young, porous sediments that covered much of North America and Asia in the earlier Cenozoic have been eroded and stripped off because of uplift of broad areas of these continents during the late Cenozoic. Furthermore, the late Cenozoic fluctuations in sea level in response to glaciation and deglaciation have resulted in modification of the continental shelves and coastal plains, offloading sediment from those regions into the continental slopes and rises and the abyssal plains. The potential groundwater reservoir that exists today may well represent the minimum that has existed for much of geologic history. In the times of abundant nonmarine and shallow carbonate deposition, the potential fluctuating global groundwater storage space could have been double the 50-m (after isostatic adjustment) estimate presented above.

A third possibility for changing the volume of groundwater lies in the possibility that the infiltration rate may change with time. It seems likely that the infiltration rate would be higher if larger areas of land were covered by relatively porous unconsolidated sediment. Significantly larger than average land areas were covered by unconsolidated nonmarine sediments in the mid-Paleozoic, late Paleozoic-early Mesozoic, and mid-Cretaceous as noted above. The potential for increased infiltration rate as well as the larger potential pore volume available for fluctuating groundwater supplies may well have constituted a

feedback system that would amplify changes in the hydrologic cycle.

Finally, there is a possibility that discharge rates from the groundwater reservoir could change with time. This is perhaps most easily envisioned as a result of canyon-cutting with concomitant exposure of aquifers as a response to either uplift or lowering of sea level. The effect would be to supply more water to the ocean, and hence raise sea level. This is a possible negative feedback that could act to damp falling sea levels. However, it seems evident that because this mechanism must of necessity affect smaller areas, it is likely to be much less effective in changing groundwater levels than any of the other three possibilities discussed above.

SIGNIFICANCE OF CHANGES IN THE GROUNDWATER TABLE FOR ACCUMULATION OF HYDROCARBONS AND FOR MINERALIZATION

Periodic filling and emptying of the pore space in aquifers implies times of relatively rapid and slow groundwater migration, which in turn should have significant implications for hydrocarbon migration, diagenesis, and mineralization. These effects could be locally very important even if the global effects were small. Assuming that when the Earth is ice free, sea-level changes are caused by filling and draining of the groundwater reservoir, a sea-level drop would imply filling of the pore space; consequently, generally higher hydrostatic heads would mean higher rates of fluid flow, and flushing of the system followed by lower concentrations of solutes. These would be times when hydrocarbons would be more likely to migrate and when secondary porosity might be produced. Sea-level rises would correspond to times of draining of the groundwater reservoirs, lowered hydrostatic head, lower rates of fluid flow, and higher concentrations of solutes. Compaction, diagenesis, and mineralization might be expected to occur during these times, with concomitant irreversible reductions in porosity. It is interesting to speculate that if such changes do occur, there is a complex feedback between the groundwater system and the ocean, which might be an important factor in concentrating natural resources. The changes in freshwater-saltwater mass balance, alluded to above, would become an important consideration, but that discussion is beyond the scope of this chapter.

SUMMARY AND CONCLUSIONS

The pore space in aquifers within the upper 1 km of average elevation of the continents is about 25×10^6 km^3, or equivalent to the volume of ice in glaciers on land today. If this pore space could be alternately filled with and emptied of water instantaneously, it would change sea level by ±76 m or ±50 m after isostatic adjustment. It seems likely that some fraction of this pore volume is subject to filling and draining as a result of climatic changes, which vary the amount of precipitation on land. The response times for the changes in the reservoir would be on the order of tens to thousands of years after a step function change in climate. Furthermore, there have been times in the geologic past (mid-Paleozoic, late Paleozoic–early Mesozoic, and mid-Cretaceous) when the pore volume of sediments residing above sea level may have been as much as twice its size today. The surficial sediments at these times were mostly young and highly porous, and may have had infiltration rates significantly greater than those of today. Clearly, changes in the global volume of groundwater with time are a possible mechanism for the changes in sea level on the order of one or a few million years as postulated by seismic stratigraphers.

In order to estimate the likely fluctuations in the groundwater reservoir more accurately, it will be necessary to determine the volumes of aquifer sediment more accurately, using more specific data for area-elevation-sediment type than are currently available. One can expect that within 5 to 10 yr, enough of the required information will become available so that it will be possible to critically assess the roles of fluctuating volume of the groundwater reservoir in effecting sea-level change.

ACKNOWLEDGMENTS

This work was supported by NSF Grant NSF OCE-8409369.

REFERENCES

Atwater, G. I., and E. E. Miller (1965). The effect of decrease in porosity with depth on the future development of oil and gas reserves in south Louisiana, *Am. Assoc. Petrol. Geol. Bull. 49*, 334.

Baldwin, B. (1971). Ways of deciphering compacted sediments, *J. Sediment. Petrol. 41*, 293–301.

Baldwin, B., and C. O. Butler (1985). Compaction curves, *Am. Assoc. Petrol. Geol. Bull. 69*, 622–626.

Blatt, H., G. Middleton, and R. Murray (1972). *Origin of Sedimentary Rocks*, Prentice-Hall, New York, 634 pp.

Budyko, M. I. (1974). *Climate and Life* (English translation by D. H. Miller, ed.), Academic Press, New York, 470 pp.

Chilingar, G. V. (1964). Relationship between porosity, permeability, and grain size distribution, in *Deltaic and Shallow Marine Deposits*, L. M. J. U. van Straaten, ed., Elsevier, Amsterdam, pp. 71–75.

Choquette, P. W., and L. C. Pray (1970). Geologic nomenclature and classification of porosity in sedimentary carbonates, *Am. Assoc. Petrol. Geol. Bull. 54*, 207–250.

Davis, S. N., and J. M. R. De Wiest (1966). *Hydrogeology*, John Wiley and Sons, New York, 463 pp.

Dickinson, G. (1953). Geological aspects of abnormal reservoir pressures in Gulf Coast Louisiana, *Am. Assoc. Petrol. Geol. Bull. 37*, 410–432.

Emery, K. O., and E. Uchupi (1972). Western North Atlantic Ocean; Topography, rocks, structure, water, life and sediments, *Am. Assoc. Petrol. Geol. Mem. 17*, 532 pp.

Garrels, R. M., and F. T. MacKenzie (1971). *Evolution of Sedimentary Rocks*, Norton, New York, 397 pp.

Gavrilenko, E. S., and V. F. Derpgol'ts (1971). *The Deep Hydrosphere of the Earth*, Naukova dumka, Kiev, 272 pp.

Gilluly, J., J. C. Reed, Jr., and W. M. Cady (1970). Sedimentary volumes and their significance, *Geol. Soc. Am. Bull. 81*, 353–376.

Hay, W. W., and C. N. Wold (1986). A 150 my cycle in erosion and sedimentation rates, *Geol. Soc. Am. Abstr. Program 18*, 632.

Korzun, V. I. (1978). *World Water Balance and Water Resources of the Earth*, UNESCO, 663 pp.

L'vovich, M. I. (1974). *World Water Resources and Their Future*, Mysl', Moscow, 447 pp.

Manger, G. E. (1963). Porosity and bulk density of sedimentary rocks, *U.S. Geol. Survey Bull. 1144-E*, 55 pp.

Maxwell, J. C. (1964). Influence of depth, temperature, and geologic age on porosity of quartzose sandstone, *Am. Assoc. Petrol. Geol. Bull. 48*, 697–709.

Morris, D. A., and A. I. Johnson (1967). Summary of hydrologic and physical properties of rock and soil materials, as analyzed by the hydrologic laboratory of the USGS 1948–1960, *U.S. Geol. Survey Water Supply Paper 1839-D*, 42 pp.

Muskat, M. (1937). *Flow of Homogenous Fluids through Porous Media*, McGraw-Hill, New York, 763 pp.

Pryor, W. A. (1973). Permeability-porosity patterns and variations in some Holocene sand bodies, *Am. Assoc. Petrol. Geol. Bull. 57*, 162–189.

Ronov, A. B. (1982). The Earth's sedimentary shell (quantitative patterns of its structure, composition, and evolution), *Int. Geol. Rev. 24*, 1313–1363, 1365–1388.

Ronov, A. B., and A. A. Yaroshevsky (1977). A new model for the chemical structure of the Earth's crust, *Geochem. Int. 13*(6), 89–121.

Schmoker, J. W., and R. B. Halley (1982). Carbonate porosity versus depth: A predictable relation for south Florida, *Am. Assoc. Petrol. Geol. Bull. 66*, 2561–2570.

Sclater, J. G., and P. A. F. Christie (1980). Continental stretching: An explanation of the post-mid-Cretaceous subsidence of the central North Sea basin, *J. Geophys. Res. 85*, 3711–3739.

Southam, J. R., and W. W. Hay (1981). Global sedimentary mass balance and sea level changes, in *The Ocean Lithosphere, The Sea, Vol. 7*, C. Emiliani, ed., John Wiley and Sons, New York, pp. 1617–1684.

Todd, D. K. (1980). *Groundwater Hydrology*, 2nd ed., John Wiley and Sons, New York, 527 pp.

Role of Land Ice in Present and Future Sea-Level Change

10

MARK F. MEIER
University of Colorado

ABSTRACT

Rising concentrations of CO_2 and other greenhouse gases in the atmosphere will cause air temperature to increase and precipitation to change. This chapter reviews our present knowledge of the response of ice sheets and glaciers to the changed climate, and the consequent effect on sea level, from the present to the next 100 yr. Glaciers other than the two existing ice sheets are currently wasting, and this has contributed about 0.46 ± 0.26 mm/yr to sea-level rise since 1900. The Greenland Ice Sheet appears to be close to a state of balance at present. The Antarctic Ice Sheet may be growing at a rate equivalent to about 0.6 mm/yr of sea-level fall; on the other hand, the rate of iceberg discharge may have been underestimated and it may be close to balance. Future growth or wastage of mountain glaciers could be calculated if the changes in climate were known, because the mass and energy balance variables are, in many regions, known functions of altitude. Calculation of future changes of the Greenland Ice Sheet, the Arctic ice caps, and the marginal areas in Antarctica requires study of the complex fluid mechanics/thermodynamics of subfreezing snow and firn subjected to water percolation. Determination of changes in iceberg calving is difficult because little is known about the rate-controlling processes: in addition, changes in the geometry of a glacier or ice stream cause changes in the rate of basal sliding, a process that is still imperfectly understood. Striking changes in calving and sliding of Columbia Glacier are occurring as it disintegrates. A warmer climate may cause warmer ocean water to intrude under the floating ice shelves of Antarctica, causing increased basal melt and ice shelf thinning. This may, in turn, reduce the back pressure on the ice streams that flow into the shelves, causing the ice streams to accelerate. This process could deplete the ice sheet, producing a sea-level rise of up to 0.3 m, or possibly more, by the year 2100. Complete disintegration of the West Antarctic Ice Sheet is not likely for many centuries or millennia. Increased accumulation on Antarctica could contribute to sea-level fall by 0.1 to 0.5 m in the next 100 years. Long-term questions include the rapid fluctuations in CO_2 and other variables observed in ice cores, the rapid deglaciation of North America at the end of the last ice age, and the possible disappearance of the West Antarctic Ice Sheet due to present-day and near-future processes.

INTRODUCTION

Could wastage of the land ice of the world account for the current rise in sea level? Will a CO_2-enhanced atmosphere cause so much ice melt in the next century that the sea will rise to cause major flooding? Thorarinsson (1940), in a comprehensive and seminal analysis, attempted to answer the first question without the benefit of good maps or sufficient data. Although more data are now available, the question is not completely answered. The second question is even more difficult, as it involves atmospheric, oceanographic, and glaciologic processes that are not completely understood and that are difficult to model.

The concentrations of CO_2 and other "greenhouse gases" in the atmosphere are currently rising (Figure 10.1), and one consequence of this will be a rise in air temperature (Carbon Dioxide Assessment Committee, 1983), which might lead to increased wastage of land-based ice. An-

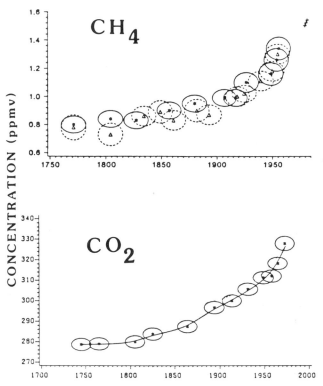

FIGURE 10.1 Data from ice cores showing the increase in two "greenhouse" gases since preindustrial times. The CO_2 results are from Neftel *et al.* (1985). The line represents a model-calculated back extrapolation assuming only CO_2 input from fossil fuel. The CH_4 results are from Stauffer *et al.* (1985). The solid ellipses represent results from melt extraction; the dashed ellipses, dry extraction. The plus signs represent atmospheric measurements. In both diagrams the vertical ellipse axis represents gas measurement error, and the horizontal axis the uncertainty in the dates of air bubble close-off.

other consequence will be changes in precipitation, which may lead to ice buildup in some areas.

In this chapter, the physical processes involved in the relation between climate, ice wastage, and runoff are emphasized; existing modern data are analyzed; and the glaciological problems involved in estimating sea-level change during the next century are discussed. Finally, a brief, longer-term perspective is presented.

Much of this chapter is derived from the results of a 1984 Workshop on *Glaciers, Ice Sheets, and Sea Level: Effects of a CO_2-Induced Climatic Change* (Committee on Glaciology, 1985), to which the reader is referred for more detail on the current state of knowledge. An earlier workshop examined what is known about the effect of CO_2-induced changes on the environment of the critical West Antarctic Ice Sheet (Committee on Glaciology, 1984). The subject of ice and sea-level change has been discussed in several recent papers, including, for instance, Grossval'd and Kotlyakov (1978), Hollin and Barry (1979), Oerlemans (1982), Revelle (1983), Barry (1984), Robin (1986), and Thomas (1987).

ICE ON EARTH AT THE PRESENT

The amount of ice on land at the present moment is huge, probably exceeding the sum of all other amounts of water substance in and on land (Table 10.1). A continuing annual loss of only 0.001 to 0.002 percent of the land ice volume would be sufficient to cause the observed change of 1 to 2 mm/yr in global sea level. But is this happening? Here we examine this question separately for small glaciers and for each of the two large ice sheets.

Glaciers and Small Ice Caps

Observational data on glacier fluctuations include measurements of advance or retreat, volume changes (surface altitude and area), and annual and seasonal mass balances (the difference between totals of mass accumulation and loss such as by melt). Mass balances can be measured directly on the glacier surface or derived from study of ice cores; mass balance sequences can be extended by use of numerical models using long-term meteorologic and hydrologic records. Most of the available glacier fluctuation data are concerned only with changes in length; these data do not provide volume change information directly and, over the short term, can be misleading in sign. Measurements of balances in cores taken at single locations in the accumulation area of a glacier are insufficient to infer the mass change of the entire glacier. Measurements of mass balance made on the glacier surface are confined to relatively short time sequences. Thus most of the inferences about the long-term effects on sea level

TABLE 10.1 Ice and Water on Earth (adapted from Meier, 1983)

	Volume ($km^3 \times 10^6$)
Oceans	1370.0
Ice on land	30.0
Groundwater	4.0[a]
Lakes, reservoirs, rivers, swamps	0.13
Water in the soil	0.06
Water in the atmosphere	0.01

[a]Values as high as 60×10^6 km^3 have been reported for water in pores deep in the Earth's crust. The amount of groundwater that can participate in the hydrologic cycle at time scales of years to centuries is probably about 4×10^6 km^3.

must come from observations of long-term volume changes, or from extensions of short-term balance measurements using long-term meteorological and hydrologic data and numerical models. When both kinds of information are brought to bear on the same glacier record for checking or calibration, the results are considered especially trustworthy (Figure 10.2). The calibration is important because the statistical relations between hydrometeorological variables (e.g., precipitation, temperature, runoff) and balance components (e.g., accumulation, melt) may not be stationary over long time intervals.

Glacier balance and volume change data for periods of record exceeding 50 yr were found for 25 glaciers in 13 regions from reports of the Permanent Service on the Fluctuations of Glaciers [IASH (1967, 1973, 1977)] and other sources. The mean period of record for those glaciers is from 1900.5 to 1961.7. The balance model sequences were pruned to the interval 1900 to 1961 to make the data set more homogeneous. Almost all glaciers showed long-term wastage, but the rates varied, depending largely on location. The unweighted average of these 25 mean balances is –0.40 m/yr (water equivalent) ± 0.25 (standard deviation), and the mean of the regional averages weighted by quality of the data is –0.38 ± 0.20 m/yr (Meier, 1984).

The few glaciers with long-term data are, with one exception, between 38° and 69°N latitude; none are at low latitudes or in the Southern Hemisphere. Thus a method is needed to transform results from this biased sample to an estimate of the mass balance of the whole Earth's cover of glaciers. Meier (1984) suggested that the magnitude of the long-term balance may be related to the magnitude of the seasonal mass fluxes (accumulation and ablation), and that this may be used as a scaling factor to derive global estimates. This way to scale results to a global average is aided by the fact that values of the annual amplitude can be calculated for most of the world's glacier areas from

data reported since 1965 and can be estimated for other areas because they are primarily functions of the climatological regime. The annual amplitudes are highest at temperate to sub-Arctic (and sub-Antarctic) latitudes and lowest near the poles, and they decrease with increasing continentality (Meier, 1985).

The 1900 to 1961 data, when averaged by region and scaled to a global average, suggest that the wastage of the world's glaciers and small ice caps contributed 0.46 ± 0.26 mm/yr to the 61-yr rise in sea level (Meier, 1984). This result is similar to some previous estimates based on many fewer data (Grossval'd and Kotlyakov, 1978; Lambeck, 1980). Unfortunately, the three major contributing regions (the mountains bordering the Gulf of Alaska, the mountains of Central Asia, and the Patagonian Ice Caps) are also regions of meager observational data. Until more observational data are obtained from these regions and the scaling by annual amplitude rigorously tested, these results will have to be considered tentative.

Greenland and Antarctic Ice Sheets

These two large ice masses gain material mainly through the accumulation of snow, and lose material through several processes that include surface melt and the runoff of meltwater, calving (discharge) of icebergs, melting of the underside of floating ice shelves, evaporation of snow, and erosion and transport of surface snow to the sea by wind. The last two processes are considered to be negligible items in the total mass balance for both ice sheets. Surface

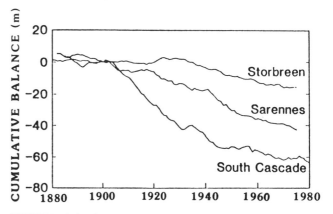

FIGURE 10.2 Cumulative mass balance, in meters of water equivalent, of Storbreen (Norway), Sarennes Glacier (France), and South Cascade Glacier (Washington state). All values are related to 1900. These curves were derived from numerical models using long-term meteorologic and hydrologic data, calibrated with several decades of mass balance observations and with long-term volume change information. (From Meier (1984), with permission of the American Association for the Advancement of Science.)

melt/runoff is also a minor item for the Antarctic Ice Sheet, but it is important to the balance of the Greenland Ice Sheet. Iceberg calving is an important loss term for both ice sheets and is the predominant one in Antarctica. Melting of the underside of ice shelves has no effect on sea level, but it does control the speed of discharge of land-based ice from Antarctica and is important for predicting the behavior of that ice sheet in the next century.

The difficulty of estimating the surface mass balance of the Greenland Ice Sheet can be illustrated by reference to the concept of snow facies (Benson, 1962; Müller, 1962), as shown on Figure 10.3. Most of the Antarctic Ice Sheet and much of the Greenland Ice Sheet are in the dry snow zone, where no melting occurs; some of the marginal areas in Antarctica and an appreciable fraction of the Greenland Ice Sheet reach the percolation zone, where surface melting occurs but the meltwater refreezes in that year's snowpack. Runoff of meltwater to the ocean cannot occur from either zone. Runoff can occur from the wet snow zone, where meltwater saturates the entire thickness of the snowpack. In order to calculate the net mass balance in this zone, the amount of runoff (which is difficult to measure) must be subtracted from the snow accumulation; alternatively, one can measure density-depth profiles through the snowpack and several underlying layers, at the beginning and again at the end of the melt season, in order to detect the amount of ice deposited by refreezing at depth. Unfortunately, the wet snow zone covers a considerable area in Greenland; so that precise determination of the net mass balance is a formidable task. Similar problems arise in the superimposed ice zone.

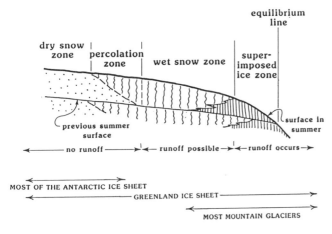

FIGURE 10.3 Diagrammatic section through a large ice mass illustrating the concept of snow facies. Dotted areas are dry (subfreezing) snow; areas with vertical wiggly lines are snow into which meltwater has percolated; and areas with vertical straight lines represent refrozen meltwater that has formed superimposed ice.

Iceberg calving is also difficult to measure. Repetitive high-altitude aerial photography or possibly satellite imagery can be used to measure calving speed as the difference between glacier speed and terminus advance, and the iceberg discharge obtained by integrating the calving speed over the area of the calving face (Brown et al., 1982). Unfortunately, aerial photography is expensive, and satellite images generally do not have sufficient resolution to measure glacier speed. Furthermore, most expeditions to measure calving do so in the summer, and so it is not known whether these data are representative of annual averages. In Antarctica, major calving events can occur infrequently in time, perhaps once in 100 yr. Major Antarctic breakoffs, unprecedented in the historic record, occurred from the Filchner and Larsen ice shelves (Jacobs et al. 1986) and the Ross Ice Shelf in 1987. Iceberg discharge can be estimated by iceberg counts coupled with estimates of iceberg lifetimes (Orheim, 1985), but the latter are difficult to obtain, the sampling problem is formidable, and the episodic nature of ice-shelf breakoffs makes it difficult to obtain a long-term average value. Thus calving data for Greenland are insufficient, and for Antarctica are extremely meager.

The difficulty of measuring the mass balance components of the ice sheets can be circumvented partly by considering the flow out of specific drainage basins. The discharge through a cross section can be determined from surface velocity and ice thickness measurements combined with models to relate surface velocity to the depth-averaged value. The surface mass balance, integrated over the drainage basin upstream from the cross section, gives the "balance flux," which can be compared with the total flux through the cross section. The difference, the "thinning flux," is a measure of the mass imbalance.

A future alternative is promising and exciting—direct measurement of the thinning flux, through the use of repetitive satellite altimeter surveys to detect the rate of change of the surface elevation. This requires a satellite in a polar orbit equipped with an altimeter; a laser altimeter would allow results to be obtained in a year or so, but a radar altimeter might require many years before results of significance were obtained. Preliminary results for that part of the Greenland Ice Sheet south of 72°N latitude, from a radar altimeter mounted in a satellite, are encouraging and suggest ice-sheet thickening (Zwally, 1985).

Estimates of the net mass balance of the Greenland and Antarctic ice sheets are presented in Table 10.2. Most of the information in this compilation (Committee on Glaciology, 1985) was obtained in the late 1970s and early 1980s, but it is difficult to assign a discrete time to these estimates, let alone to determine the time rate of change.

In spite of wide error bands, the results suggest that the current rise in global sea level is not due to the melting of

TABLE 10.2 Estimated Balance of Glaciers and Ice Sheets at the Present Time (from Committee on Glaciology, 1985)[a]

Ice Mass	Period of Observation	Area (10^6 km²)	Average Mass Balance (water equivalent) (m/yr)	Effect on Sea Level (mm/yr)
Glaciers and small ice caps	1900 to 1960	0.54	-1.2 ± 0.7	$+0.5 \pm 0.3$
Greenland Ice Sheet	1929 to 1984[b]	1.73	$+0.02 \pm 0.08$[c]	-0.1 ± 0.4[c]
Antarctic Ice Sheet	1970 to 1984[d]	11.97	$+0.02 \pm 0.02$	-0.6 ± 0.6

[a]Error limits represent approximate bounds of estimation and cannot be defined statistically.

[b]Observations scattered in both time and space.

[c]Combination of (a) historical estimates, (b) extrapolation of modern ablation data from West Greenland and accumulation data from Central Greenland, and (c) extrapolation of thickness change data from Central Greenland..

[d]Some observations taken earlier are included.

the polar ice sheets. In fact, these published results suggest that current growth of the Antarctic Ice Sheet is, at least in part, canceling out the effect of small glacier wastage, leaving even more of the sea-level rise to be explained by ocean warming, loss of volume of the ocean basins, or other such mechanisms, an alternative that seems unlikely.

One way to test the meager glaciological results on long-term mass balance is through study of the effect that mass balance changes might have on the rate of Earth rotation and the relative motion of the rotational pole (Peltier, 1985, and Chapter 4, this volume). Peltier (1988) applied Meier's (1984) model of small glacier wastage to Earth rotation models, assuming no current change in the major ice sheets and that a major ice sheet had existed on the Barents Sea platform during the last glacial period (a somewhat controversial point). His results correctly predict the observed changes in polar motion or rotation rate. This further suggests that the Greenland and Antarctic ice sheets are currently close to balance, and that the long-term iceberg discharge has been underestimated (Orheim, 1985).

CHANGES IN ICE IN THE NEXT CENTURY

The increase in air temperature caused by a rise in concentration of CO_2 may cause increased snow and ice melting, and some of this meltwater may run off to the oceans, causing a rise in sea level. Increased air temperature and/or meltwater production may also cause some outlet glaciers and ice streams to flow faster, transferring land-based ice to the ocean and causing a further rise in

sea level. On the other hand, a rise in CO_2 concentration may, in some regions, lead to increased snow precipitation on glaciers and ice sheets, which will have the opposite effect on sea level. Predicting the effects of climate change on ice growth and wastage is a complex problem because several different, interacting processes must be considered.

Here we discuss five important processes that are involved in potential impacts on sea level of future changes in glaciers and ice sheets: (1) the variation of energy and mass balance components with altitude, (2) the dynamic response of ice masses to changes in thickness, (3) the warming of cold firn to allow meltwater runoff, (4) increased flow and iceberg calving of tidal glaciers due to increased meltwater, and (5) the stability of ice-sheet/ice-stream/ice-shelf systems. We restrict the time frame to the next 100 yr, approximately the time of doubling of the present level of CO_2.

Energy and Mass Balances as a Function of Altitude

Mountain glaciers continuously move ice from an area of high altitude where, on an annual basis, snow accumulation exceeds snow and ice wastage, to an area at a lower altitude where snow and ice ablation exceeds snow accumulation (Figure 10.4). Thus the magnitude of these processes is dependent on altitude, and consideration of the altitudinal gradients in the mass and energy fluxes allows one to generalize from measurements in one restricted area to large regions (Kuhn, 1981). Mass balance components also vary, but more weakly, with latitude and with distance along air mass trajectories as the airmass

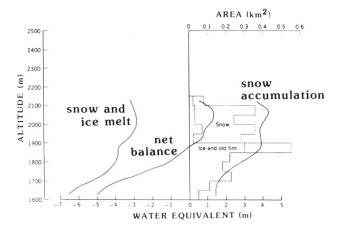

FIGURE 10.4 Snow accumulation, snow and ice melt, and net balance for South Cascade Glacier, Washington, for the 1964–1965 balance year. Also shown is the area at each altitude increment. From Meier *et al.* (1971).

moves from the sea across mountain ranges. The mass balance of the Greenland Ice Sheet can be analyzed by considering the altitudinal gradient as primary, although the latitude and distance from the sea are also important (Ambach, 1985). Such an approach is less useful in Antarctica, where iceberg discharge and subshelf melting are the dominant processes causing ice loss to the sea.

Many of the energy fluxes that cause melting are known or observable functions of altitude. The most sensitive of these are the absorbed solar radiation and the precipitation as snow. The first of these is critically dependent on the albedo (reflectivity) of the surface, which in turn depends on how long the surface is covered with high-albedo snow during the course of the melt season. The persistence of the snow cover depends, of course, on the amount of snow and the intensity of melt processes, both of which also depend on altitude. Kuhn (1985) and Ambach and Kuhn (1985) estimated the altitudinal gradients in energy balance components for an Alpine mountain glacier and for the central western portion of the Greenland Ice Sheet, respectively. Bindschadler (1985) used such information to calculate the static response of the Greenland Ice Sheet to doubled CO_2. Such analyses should be done in other parts of the world, as this information leads to a relatively simple way to relate climate change due to a CO_2 perturbation to a change in the mass balance of an ice mass. A rise in the annual melt rate at the equilibrium line leads to a rise in the altitude of that equilibrium line, displacing the ablation/accumulation rates as functions of altitude, and changing the size of the ablation and accumulation areas.

The altitudinal dependence of the mass and energy fluxes leads to a potential instability, which is one of the reasons for the sensitivity of glaciers to slight climate changes. A rise in melting causes a lowering of the ice surface, which

in turn may cause a further increase in melting or decrease in snow accumulation, leading to further changes accentuating the melting and vice versa (Bodvarsson, 1955). This sensitivity/instability thus requires improved understanding of the exact magnitude of annual and seasonal changes in precipitation, surface air temperature, and other variables, on a regional basis, before reliable estimates can be made of the changes to be expected as a consequence of a rise in CO_2 concentration. This is a formidable task indeed.

Time Scales of Ice Wastage and Dynamic Response

Studies in the Alps (Kuhn, 1985) suggest that a 4°C rise in temperature, caused by a doubling of CO_2 in the next 100 yr, would lead to a negative mass balance change from 0 to about 0.3 m/yr. If this wastage were to increase linearly with time, the effect over time could be estimated for a glacier of known initial dimensions. Assuming a nonflowing, wedge-shaped glacier of length $L(t)$, surface slope β resting on a bed of slope β, and having a constant width W, the amount of thinning H_1 during the period 0 to t_1, for a mass balance scenario $b(t)$ is

$$H_1 = \int_0^{t_1} b(t)\,dt .   \quad (10.1)$$

Assuming a linear decrease (more negative) with time in the balance, $b(t) = kt$,

$$H(t) = 0.5kt^2.   \quad (10.2)$$

The glacial length $L(t)$ is thus

$$L(t) = L_0 - \frac{kt^2}{2(\tan\alpha - \tan\beta)},   \quad (10.3)$$

and the loss of ice to increased runoff (and thus to the ocean) is

$$R(t) = WL(t)b(t) = WL_0 kt - \frac{Wk^2 t^3}{2(\tan\alpha - \tan\beta)}.   \quad (10.4)$$

$R(t)$ reaches a maximum at the time

$$t_{max} = [2L_0(\tan\alpha - \tan\beta)/3k]^{0.5},   \quad (10.5)$$

which is of the order of decades to centuries for typical mountain glaciers. Under these assumptions, South Cascade Glacier, Washington, will reach a maximum rate of ice loss in about 70 yr, and be gone completely in about 130 yr, assuming a constant increase in the rate of wastage of 0.03 m/yr² (Figure 10.5).

This simplistic analysis needs to be modified, not only to reflect a more realistic geometry, but also to incorporate glacier dynamics and the change of balance with altitude. The effect of thinning will be to decrease the rate of flow, causing a steepening of the longitudinal profile. The normal altitudinal balance gradient will tend to accelerate the ice loss as the surface profile lowers, but the dynamic steepening and the loss of ice at the lowest altitudes will, in part, reduce this acceleration in wastage. For a mountain glacier, the characteristic time for dynamic response is of the same order of magnitude as the time for glacier disintegration (Figure 10.6).

On the other hand, for the Greenland and Antarctic ice sheets, the characteristic time for dynamic response is one or more orders of magnitude longer (Alley and Whillans, 1984; Bindschadler, 1985), with the possible exception of ice-stream/ice-shelf interactions discussed later. Except for this last possibility, the dynamic response to changes in the two existing ice sheets due to altered climate in the next century can be ignored.

The Warming of Cold Snow and Firn

Figure 10.3 illustrates the fact that most of the Antarctic Ice Sheet and much of the Greenland Ice Sheet are not likely to be affected by small rises in air temperature, such

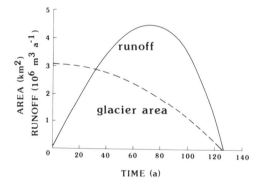

FIGURE 10.5 Excess runoff from South Cascade Glacier produced a linearly increasing wastage rate of 0.03 m/yr², assuming a nonflowing glacier of simple wedge shape. Data on South Cascade Glacier taken from Meier and Tangborn (1965).

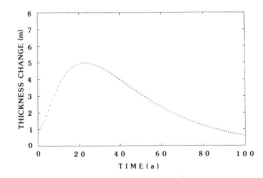

FIGURE 10.6 Thickness change at the terminus of South Cascade Glacier in response to a 1-m balance perturbation for 1 yr. From Nye (1965).

as those that might be caused by an increased CO_2 concentration in the next century. Increases in summer air temperature cause no melting unless the increases exceed the present (negative) surface air temperature. The more interesting question, however, concerns the situation where an increase of the summer air temperature brings the surface snow up to the melting point. With an increase of mean annual temperature of 5° to 10°C in the polar regions, as suggested by many atmospheric circulation models for a twofold enhancement of CO_2, there would be a large shift of the boundary between the dry snow, percolation, and wet snow facies on the two ice sheets. But how long would it take to develop enhanced runoff to the sea?

The transient response of infiltration into and warming of snow and firn (consolidated snow that has survived at least one summer season and has not yet become impermeable ice) is complex. The equations of flow through a porous medium are complicated by heat flow considerations and the physics of the freezing interstitial water. Research on this problem is now under way (Illangasekare *et al.*, 1988; Meier *et al.*, 1988).

Flow and Iceberg Calving of Tidewater Glaciers

Much of the transfer of water mass from land to the ocean is accomplished by the breakoff (calving) of icebergs. This is the dominant process in Antarctica and is also important in Greenland, the Arctic Islands, Alaska, and Chile. Unfortunately, our understanding of calving glacier dynamics is insufficient to predict how these glaciers will react to an altered climate in the future.

The length of a calving glacier depends on the balance between ice flow to the calving face and the discharge of icebergs from the face,

$$\frac{dL}{dt} = V - V_c , \qquad (10.6)$$

where L is the glacier length, t is time, V is the speed of the glacier at the terminus, and V_c the calving speed. Both V and V_c are averages over the glacier thickness and width. The calving speed (V_c) can be considered as the iceberg discharge (in units of volume per time) divided by the area of the calving face. Both V and V_c can vary, causing changes in the length and volume of the glacier. These changes in length may then affect V and V_c, and the ensuing feedback can cause rapid disintegration of a calving glacier (Post, 1975; Meier and Post, 1987).

Studies of Columbia Glacier, Alaska, and other grounded calving glaciers in Alaska suggest that

$$V_c = k'h_w, \qquad (10.7)$$

where h_w is water depth at the terminus and k' is an empirical coefficient (Brown *et al.*, 1982), or,

$$V_c = \left(\frac{k'R}{h_u^2}\right)^{0.5}, \qquad (10.8)$$

where R is the liquid water runoff and h_u is the thickness of ice unsupported by buoyancy at the terminus (Sikonia, 1982). Although these two equations are very different in form, they produce similar values of V_c over the range of h_w, h_u, and R as measured at Columbia Glacier during the period 1977 to 1980. Unfortunately, the generality of these formulae for a wider range of variables is not yet established.

No calving law has been established for a floating, calving glacier, although Reeh (1968) and others investigated the stresses in a floating ice tongue, and Fastook and Schmidt (1982) analyzed the role of fracture in the calving process. Obviously, a calving law for floating ice is needed in order to predict future changes in glaciers and ice sheets.

Most calving glaciers flow rapidly, and most of this flow is by basal sliding. No sliding law has been generally accepted, but a relation of the form

$$V_b = \frac{k'\tau^n}{(P_i - P_w)}, \qquad (10.9)$$

where τ is the shear stress on the bed, n is a constant (≈ 3), and P_i and P_w are the hydrostatic pressures in the ice at the bed and in the subglacial water (Budd et al., 1979; Bindschadler, 1983), is commonly used for glaciers with $P_i > P_w$ (not floating). At the terminus, h_u and $P_i - P_w$ are proportional. A grounded calving glacier with its bed well below sea level slides faster than a similar glacier on land, because P_w approaches P_i due to the pressurization of the subglacial water system by the saltwater column at the terminus.

The effect of CO_2-induced climate change on calving glaciers is complex; increased melt due to warmer temperature might cause thinning of the ice. This thinning would decrease t and P_i; if $P_i - P_w$ were small initially, the decrease in $P_i - P_w$ could predominate and V_b would increase. Because h_u would decrease, V_c would increase, if the glacier were grounded.

Increased flow would cause further thinning, and an unstable disintegration would ensue. Increased melt or increased rainfall also might increase P_w, leading to more rapid flow, and the increase in R (for a grounded glacier) would lead to more rapid calving. Thus calving glaciers could respond to slight increases of melting and rainfall by rapid and dramatic disintegration (Meier and Post, 1987), but it is difficult to calculate the rate of change because of our insufficient knowledge of subglacial hydraulics, basal sliding, and calving.

The rapidity of change in the dynamics of a calving glacier is illustrated by the current disintegration of Columbia Glacier, Alaska (Figure 10.7).

Stability of Ice Streams and Ice Shelves

Mercer (1978) warned that a climatic change due to increased CO_2 in the atmosphere could lead to disintegration of the West Antarctic Ice Sheet, causing a 5-m rise in global sea level. The discharge of ice from this ice sheet is mainly through rapidly moving ice streams, which flow into floating ice shelves (Figure 10.8). Hughes (1973) and Weertman (1974) pointed out that an ice sheet that rests on a flat bed situated below sea level can be inherently unstable. If the climate in the future were to become warmer and this caused warmer ocean water to intrude under the ice shelves causing increased melting under the shelves, then the back pressure exerted on the ice streams by the shelves would be reduced and the ice streams would accelerate, draining the ice sheet itself. But how quickly could this happen?

The interaction between an ice stream and an ice shelf is illustrated in Figure 10.9. The ice stream pushes the ice shelf past its margins and around shoals (ice rises). This results in a compressive force (back pressure) F on the ice stream at the grounding line. The weight-induced spreading of the ice stream is resisted by F and by the shear along the sides and base of the ice stream. If F is reduced due to ice shelf thinning, the rate of extension of the ice stream increases; the upper end flows at a very slow speed, and the increasing extension rate is manifest in an increasing velocity at the grounding line, and ice is moved from land to sea at an increasing rate.

The critical questions then are (1) how much additional heat will be delivered to the underside of the ice shelves in an altered climate, (2) how will the changed conditions affect calving rates and thus the dimensions of ice shelves, and (3) how rapidly will the ice streams react to changes in the back pressure?

The current basal melt rate for the ice shelves of Antarctica has been estimated at about 0.4 m/yr (Jacobs et al., 1985). However, nearly undiluted circumpolar water, which is several degrees above the in situ melting temperature, intrudes onto the continental shelf of the Bellingshausen Sea and causes about 2 m/yr of melt from the base of the George VI Ice Shelf (Potter and Paren, 1985). One coupled atmosphere-ocean circulation model (Schlesinger et al., 1985) suggests that the ocean water 250 to 750 m below the surface just north of the Ross Ice Shelf will be warmed by 0.5° to 1.5°C, as a consequence of a doubled concentration of CO_2. Such results, however, cannot be translated into terms of ice-shelf melt, because our knowledge of circulation under ice shelves is still incomplete (MacAyeal, 1984a,b; Jacobs, 1985). It seems possible that, by the time of doubled CO_2, the basal ice shelf melt rate could increase by as much as 1 to 3 m/yr (Committee on Glaciology, 1985).

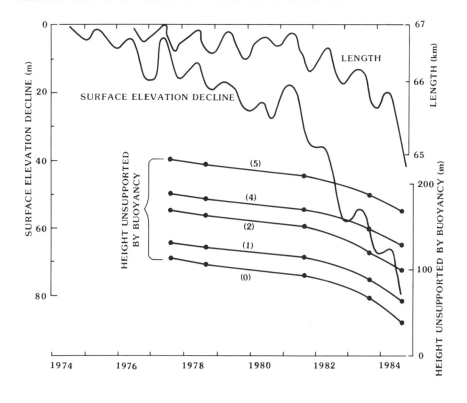

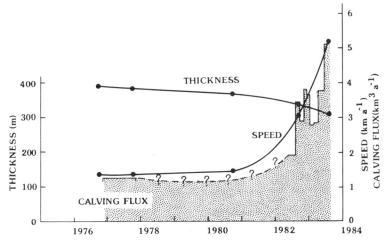

FIGURE 10.7 Changes in Columbia Glacier, Alaska, as disintegration begins. This grounded, calving glacier resembles a small ice stream, but it does not terminate in a floating ice shelf. The upper diagram shows the acceleration in retreat, the decline in ice-surface altitude, and the decline in the ice thickness unsupported by buoyancy (h_u); the values in parentheses represent the distances above the 1984 terminus in kilometers. The lower diagram shows the declining ice thickness and accelerating velocity, as measured at a point near the 1984 terminus, and the rise in calving flux. From Meier *et al.* (1985).

Unfortunately, almost nothing is known on how the calving rate, and thus ice-shelf dimensions, will be affected by a warmer atmosphere and ocean. At this stage, one can only make arbitrary guesses. This does not invalidate but certainly limits the confidence one can have in the results of ice-stream/ice-shelf modeling.

Ice-stream/ice-shelf interactions resulting from a CO_2-enhanced climate change have been explored by several authors. Lingle (1984, 1985) modeled the evolution of Ice Stream E, by assuming changes in the back pressure and assuming no change in the ice-shelf dimensions. His results show a slight and very slow response to a 10 percent

reduction in back pressure, but a dramatic and rapid disintegration in response to a 50 percent reduction (Figure 10.10). Thomas (1985) made a simple calculation for Ice Stream B in which various combinations of basal melt rates and calving (change in ice-shelf dimensions) were assumed.

Assuming that the model results from Ice Stream E are representative of the whole West Antarctic Ice Sheet, Lingle (1985) estimated that the contribution of this ice sheet to sea-level rise by the year 2100 would be only 0.03 to 0.05 m. Thomas (1985), however, pointed out that most of the discharge from Antarctica is through ice streams. He

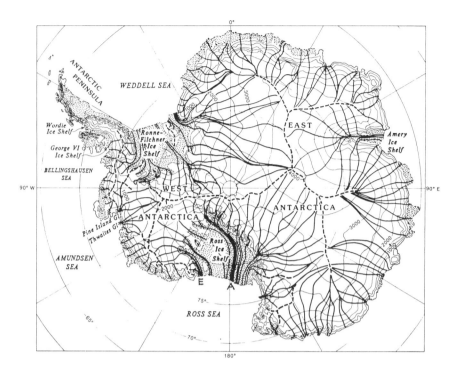

FIGURE 10.8 Map of the Antarctic Ice Sheet, with generalized flow lines (solid lines) and ice divides (dashed lines). Ice streams A and E in West Antarctica are shown; floating ice shelves are stippled.

extrapolates the results from Ice Stream B to all of Antarctica, and with his upper limit scenarios, predicts a probable contribution to sea-level rise by 2100 of 0.2 to 0.8 m. This represents a reasonable limit to the response, as there are several processes that exert negative feedback that were not taken into account. Bentley (1984) summarized a number of reasons why disintegration of the West Antarctic Ice Sheet will not happen in a few decades or centuries.

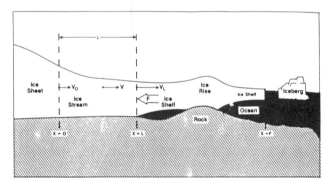

FIGURE 10.9 A typical Antarctic ice-drainage system. The ice stream pushes the ice shelf seaward past its margins and around the grounded ice rises, and there is a compressive force (F) at the grounding line ($X = L$) between ice stream and ice shelf. This compressive force is transmitted for some distance (L) upstream of the grounding line, and there is an additional force due to shear between the ice stream and its sides and bed. These forces resist the extending flow of the ice stream. From Thomas (1985).

Clearly, sea level will not rise catastrophically in the near future due to the demise of this ice sheet; see Table 10.3.

Increased Accumulation on Antarctica

The Antarctic continent is large in area, but it is a desert, with average annual precipitation of only 15 to 17 mm/yr, with a strong concentration of precipitation along the coast. Greenhouse warming could lead to increased precipitation because of two mechanisms: (1) reduction of the surrounding sea ice allowing increased water vapor exchange and transport to the continent, and (2) increased water vapor content of the air due to increased temperature.

Oerlemans (1982) estimated a precipitation increase of 12 to 30 percent based on a temperature increase of 3° to 8°C; over a 100-yr period this would produce a sea-level fall of 0.08 to 0.20 m. Warren and Frankenstein (in press), using a simple relation of saturation vapor pressure to the temperature of the inversion layer, suggest that a temperature rise of 5°C could lead to an increase in moisture of 60 to 90 percent, leading in 100 yr to a sea-level fall of 0.5 m. They point out that the simple vapor pressure/temperature model appears to be confirmed by the change in precipitation observed at Dome C on the Antarctic plateau since the last glacial period, but also that such a simple model might not be valid when applied to the coastal regions.

The Committee on Glaciology (1985) report ascribed a potential sea-level fall in their estimated range of only 0.1

(a)

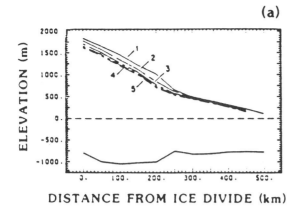

DISTANCE FROM ICE DIVIDE (km)

(b)

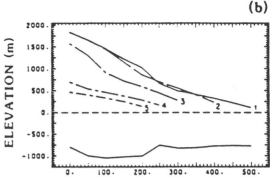

DISTANCE FROM ICE DIVIDE (km)

FIGURE 10.10 Thickness profiles for numerical models of Ice Stream E. In (a) the back pressure is reduced by 10 percent from equilibrium back pressure; the profiles 1 through 5 are 0, 500, 1000, 1500, and 2000 yr, respectively. In (b) the back pressure is reduced by 50 percent; the profiles 1 through 5 are 0, 39, 251, 487, and 664 yr. From Lingle (1985).

m in the next 100 yr (Table 10.3). The study by Warren and Frankenstein suggests that this may be seriously underestimated. This is obviously a topic that deserves further study.

A LONGER-TERM PERSPECTIVE

Fluctuations in CO$_2$ Observed in Ice Cores

Recent studies of deep ice cores have shown rapid fluctuations in climate during the last glacial period (Dansgaard *et al.*, 1982). These events, at time scales of decades to a few centuries, appear to be paralleled by similar changes in the CO$_2$ content of the entrapped air (Oeschger *et al.*, 1984, 1985). It is also clear that the CO$_2$ fluctuations do not lag the changes in climate; they may lead them, but this is not well established. Although none of these events has occurred so far in the Holocene (since the last glacial age), they do indicate that a mechanism involving CO$_2$ exists that can cause rapid switches in climate over large regions, perhaps from one quasi-stable mode to another (Broecker *et al.*, 1985).

These results prove that changes in CO$_2$, climate, and ice are directly related. The ice core results do not indicate directly how rapidly the ice masses responded to climate changes. But these fluctuations represent changes in the complex system involving the ocean, the ice, the atmosphere, and the biosphere (Broecker *et al.*, 1985); so some sort of rapid fluctuation in ice mass must have happened.

Rapid Deglaciation in North America

The Laurentide Ice Sheet in North America covered most of Canada and part of the United States as late as 13,000 yr ago, but 7000 yr ago it was reduced to isolated ice caps and Hudson Bay was opened (Andrews and Falconer, 1969; Denton and Hughes, 1981; Andrews, 1988). Field evidence suggests that this retreat was not a smooth, continuous meltback, but was accomplished by episodes of rapid retreat, surrounded by periods of slower retreat or advance. Field evidence (e.g., Prest, 1969) is not yet sufficient to define the rates of retreat precisely, but the evidence points to rates that can only be explained by rapid calving disintegration. The rapid retreat in some areas appears to have triggered rapid advances (surges?)

TABLE 10.3 Estimates of the Contribution to Sea-Level Rise by Ice Wastage in a CO$_2$-Enhanced Environment (from Committee on Glaciology, 1985)

Ice Mass	Annual Probable Contribution to Sea Level with Steady-State 2 × CO$_2$ Atmosphere (mm/yr)	Range of Estimated Contribution to Total Sea-Level Change to Year 2100 (m)
Glaciers and small ice caps	2 to 5	0.1 to 0.3
Greenland Ice Sheet	1 to 4	0.1 to 0.3
Antarctic Ice Sheet	–3 to 10	–0.1 to 1[a]

Note: Thermal expansion of the oceans and other nonglacial processes that might cause additional sea-level rise are not included here.

[a]Values in the range of 0 to 0.3 are considered most likely.

of ice into the recently deglaciated areas such as the Cochrane surges of southern Hudson Bay (Prest, 1970). There is evidence that such rapid deglaciations occurred about 40,000 and 80,000 years ago as well as 8000 years ago (Shilts, 1982; Andrews *et al.*, 1983).

The rapid and complex deglaciation at the end of the last ice age has obvious implications to understanding and predicting the future of the marine-based portions of the Antarctic Ice Sheet. The problem now is to obtain additional datable evidence of retreat, and to use these data in ice-sheet models so that ice-sheet stability can be defined more precisely.

Did the West Antarctic Ice Sheet Disappear During the Last Interglacial?

During the last interglacial, global sea level stood, on the average, 5 to 7 m higher than today. If the West Antarctic Ice Sheet were to disintegrate, sea level would rise 5 to 7 m. Mercer (1968) first suggested the connection. If this ice sheet disintegrated in the last interglacial, then one would expect that it could do the same in the forthcoming CO_2-enhanced "super interglacial."

Soviet engineers have drilled an ice core from Vostok in East Antarctica that penetrates ice from the last interglacial, 116,000 to 140,000 yr ago. The results indicate that the surface air temperature during the interglacial was about 3°C warmer than during the Holocene (Lorius *et al.*, 1985). But does this mean that the regional climate in Antarctica was warmer during the last interglacial?

Robin (1985) suggested that the warmth was due to a lower ice surface at that time, a suggestion that is supported by other evidence as well. Also, there is evidence from other parts of the world that the climate of the last interglacial was similar to that of the present. So, if the climate were the same as that of today, the Vostok results imply that the ice surface must have been 300 to 350 m lower than today. If this were true over most of central Antarctica, the lower ice volume would cause a higher sea level during the last interglacial similar to that observed. So the question is not yet resolved. A deep ice core in West Antarctica penetrating interglacial ice, or penetrating the last glacial but showing no interglacial ice below, is vitally needed if we are to evaluate the hypothesis of rapid collapse of the West Antarctic Ice Sheet and the dramatic effect on sea level that would ensue.

ACKNOWLEDGMENTS

I thank J. T. Andrews, C. R. Bentley, and T. J. Hughes for helpful comments on the manuscript.

REFERENCES

Alley, R. B., and I. M. Whillans (1984). Response of the East Antarctica Ice Sheet to sea-level rise, *J. Geophys. Res. 89*, 6487–6493.

Ambach, W. (1985). Characteristics of the heat balance of the Greenland Ice Sheet for Modelling, *J. Glaciol. 31*, 3–12.

Ambach, W., and M. Kuhn (1985). The shift of equilibrium-line altitude on the Greenland Ice Sheet following climatic changes, in *Glaciers, Ice Sheets, and Sea Level: Effects of a CO_2-Induced Climatic Change*, Committee on Glaciology, National Academy Press, Washington, D.C., pp. 255–257.

Andrews, J. T. (1988). Climatic evolution of the eastern Canadian Arctic and Baffin Bay during the past three million years, *Philos. Trans. R. Soc. London B318*, 645–660.

Andrews, J. T., and G. Falconer (1969). Late glacial and postglacial history and emergence of the Ottawa Island, Hudson Bay, N.W.T.: Evidence on the deglaciation of Hudson Bay, *Can. J. Earth Sci. 6*, 1263–1276.

Andrews, J. T., W. W. Shilts, and G. H. Miller (1983). Multiple deglaciations of the Hudson Bay lowlands, Canada, since deposition of the Missinaibi (last-interglacial?) formation, *Quat. Res. 19*, 18–37.

Barry, R. G. (1984). Possible CO_2-induced warming effects on the cryosphere, in *Climatic Changes on a Yearly to Millennial Basis*, N.-A. Mörner and W. Karlén, eds., D. Reidel, Dordrecht, Holland, pp. 571–604.

Benson, C. S. (1962). Stratigraphic studies in the snow and firn of the Greenland Ice Sheet, SIPRE Research Report 70.

Bentley, C. R. (1984). Some aspects of the cryosphere and its role in climatic change, in *Climate Processes and Climate Sensitivity*, J. E. Hansen and T. Takahashi, eds., Geophysical Monograph 29, Maurice Ewing Vol. 5, American Geophysical Union, Washington, D.C., pp. 207–220.

Bindschadler, R. (1983). The importance of pressurized subglacial water in separation and sliding at the glacier bed, *J. Glaciol. 29*, 3–19.

Bindschadler, R. (1985). Contribution of the Greenland Ice Cap to changing sea level present and future, in *Glaciers, Ice Sheets, and Sea Level: Effects of a CO_2-Induced Climatic Change*, Committee on Glaciology, National Academy Press, Washington, D.C., pp. 258–266.

Bodvarsson, G. (1955). On the flow of ice-sheets and glaciers, *Jokull 5*, 1–8.

Broecker, W. S., D. M. Peteet, and D. Rind (1985). Does the ocean-atmosphere system have more than one stable mode of operation?, *Nature 315*, 21–26.

Brown, C. S., M. F. Meier, and A. Post (1982). *Calving Speed of Alaska Tidewater Glaciers, with Application to Columbia Glacier*, U.S. Geol. Surv. Prof. Paper 1258-C, 13 pp.

Budd, W. F., P. L. Keage, and N. A. Blundy (1979). Empirical studies of ice sliding, *J. Glaciol. 23*, 157–170.

Carbon Dioxide Assessment Committee (1983). *Changing Climate*, National Academy Press, Washington, D.C., 496 pp.

Committee on Glaciology (1984). *Environment of West Antarctica, Potential CO_2-Induced Changes*, National Academy Press, Washington, D.C., 236 pp.

Committee on Glaciology (1985). *Glaciers, Ice Sheets, and Sea Level: Effects of a CO₂-Induced Climatic Change*, National Academy Press, Washington, D.C., 330 pp.

Dansgaard, W., H. B. Clausen, N. Gundestrup, C. U. Hammer, S. F. Johnson, P. M. Kristinsdottier, and N. Reeh (1982). A new Greenland deep ice core, *Science 218*, 1273–1277.

Denton, G. H., and T. J. Hughes (1981). *The Last Great Ice Sheets*, John Wiley, New York, 484 pp.

Fastook, J. L., and W. Schmidt (1982). Finite-element analysis of calving from ice fronts, *Ann. Glaciol. 3*, 103–106.

Grossval'd, M. S., and V. M. Kotlyakov (1978). Impending climatic change and the fate of glaciers, *Izv. Akad. Nauk SSSR, Ser. Geogr. 6*, 21–32.

Hollin, J. T., and R. G. Barry (1979). Empirical and theoretical evidence concerning the response of the Earth's ice and snow cover to a global temperature increase, in *Environment International 2*, Pergamon Press, pp. 437–444.

Hughes, T. J. (1973). Is the West Antarctic Ice Sheet disintegrating? *J. Geophys. Res. 78*, 7884–7910.

IASH (1967). *Fluctuations of Glaciers 1959–65*, P. Kasser, compiler, International Association of Scientific Hydrology/Unesco, 52 pp.

IASH (1973). *Fluctuations of Glaciers 1965–70*, P. Kasser, compiler, International Association of Scientific Hydrology/Unesco, 357 pp.

IASH (1977). *Fluctuations of Glaciers 1970–75*, F. Müller, compiler, International Association of Scientific Hydrology/Unesco, 269 pp.

Illangasekare, T. H., M. F. Meier, R. J. Walter, and W. T. Pfeffer (1988). Modeling of water flow in sub-frozen snow to study future sea level changes due to ice wastage, *EOS 69*, 366.

Jacobs, S. S. (1985). Oceanographic evidence for land ice/ocean interactions in the Southern Ocean, in *Glaciers, Ice Sheets, and Sea Level: Effects of a CO₂-Induced Climatic Change*, Committee on Glaciology, National Academy Press, Washington. D.C., pp. 116–128.

Jacobs, S. S., R. G. Fairbanks, and Y. Horibe (1985). Origin and evolution of water masses near the Antarctic continental margin: Evidence from H₂¹⁸O/H₂¹⁶O ratios, in *Oceanology of the Antarctic Continental Shelf*, Ant. Res. Ser. 43, American Geophysical Union, Washington, D.C.

Jacobs, S. S., D. R. MacAyeal, and J. L. Ardai, Jr. (1986). Recent advance of the Ross Ice Shelf, Antarctica, *J. Glaciol. 32*, 464–474.

Kuhn, M. (1981). Climate and glaciers, sea level, ice and climatic change, in *Proceedings of the Canberra Symposium, December 1979*, I. Allison, ed., IASH Publ. 131, pp. 3–20.

Kuhn, M. (1985). Reactions of mid-latitude glacier mass balance to predicted climatic changes, in *Glaciers, Ice Sheets, and Sea Level: Effects of a CO₂-Induced Climatic Change*, Committee on Glaciology, National Academy Press, Washington, D.C., pp. 248–254.

Lambeck, K. (1980). *The Earth's Variable Rotation, Geophysical Causes and Consequences*, Cambridge University Press, 449 pp.

Lingle, C. (1984). A numerical model of interactions between a polar ice stream and the ocean: Application to Ice Stream E, West Antarctica, *J. Geophys. Res. 89*, 3523–3549.

Lingle, C. (1985). A model of a polar ice stream, and future sea-level rise due to possible drastic retreat of the West Antarctic Ice Sheet, in *Glaciers, Ice Sheets, and Sea Level: Effects of a CO₂-Induced Climatic Change*, Committee on Glaciology, National Academy Press, Washington, D.C., pp. 317–330.

Lorius, C., J. Jouzel, C. Ritz, L. Merlivat, N. I. Barkov, Y. S. Korotkevich, and V. M. Kotlyakov (1985). A 150,000-year climate record from Antarctic ice, *Nature 316*, 591–596.

MacAyeal, D. R. (1984a). Potential effect of CO₂ warming on sub–ice–shelf circulation and basal melting, in *Environment of West Antarctica: Potential CO₂-Induced Changes*, Committee on Glaciology, National Academy Press, Washington, D.C., pp. 212–221.

MacAyeal, D. R. (1984b). Thermohaline circulation below the Ross Ice Shelf: A consequence of tidally induced vertical mixing and basal melting, *J. Geophys. Res. 89*, 597–606.

Meier, M. F. (1983). Snow and ice in a changing hydrological world, *Hydrolog. Sci. J. 28*, 3–22.

Meier, M. F. (1984). Contribution of small glaciers to global sea level, *Science 226*, 1418–1421.

Meier, M. F. (1985). Mass balance of the glaciers and small ice caps of the world, in *Glaciers, Ice Sheets, and Sea Level: Effects of a CO₂-Induced Climatic Change*, Committee on Glaciology, National Academy Press, Washington, D.C., pp. 139–144.

Meier, M. F., and A. Post (1987). Fast tidewater glaciers, *J. Geophys. Res. 92*, 9051–9058.

Meier, M. F., and W. V. Tangborn (1965). Net budget and flow of South Cascade Glacier, Washington, *J. Glaciol. 5*, 547–566.

Meier, M. F., W. V. Tangborn, L. R. Mayo, and A. Post (1971). *Combined Ice and Water Balances of Gulkana and Wolverine Glaciers, Alaska, and South Cascade Glacier, Washington, 1965 and 1966 Hydrologic Years*, U.S. Geol. Surv. Prof. Paper 715-A, 23 pp.

Meier, M. F., L. A. Rasmussen, and D. S. Miller (1985). Columbia Glacier in 1984, Disintegration Underway, *U.S. Geol. Surv. Open File Report 85-81*, 17 pp.

Meier, M. F., T. Pfeffer, and T. Illangasekare (1988). The behavior of liquid water in cold ice caps—Initial results, *EOS 69*, 1211–1212.

Mercer, J. H. (1968). Antarctic ice and Sangamon sea level, *Int. Assoc. Sci. Hydrol. 79*, 217–225.

Mercer, J. H. (1978). West Antarctic Ice Sheet and CO₂ greenhouse effect: A threat of disaster, *Nature 271*, 321–325.

Müller, F. (1962). Zonation in the accumulation area of the glaciers of Axel Heiberg Island, N.W.T., Canada, *J. Glaciol. 4*, 302–311.

Neftel, A., E. Moor, H. Oeschger, and B. Stauffer (1985). Evidence from polar ice cores for the increase in atmospheric CO₂ in the past two centuries, *Nature 315*, 45–47.

Nye, J. F. (1965). A numerical method of inferring the budget of history of a glacier from its advance and retreat, *J. Glaciol. 5*, 589–607.

Oerlemans, J. (1982). Response of the Antarctic Ice Sheet to a climatic warming: A model study, *J. Climatology 2*, 1–11.

Oeschger, H. *et al.* (1984). Late glacial climate history from ice cores, in *Climate Processes and Climate Sensitivity*, J. E. Hansen and T. Takahashi, eds., Geophysical Monograph 29,

Maurice Ewing Vol. 5, American Geophysical Union, Washington, D.C., pp. 299–306.

Oeschger, H. *et al.* (1985). Variations of the CO_2 concentration of occluded air and of anions and dust in polar ice cores, in *The Carbon Cycle and Atmospheric CO_2: Natural Variations Archean to Present*, E. T. Sundquist and W. S. Broecker, eds., Geophysical Monograph 32, American Geophysical Union, Washington, D.C., pp. 132–142.

Orheim, O. (1985). Iceberg discharge and the mass balance of Antarctica, in *Glaciers, Ice Sheets, and Sea Level: Effects of a CO_2-Induced Climatic Change*, Committee on Glaciology, National Academy Press, Washington, D.C., pp. 210–215.

Peltier, W. R. (1985). Climatic implications of isostatic adjustment constraints on current variations of eustatic sea level, in *Glaciers, Ice Sheets, and Sea Level: Effects of a CO_2-Induced Climatic Change*, Committee on Glaciology, National Academy Press, Washington, D.C., pp. 92–103.

Peltier, W. R. (1988). Global sea level and Earth rotation, *Science 240*, 895–901.

Post, A. (1975). Preliminary hydrography and historic terminal changes of Columbia Glacier, Alaska, *U.S. Geol. Surv. Hydrol. Invest. Atlas 559*.

Potter, J. R., and J. G. Paren (1985). Interaction between ice shelf and ocean in George VI Sound, Antarctica, in *Oceanology of the Antarctic Continental Shelf*, Antarct. Res. Ser. 43, American Geophysical Union, Washington, D.C., 1985.

Prest, V. K. (1969). Retreat of Wisconsin and Recent Ice in North America, *Geol. Surv. Canada Map 1257A*.

Prest, V. K. (1970). Quaternary geology in Canada, in *Geology and Economic Minerals in Canada*, 5th edition, R. J. Douglas, ed., Department of Energy, Mines, and Resources, Ottawa, Canada, pp. 676–764.

Reeh, N. (1968). On the calving of ice from floating glaciers and ice shelves, *J. Glaciol. 7*, 215–234.

Revelle, R. R. (1983). Probable future changes in sea level resulting from increased atmospheric carbon dioxide, in *Changing Climate*, Carbon Dioxide Assessment Committee, National Academy Press, Washington, D.C., pp. 433–448.

Robin, G. de Q. (1985). Contrasts in Vostok core—changes in climate or ice volume? *Nature 316*, 578–579.

Robin, G. de Q. (1986). Changing the sea level, in *The Greenhouse Effect, Climatic Changes, Ecosystems*, B. Bolin, B. R. Döös, J. Jäger, and R. A. Warrick, eds., SCOPE 29, Wiley, Chichester, Mass., pp. 323–359.

Schlesinger, M. E., *et al.* (1985). The role of the ocean in CO_2-induced climate change: Preliminary results from the OSU coupled atmosphere-ocean general circulation model, in *Coupled Ocean—Atmosphere Models*, J. C. J. Jihoul, ed., Elsevier, Amsterdam, pp. 447–478.

Shilts, W. W. (1982). Quaternary evolution of the Hudson/James Bay region, in University of Guelph Symposium on Hudson Bay, *Can. Natur. 109*, 309–332.

Sikonia, W. G. (1982). *Finite Element Glacier Dynamics Model Applied to Columbia Glacier, Alaska*, U.S. Geol. Surv. Prof. Paper 1258-B, 74 pp.

Stauffer, B., G. Fischer, A. Neftel, and H. Oeschger (1985). Increase of atmospheric methane recorded in Antarctic ice core, *Science 229*, 1386–1388.

Thomas, R. H. (1985). Responses of the polar ice sheets to climatic warming, in *Glaciers, Ice Sheets, and Sea Level: Effects of a CO_2-Induced Climatic Change*, Committee on Glaciology, National Academy Press, Washington, D.C., pp. 301–316.

Thomas, R. H. (1987). Future sea-level rise and early detection by satellite remote sensing, *Prog. Oceanog. 18*, 23–40.

Thorarinsson, S. (1940). Present glacier shrinkage and eustatic changes of sea-level, *Geogr. Ann. 22*, 131–159.

Weertman, J. (1974). Stability of the junction of an ice sheet and an ice shelf, *J. Glaciol. 13*, 3–11.

Warren, S., and S. Frankenstein (in press). Increased accumulation on the Antarctic Ice Sheet due to climatic warming, abstract for symposium on Ice and Climate, International Glaciol. Soc., to appear in *Ann. Glaciol.*

Zwally, H. J. (1985). Mapping ice sheet growth by satellite altimetry, paper presented to Symposium on Glacier Mapping and Surveying, International Glaciological Society, Reykjavik, Iceland, Aug. 26, 1985.

Sea Level and Climate Change

11

ERIC J. BARRON
Pennsylvania State University

STARLEY L. THOMPSON
National Center for Atmospheric Research

INTRODUCTION

It has long been suggested that a change in global sea level produces a change in climate. This hypothesis receives substantial support from the geologic record of sea-level change. Over a broad range of time scales (10^4 to 10^7 yr), eustatic sea level and continental flooding are well correlated with paleoclimatic data, particularly during the past 100 million yr (m.y.).

The relationship between sea level and climate has three components. First, climate directly influences sea-level variations largely through the processes that control the growth and decay of continental ice. At peak glaciation 18,000 yr ago, approximately one-sixth of the planet was covered by ice and continental glaciation, and sea level was lower by 85 to 130 m than it is at present (CLIMAP, 1976).

Second, a variety of physical mechanisms exist whereby sea-level changes can directly affect global climate. By changing the nature of the atmosphere-surface interface, sea-level changes can alter the transfer of heat, moisture, and momentum between the surface and the atmosphere. In addition, sea level can alter ocean currents by introducing or removing geographic barriers. Ocean currents are a principal means of transferring heat from the tropics to polar regions and play a crucial role in controlling some regional climates of the present. Other potential direct influences of sea level on climate include effects on ice sheet size, ocean chemistry, and atmospheric content of carbon dioxide and other trace gases.

Third, the correlation between sea level and climate during Earth history may be the product of an indirect association. One plausible indirect link is a relationship between increased volcanism, higher atmospheric carbon dioxide levels, and high global sea level caused by rapid sea-floor spreading (Berner *et al.*, 1983). In this case, climate change would be directly related to changes in carbon dioxide levels, and the correlation of climate change with sea level would not be indicative of a direct cause-and-effect relationship.

Only the direct influence of sea level on climate and the indirect association between sea level and climate are considered here. The direct influences of climate on sea level, through land ice volume or ocean temperature change, are considered in Chapters 10 and 13 (this volume), respectively. Simulations with global dynamical climate models and examples from the paleoclimatic record provide insight into the importance of various physical mechanisms and the explanations of correlation between climate and sea-level variations during earth history.

DIRECT EFFECTS OF SEA LEVEL ON CLIMATE

The potential direct effects of sea level on climate can arise from physical mechanisms that fall into at least five categories: (1) regional changes in atmosphere-surface coupling, (2) changes in ocean circulation, (3) changes in surface heat capacity, (4) changes at ice sheet-ocean margins, and (5) changes in ocean-atmosphere chemical composition. The mechanisms are listed roughly in order of increasing uncertainty, although insufficient work has been completed in the investigation of the climatic effect of sea-level changes in every category. Hence the discussion that follows will necessarily rely heavily on work not originally conducted with an eye to sea-level change. In some cases, educated speculation will have to suffice until more research specific to the problem of sea level and climatic change is conducted.

More is known about atmosphere-surface coupling than the other four categories, perhaps because atmospheric and oceanographic scientists have studied and modeled the general problem for many years. Ocean circulation modeling has also been conducted for many years, but the level of detail required for comprehensive study of the global effects of sea-level change has exceeded the capability of even the largest computers. [Current global ocean models use about a 500-km grid. Typical sea-level changes (over geologic time) produce coast line changes on the order of 100 to 1000 km. Thus sea-level changes are either subgrid scale or, at best, barely resolvable (hence poorly treated) by current global ocean models.] Changes in ice sheet volume are usually thought of as influencing sea level; interest in the influence of sea level on ice sheets is relatively recent. Likewise, it is uncertain how sea-level changes can have a direct impact on ocean-atmosphere chemical composition, but such composition changes, if they exist, could cause substantial changes in global climate by altering the "greenhouse" effect.

Atmosphere-Surface Coupling

Virtually every facet of Earth's climate is controlled or affected by the transfer of heat, moisture, and momentum between the atmosphere and the underlying surface. For example, most of the heat that drives the atmospheric circulation is first absorbed as solar energy at the surface and then made available as sensible heat (direct temperature change) and latent heat of condensation (heat released when evaporated moisture condenses to form rain or snow). Furthermore, the long-term circulation of the atmosphere is constrained to conserve absolute angular momentum. Momentum transferred to the Earth when the atmosphere "rubs" against the surface must be balanced by a reverse momentum transfer elsewhere arising from changes in wind speed and direction. Surface easterly winds represent absolute momentum transfer from the Earth to the atmosphere, and surface westerlies vice versa. Changes in the regional distribution of surface "roughness" can di-

rectly affect local surface winds and, owing to the momentum conservation constraint, global wind patterns as well.

Surface Heat and Moisture Balance   Heat balances typical of three surface types are shown in Figure 11.1. In general, the surface of the Earth receives a net positive flux of radiant energy (a gain of solar energy minus a loss of infrared energy) that is balanced by sensible heat and evaporative losses. Land in the subtropics is relatively dry and has a relatively high albedo, that is, reflectivity for solar energy. Most of the net radiative gain is balanced by sensible heat loss owing to the lack of moisture to be evaporated. In contrast, the adjacent subtropical oceans have low albedo and lose heat almost exclusively by evaporation. Changing subtropical land to ocean, e.g., by a sea-level increase, would produce a major change in the amount of heat and moisture received by the overlying atmosphere, perhaps producing a change in regional climate. As Figure 11.1 shows, near equatorial land is qualitatively similar to ocean because abundant vegetation makes it relatively dark and moist. Thus changing equatorial-type land to ocean would produce a smaller climatic effect than changing subtropical land to ocean.

A local change in albedo of the surface can produce a large effect on the local surface energy balance, but what effect can it have on global-scale climate? A simple analy-

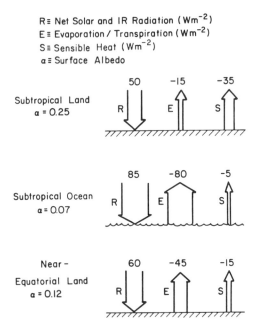

FIGURE 11.1   Heat balances for some representative surface types. On average, net radiative heating of the surface is balanced by evaporative and sensible heat loss. Evaporation is the preferred mode of heat loss at warm temperature; thus subtropical oceans lose heat mainly through evaporation.

sis based on our current understanding of global climate theory suggests that a global change in surface albedo of 0.01 will produce about a 1°C change in surface temperature. The effect of less than global-scale changes in surface albedo will be smaller in general, but requires detailed modeling calculations to determine the actual sensitivity.

We can now make a rough estimate of the effect of sea-level change on surface albedo and temperature. Suppose that in a latitude zone with an initial land fraction of 0.4, a sea-level rise reduces the land area fraction to 0.3 (a rather large change, probably requiring a sea-level increase well in excess of 100 m). Assume also that the difference in surface albedos between the covered land and the ocean is 0.1. The average change in surface albedo in the zone is thus $(0.4 - 0.3) \times 0.1 = 0.01$. The predicted rise in surface temperature of the zone (assuming no interaction with adjacent zones or other effects) is about 1°C. Thus a major change in sea level is not likely to cause a large change in global-scale surface temperature from an albedo effect only. This approximate result is confirmed by more detailed calculations by Thompson and Barron (1981) that examined the effect on albedo of shifting the continents to the positions they had 100 m.y. ago (Ma). Factor of 2 changes in land area in the subtropics produced a change of only a few watts per square meter (W/m^2) in the surface energy budget of the zone.

Referring again to Figure 11.1, we see that plausible changes in surface wetness can produce much larger changes in the surface energy balance than can plausible changes in albedo. Sea-level change provides an ideal way to produce an extreme change in surface wetness. No calculations have been done with global climate models in which sea level has been changed and only surface wetness has been allowed to vary (i.e., no albedo changes or other effects). Thus it is difficult to say just how important changes in wetness are compared to albedo changes, but we can get some indication from simulation studies that have changed land surface properties to investigate other problems. Shukla and Mintz (1982) performed two simulations with a global climate model: one in which all land was assumed to be completely dry (no evaporation allowed) and the other assuming land to be completely wet (like the ocean). Thus the simulations can give an indication of what might happen if very dry desert were changed to ocean without any albedo change. The calculations showed the dry land case producing summer land surface temperatures 10° to 20°C higher than the wet case. Clearly, evaporation from the surface can have a powerful cooling effect.

Moisture flux from the surface not only transports heat but causes changes in atmospheric water vapor content that can affect cloudiness and precipitation. Yeh *et al.*

(1984) have done simulations with a global climate model in which soil moisture was perturbed to see the effect on surface temperature and rainfall. They found, in general, that increased surface evaporation led to increased precipitation, but not necessarily in the same region as the increased evaporation. Increasing evaporation in a surface divergence region (a region of downward air motion) led to increased precipitation in adjacent convergence regions, but not locally. The explanation of the effect is that it is difficult to form precipitation in regions of downward motion regardless of the amount of water vapor in the air. Excess humidity in subsidence regions is moved by winds to regions of upward motion and then rained out. More recently, Gordon and Hunt (1987) have further emphasized the role of simulated surface hydrology and precipitation patterns.

From the above discussion, it can be concluded, for example, that replacing land with ocean in the normally dry subtropics might not lead to any local rainfall increase, but might instead produce increased equatorial rainfall. Aside from noting these potential local and nonlocal mechanisms, more specific statements cannot be made about the effects of sea-level change on rainfall without more-detailed global climate model calculations.

What happens if surface albedo and moisture availability are changed simultaneously? Henderson-Sellers and Gornitz (1984) performed climate model experiments in which the effects of Amazon deforestation were examined. The deforestation was simulated by changing the surface albedo from 0.11 to 0.19 and by greatly decreasing the potential for surface evaporation and transpiration. These changes are similar to what one would expect from a sea-level drop, i.e., replacing sea with land. The researchers found that there was no net surface temperature change in the deforested area. The surface albedo and wetness changes for this particular case produced nearly complete compensating effects. There was, however, a local decrease in rainfall of about 10 percent over the deforested area, as expected from the discussion above. There were no global-scale climate effects that could be reliably detected above the normal model variability. Even though these results are only suggestive, they do indicate that the direct climate effect of large sea-level changes may not be large even on regional scales.

Surface Roughness The transfer of heat, moisture, and momentum between the atmosphere and surface is proportional to the surface drag coefficient (C_D), a nondimensional quantity that expresses the proportionality between the friction force per unit area at the surface and the square of the surface wind speed. The value of C_D is strongly dependent on the "roughness" of the surface as measured by the characteristic height of surface protrusions such as

188   ERIC J. BARRON AND STARLEY L. THOMPSON

TABLE 11.1 Typical Values
of Surface Drag Coefficient,
C_D ($\times 10^{-3}$) (from Garratt 1977)

	C_D
Ocean	
Wind speed 5 m/s	1
Wind speed 20 m/s	2
Land	
Desert	3
Tropical forest	25
Typical	10

rocks or plants. Typical values of C_D for land and ocean are given in Table 11.1. In general, the drag coefficient over ocean is an order of magnitude smaller than it is over land. This has the effect of (a) tending to produce decreased surface fluxes over ocean and (b) producing a higher surface wind speed over ocean by way of compensation. [Substantial momentum transfer between the Earth and the atmosphere is also created by large-scale orography, the so-called "mountain pressure torque" (see Holton, 1979, pp. 264–265). It is unlikely that reasonable sea-level changes alone would produce climatologically important changes in topographic heights relative to sea level.]

It is not clear what effect a change in sea level could have on large-scale climate through a change in C_D. Some sensitivity studies with climate models (e.g., Hansen *et al.*, 1983) have shown that the simulated climates are not very sensitive to C_D. On the other hand, it is well known that changes in momentum exchange at the surface can have a substantial effect on atmospheric circulation systems (Holton, 1979, p. 260). It is not possible at this time to say how important sea-level-induced changes in surface roughness are in relation to changes in albedo or wetness.

Combined Surface Effects Barron and Washington (1984) performed two climate model simulations for continental positions of 100 Ma. One case was for present day sea levels and the other included a sea level estimated to be 330 m higher than the present that is thought to have occurred during the Cretaceous period. This is a massive sea-level increase that would have caused a 20 percent decrease in global land area. Barron and Washington found that the sea-level change produced systematic local climate changes only in the subtropics and tropics. In these regions, changing land to ocean produced local cooling (the evaporation increase outweighed the albedo decrease) and slight warming elsewhere. Zonally averaged temperatures changed by less than 2°C at any latitude (see Figure 11.2). The global net effect on surface tem-

perature was small. The postulated change in Cretaceous sea level produced a smaller global climate effect than did correspondingly large changes in topography and continental positions.

Ocean Circulation

In a number of instances, sea-level changes have been invoked as a mechanism of modifying the ocean surface and thermohaline circulations. The sea-level ocean circulation mechanism involves bathymetric control of current directions, importance of barriers in basin-basin communication, and the role of shallow epicontinental seas and marginal basins in the formation of deep water. Each example cited below illustrates the potential for sea-level changes to influence climate, but none of the studies to date provides quantitative links between sea-level-induced circulation changes and climate change.

The high-velocity core of the Gulf Stream scours the Blake Plateau, producing a 50- to 75-km-wide band of erosional topography. Pinet and Popenoe (1982) used seismic stratigraphy and a series of drill holes to show that the axis of the Gulf Stream has shifted hundreds of kilometers repeatedly during the past 20 m.y. The position of the Gulf Stream axis is well correlated with sea-level

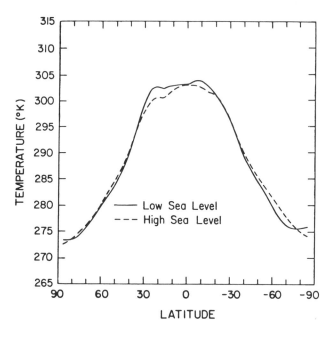

FIGURE 11.2 A comparison of the zonally averaged surface temperature for a general circulation model experiment with Cretaceous continental positions with low sea level (present-day total area) and no topography and an experiment with Cretaceous continental positions with high sea level (about 20 percent of continental area flooded) and no topography.

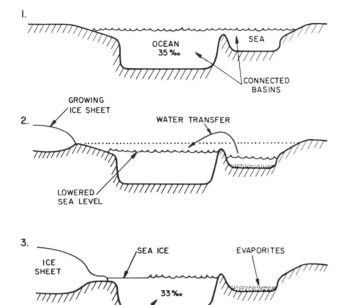

FIGURE 11.3 A hypothetical example of the potential for sea-level change to affect climate through a mechanism that relies on isolation of ocean basins. (1) The world ocean and a subtropical sea communicate across a shallow strait. (2) A global climate change induces ice sheets to grow. Lowered sea level isolates the subtropical sea, which begins to evaporate. Fresh water is transferred to the world ocean by precipitation. (3) Lower salinity of the world ocean promotes a sea ice increase, which acts as a positive feedback to the global cooling.

variations. During lowstands of sea level, a broad bathymetric bulge, called the Charleston bump, deflects the Gulf Stream offshore to a position across the Blake Plateau and along the steep continental slope just north of the Blake Spur. During sea-level highstands, the Gulf Stream parallels the Florida–Hatteras Slope. The Gulf Stream has a notable impact on regional climates, thus the potential for climate change. However, the extent of possible climatic change due to such movements in current direction in response to sea-level variation is unknown.

The elevation of a barrier above sea level or subsidence of a ridge below sea level has been cited as a mechanism of climatic change in numerous instances. The subsidence of the Greenland–Iceland–Faeroe Ridge (about 38 Ma) probably resulted in a significant interaction between the Arctic and Atlantic oceans; subsidence of the Walvis Ridge off South Africa (about 20 Ma) greatly influenced the Bengula current system; and formation of the Isthmus of Panama (about 2.3 Ma) eliminated equatorial Atlantic–Pacific flow (Berggren, 1982). Berggren stressed the role of ocean gateways in modifying heat transport by the oceans. Sea-level variations have the potential to have

breached or created surface circulation barriers during Earth history, but few instances related specifically to sea level are documented, and in every case the climatic implications are qualitative.

Sea-level change can isolate or reconnect small basins. Particularly in subtropical high-evaporation zones or regions where precipitation greatly exceeds evaporation, isolation of a basin from the main ocean can result in large density contrasts (i.e., hypersalinity or freshening of the basin). Thierstein and Berger (1978) hypothesized that reconnection of a temporarily isolated basin can result in an injection event that favors either abyssal stratification or surface stratification depending on the salinity characteristics. Brass et al. (1982) showed that deep water can form in marginal seas in the subtropics due to high evaporation (warm, salty bottom water). These authors suggest that changes in the size and configuration of marginal seas in the subtropics owing to sea-level change may have controlled deep-water formation during different periods of earth history. Deep water tends to form in semirestricted basins or marginal seas because the isolation allows the water mass to obtain different density characteristics, through atmospheric interaction, in comparison with the main ocean. These marginal seas are certainly susceptible to sea-level variations that may modify their size and configuration. Basin isolation and induced salinity variations may also result in climate change because fresh water freezes more readily (see hypothetical example in Figure 11.3). The potential of this mechanism to modify climate is largely unexplored.

Annual Temperature Cycle

The heat capacity or thermal inertia of the surface is a strong determinant of the annual cycle of temperature at a given latitude. The thermal inertia of the ocean is dependent on the depth of the mixed layer, ranging from tens of meters to in excess of 200 m. In contrast, the thermal inertia of land is roughly equivalent to a 1.0-m depth of water. Extensive continental flooding is likely to modify the local thermal inertia, and hence influence the annual cycle of temperature.

In experiments using the seasonal zonal energy balance climate model of Thompson and Schneider (1979) to investigate the contrast between the climate of the Cretaceous and that of the present (Barron et al., 1981), the changes in land fraction associated with Cretaceous geography resulted in a 3° to 5°C reduction in the amplitude of the annual cycle of surface temperature in the Northern Hemisphere mid-latitudes (i.e., the summers were cooler and winters warmer than present). The mid-latitude ocean fraction increased by approximately 20 percent in the Cretaceous case. In this model the zonal thermal inertia was computed from an area-weighted harmonic mean assuming no land–sea heat transfer within a zone.

More recently, the general circulation model experiments with Cretaceous geography and mean annual isolation of Barron and Washington (1984) have been extended to a seasonal mixed-layer experiment. The zonally averaged amplitude of the annual cycle of surface temperature in the Northern Hemisphere mid-latitudes was reduced 4° to 8°C in comparison with a present-day control experiment. The largest component of this zonally averaged difference is regions that were land initially but become oceanic when the sea level was increased. If this component of the zonal average is removed, the change in amplitude is 1° to 3°C.

The above two simulations illustrate the potential for sea-level fluctuations to produce changes in seasonality. However, each experiment described above includes a number of variables that could influence seasonality (e.g., ice-albedo feedback), and consequently the importance of thermal inertia has not been isolated.

Ice Sheets

The influence of sea level on ice sheets became a topic of interest when the potential instability of the present West Antarctic Ice Sheet became widely known. This ice sheet is the smaller of two major ice masses covering Antarctica. Unlike the East Antarctic Ice Sheet, the western ice sheet is mostly grounded below where sea level would be if the ice were removed. Apparently, the ice sheet exists only because floating ice shelves at its edge act as buttresses to prevent the ice sheet from quickly flowing out into the ocean. These ice shelves are themselves stabilized by friction against protruding islands, underwater rises, and the sides of the bays in which the shelves sit. Such marine ice sheet-shelf systems have been shown to be potentially unstable to perturbations such as ocean warming or sea-level rise (Thomas *et al.*, 1979).

Evidence and modeling studies (Thomas and Bentley, 1978) indicated that the West Antarctic Ice Sheet was much larger 18,000 yr ago during the height of the last glacial episode. This increased size was made possible by the almost 100-m lower sea level at that time. The hypothesis is that the rise in sea level associated with the melting of the great Northern Hemisphere ice sheets caused a substantial collapse of the large West Antarctic Ice Sheet until it stabilized at its present size. A further sea-level rise could presumably act to unpin the buttressing ice shelves that allow the remnant ice sheet to exist.

Although the massive East Antarctic Ice Sheet is in no danger of total collapse, it too has undergone fluctuations in size associated with sea-level changes. During the last glacial maximum this ice sheet probably extended 75 to 90

km farther onto the continental shelf than at present (Alley and Whillans, 1984). The rise in sea level at the end of the Ice Age produced a rapid retreat response at the ice sheet edge that propagated as a wavelike reduction of ice sheet thickness to the interior of the ice sheet.

Although the basic link between large sea-level changes and ice sheet growth and decay is fairly clear, it is not obvious that small sea-level changes in isolation have a substantial effect on global climate. The principal effect on climate would probably arise from the change in area of the ice sheet and associated sea ice. A further regional climatic effect would come from changes in ice sheet topography. If large glacio-climatic changes in sea level are needed to have significant effects on ice sheets, then any sea-level-induced ice sheet changes would act only as positive climatic feedbacks rather than actual driving forces for large climatic changes. The level of positive feedback could be quite small, e.g., a 50-cm rise in sea level due to ocean thermal expansion could cause such a small reduction in ice sheet size that any positive feedback on ocean temperature would be negligible.

Ocean Chemistry

The most uncertain potential direct effect of sea level on climate involves the conjectured relation between sea level, ocean–atmosphere chemistry, and the greenhouse effect. Studies of cores drilled in ice sheets have shown that atmospheric CO_2 concentration was about two thirds of its present value during the last glacial maximum. Since CO_2 is a greenhouse gas, the change in atmospheric composition apparently acted to substantially augment the postglacial warming (Thompson and Schneider, 1981).

It is not clear what caused the atmospheric CO_2 increase at the time of deglaciation, but it is known that the changes must have originated in the oceans (Broecker, 1984). An initial hypothesis (Broecker, 1982) was that increased sea level due to melting Northern Hemisphere ice sheets covered previously exposed continental shelf. Increased sedimentation of organic matter onto the shelf would have removed phosphorous from the ocean. Since phosphorous is a limiting nutrient, such a loss would decrease ocean phytoplankton productivity, reduce the uptake of CO_2 in surface water, and hence produce an atmospheric CO_2 increase. A second hypothesis (Berger, 1982) was that the postglacial sea-level rise encouraged deposition of carbonate sediments on the shelf, an action that would have generated an increased CO_2 content in the surface water and the atmosphere. The source of carbonate to be deposited, and hence CO_2, would have been dissolution of deep-sea sediments. However, more recent hypotheses of the glacial to interglacial CO_2 change have not directly involved sea-level changes (Broecker, 1984).

Whether sea-level changes can directly affect atmospheric CO_2 concentration is an unresolved question.

INDIRECT SEA-LEVEL AND CLIMATE ASSOCIATIONS

The global cooling trend over the past 70 m.y. is well correlated with a gradual decline in global sea level or increase in global land area. The global cooling from the Cretaceous to the present day has been estimated to be in the range of 6° to 12°C in globally averaged surface temperature (Barron, 1983). The decrease in global sea level is estimated to be as much as 300 to 400 m (Hardenbol *et al.*, 1982) and an increase in total land area of 20 percent (Barron *et al.*, 1980). Early comparisons between sea-level variations associated with this long-term trend and paleotemperature (e.g., Damon, 1968) have been used to infer a strong causal relationship between sea level and climate over the past 100 m.y.

Barron and Washington (1984) performed an extreme sea-level sensitivity experiment with a general circulation model of the atmosphere. This experiment compared mean annual simulations for Cretaceous geography with flooded continents and with present land area (nonflooded continents), which were described earlier. The globally averaged surface temperature response to increased sea level and decreased land area in the model was –0.2°C (Figure 11.2). In the subtropics, where the majority of the flooding occurred, increased evaporative cooling (about 60 W/m^2) and a small increase in cloud cover (2.5 percent) compensated for the surface albedo change.

The results of the sea-level climate model sensitivity experiment bring into question the direct explanation of the global cooling trend as a function of sea-level and surface-albedo variations. The alternative possibilities are (1) inadequate model sensitivity, most likely because of a lack of a seasonal cycle or a fully resolved coupled ocean model, or (2) an indirect association of sea level and paleotemperatures. Point (1) is discussed in detail by Barron and Washington (1984), and here we will introduce only one plausible indirect association between sea level and paleotemperatures.

Berner *et al.* (1983) and Lasaga *et al.* (1985) performed calculations with a geochemical model based on the carbonate–silicate geochemical cycle to test the possibility of atmospheric CO_2 concentration changes on time scales of 1 m.y. over the past 90 m.y. On this time scale, CO_2 removal from the atmosphere is largely dependent on weathering of exposed silicate rocks (with higher sea level this area decreases). One explanation of higher global sea levels is increased rates of seafloor spreading (see Harrison, Chapter 8, this volume). If seafloor spreading rates are high, then a logical implication is greater volcanism

and hence higher CO_2 input into the atmosphere (see Arthur *et al.*, 1985, for a discussion on substantially higher volcanic input estimates for the Cretaceous). Berner *et al.* (1983) suggested that this substantially higher volcanic input of CO_2 cannot be compensated by increased rate of weathering of exposed silicate rocks. In this scenario, high paleotemperatures would result from a CO_2-induced warming that would be associated with high sea level, but the warming would not be a direct response to sea-level variations.

The problems with the atmospheric CO_2 model of Berner *et al.* (1983) include estimating actual rates of CO_2 degassing through time, estimating actual seafloor spreading rates (Berner *et al.* used four different estimates of seafloor spreading variations), and determination of how directly weathering of silicate and carbonate rocks responds to the level of atmospheric CO_2 concentrations (Berner and Barron, 1984). In addition, Shackleton (1985) suggested that the carbon isotope record may limit the potential variability of atmospheric CO_2. Despite some unresolved problems, the CO_2 model of Berner *et al.* (1983) and Lasaga *et al.* (1985) would solve the problem of explaining the long-term correlation between sea level and paleotemperatures in light of small climate model sensitivity to large-scale continental flooding, and is a good example of a potential indirect association between sea level and climate.

SUMMARY

The direct effects of sea level on climate include changes in atmosphere–surface coupling, ocean circulation, thermal inertia, ice sheet-ocean interactions, and changes in ocean–atmosphere chemical composition. In only a few cases has the direct effect of a specific variable been isolated. In the majority of the climate model simulations, the model experiments provide insight into the importance of various physical mechanisms, but a series of sensitivity experiments should be conducted to isolate the importance of variables such as surface albedo, surface wetness, surface roughness, and thermal inertia. Many of the arguments presented for the importance of various mechanisms (e.g., ocean gateways that influence oceanic heat transport) are purely qualitative. The first steps toward placing these concepts on a more firm physical foundation must be taken if we are to demonstrate the potential role of sea level in climate change.

Further, correlations between sea level and climate in the geologic record may not be a product of the direct mechanisms described here. The possibility exists, for example, that the causes of sea-level change may influence climate (e.g., tectonic control on sea level and volcanic CO_2 emissions), and the paleoclimatic associations with sea level may be indirect.

REFERENCES

Alley, R. B., and I. M. Whillans (1984). Response of the East Antarctic Ice Sheet to sea-level rise, *J. Geophys. Res. 89*, 6487–6493.

Arthur, M. A., W. E. Dean, and S. O. Schlanger (1985). Changes in the global carbon cycle and climate during the mid-Cretaceous: Their relationship to volcanism and possible changes in atmospheric CO_2, in *The Carbon Cycle and Atmospheric CO_2: Natural Variations Archean to Present*, E. T. Sundquist and W. S. Broecker, eds., Geophys. Monogr. Ser. 32, American Geophysical Union, Washington, D.C.

Barron, E. J. (1983). A warm equable Cretaceous: The nature of the problem, *Earth Sci. Rev. 19*, 305–338.

Barron, E. J., and W. M. Washington (1984). The role of geographic variables in explaining paleoclimates: Results from Cretaceous climate model sensitivity studies, *J. Geophys. Res. 89*, 1267–1279.

Barron, E. J., J. L. Sloan, and C. G. A. Harrison (1980). Potential significance of land–sea distribution and surface albedo variations as a climatic forcing factor; 180 m.y. to the present, *Palaeogeogr. Palaeoclim. Palaeoecol. 30*, 17–40.

Barron, E. J., S. L. Thompson, and S. H. Schneider (1981). An ice-free Cretaceous? Results from climate model simulations, *Science 212*, 501–508.

Berger, W. H. (1982). Increase of carbon dioxide in the atmosphere during deglaciations: The coral reef hypothesis, *Naturwissenchaften 69*, 87–88.

Berggren, W. (1982). Role of ocean gateways in climatic change, in *Climate in Earth History*, Studies in Geophysics, National Academy Press, Washington, D.C., pp. 118–125.

Berner, R. A., and E. J. Barron (1984). Comments on the BLAG model: Factors affecting atmospheric CO_2 and temperature over the past 100 million years, *Am. J. Sci. 284*, 1183–1192.

Berner, R. A., A. C. Lasaga, and R. M. Garrels (1983). The carbonate–silicate geochemical cycle and its effect on atmospheric carbon dioxide over the last 100 million years, *Am. J. Sci. 283*, 641–683.

Brass, G., E. Saltzman, J. Sloan, J. Southam, W. Hay, W. Holzer, and W. Peterson (1982). Ocean circulation, plate tectonics, and climate, in *Climate in Earth History*, Studies in Geophysics, National Academy Press, Washington, D.C., pp. 118–125.

Broecker, W. S. (1982). Glacial to interglacial changes in ocean chemistry, *Prog. Oceanogr. 11*, 151–197.

Broecker, W. S. (1984). Carbon dioxide circulation through the ocean and atmosphere, *Nature 308*, 602.

CLIMAP (1976). The surface of the ice-age Earth, *Science 191*, 1131–1137.

Damon, P. (1968). The relationship between terrestrial factors and climate, *Meteorol. Monogr. 8*, 106–111.

Garratt, J. R. (1977). Review of drag coefficients over oceans and continents, *Mon. Weather Rev. 105*, 915–929.

Gordon, H. B., and B. G. Hunt (1987). Interannual variability of simulated hydrology in a climate model: Implications for drought, *Climate Dynamics 1*, 113–130.

Hansen, J., G. Russell, D. Rind, P. Stone, A. Lacis, S. Lebedeff, R. Ruedy, and L. Travis (1983). Efficient three-dimensional global models for climate studies: Models I and II, *Mon. Weather Rev. 111*, 609–662.

Hardenbol, J., P. R. Vail, and J. Ferrer (1982). Interpreting paleoenvironments, subsidence history, and sea-level changes of passive margins from seismic and biostratigraphy, in *Climate in Earth History*, Studies in Geophysics, National Academy Press, Washington, D.C., pp. 139–153.

Henderson-Sellers, A., and V. Gornitz (1984). Possible climatic impacts of land cover transformations, with particular emphasis on tropical deforestation, *Climatic Change 6*, 231–257.

Holton, J. R. (1979). *An Introduction to Dynamic Meteorology*, 2nd ed., Academic Press, New York, 391 pp.

Lasaga, A. C., R. A. Berner, and R. M. Garrels (1985). An improved geochemical model of atmospheric CO_2 fluctuations over the past 100 million years, in *The Carbon Cycle and Atmospheric CO_2: Natural Variations Archean to Present*, E. T. Sundquist and W. S. Broecker, eds., Geophys. Monogr. Ser. 32, American Geophysical Union, Washington, D.C., pp. 397–411.

Pinet, P. R., and P. Popenoe (1982). Blake Plateau: Control of Miocene sedimentation patterns by large-scale shifts of the Gulf Stream axis, *Geology 10*, 257–259.

Shackleton, N. (1985). The ocean carbon isotope record: Implications for the history of global carbon cycling, in *The Carbon Cycle and Atmospheric CO_2: Natural Variations Archean to Present*, E. T. Sundquist and W. S. Broecker, eds., Geophys. Monogr. Ser. 32, American Geophysical Union, Washington, D.C.

Shukla, J., and Y. Mintz (1982). Influence of land-surface evapotranspiration on the Earth's climate, *Science 215*, 1498–1501.

Thierstein, H. R., and W. H. Berger (1978). Injection events in ocean history, *Nature 276*, 461–466.

Thomas, R. H., and C. R. Bentley (1978). A model for Holocene retreat of the West Antarctic Ice Sheet, *Quat. Res. 10*, 150–170.

Thomas, R. H., T. J. O. Sanderson, and K. E. Rose (1979). Effect of climatic warming on the West Antarctic Ice Sheet, *Nature 277*, 355–358.

Thompson, S. L., and E. J. Barron (1981). Comparison of Cretaceous and present Earth albedos: Implications for the causes of paleoclimates, *J. Geol. 89*, 143–167.

Thompson, S. L., and S. H. Schneider (1979). A seasonal zonal energy balance climate model with an interactive lower layer, *J. Geophys. Res. 84*, 2401–2414.

Thompson, S. L., and S. H. Schneider (1981). Carbon dioxide and climate: Ice and ocean, *Nature 290*, 9–10.

Yeh, T.-C., R. T. Wetherald, and S. Manabe (1984). The effect of soil moisture on the short-term climate and hydrology change—A numerical experiment, *Mon. Weather Rev. 112*, 474–490.

Long-Term Aspects of Future Atmospheric CO_2 and Sea-Level Changes

12

ERIC T. SUNDQUIST

U.S. Geological Survey, Woods Hole

INTRODUCTION

The primary motivation for recent concern about future sea-level change is the anthropogenic production of carbon dioxide and other infrared-absorbing trace gases. Climate models predict a rise of about 1.5 to 4.5°C in mean global temperatures for a doubling of atmospheric CO_2 levels, expected to occur during the next century (National Research Council, 1983). More extreme warming is predicted for higher latitudes. Analyses of air trapped in polar ice have shown that the warming that marked the end of the most recent ice age was accompanied by a rise of atmospheric CO_2 from about 200 to 280 ppm (Berner *et al.*, 1980; Delmas *et al.*, 1980; Neftel *et al.*, 1982). That event was accompanied by the melting of enough polar ice to cause a sea-level rise of about 100 m (see Matthews, Chapter 5, this volume). Thus, both climate theory and history strongly suggest a close interconnection among CO_2, climate, and sea level.

However, the ice core data do not tell us whether CO_2 caused climate and sea level to change. The most viable hypotheses to explain the CO_2 changes observed in ice cores call on climate-induced redistributions of carbon within the ocean–atmosphere system (for example, see the section entitled "The Last Deglaciation" in Sundquist and Broecker, 1985). On the other hand, some manifestations of CO_2 change appear to lead climate-sensitive oxygen

isotope changes in the deep-sea sedimentary record (Shackleton and Pisias, 1985). The sequence of events and the details of cause and effect are still vague.

Such uncertainties will probably persist for a long time to come. Perhaps, rather than trying to discern whether CO_2 or climate has been cause or effect, we would do better to work toward models in which climate and the carbon cycle are considered parts of the same system. This approach was suggested long ago by Chamberlin (1898) and has recently reappeared in the work of Walker and Hays (1981) and Berner *et al.* (1983). The geochemical model of Berner *et al.* suggests that very high CO_2 concentrations were associated with high volcanic activity 100 million years ago (Ma). To simulate the reduction in CO_2 concentrations to present levels, the model connects CO_2 level to climate by including equations for the increase in chemical weathering rates caused by the temperatures associated with high CO_2 concentrations. An updated version of this model, incorporating the effects of sulfur and organic carbon cycling, suggests that late Cretaceous atmospheric CO_2 concentrations were 13 times their present level (Lasaga *et al.*, 1985).

One of the key problems in further developing this approach is to sort out which processes and interrelationships are important to which time scales. Sea-level change encompasses a broad range of time scales, with different mechanisms associated with change over

193

different time scales. This volume includes discussion of sea-level records ranging over seasons to hundreds of millions of years. The oceanographer concerned with storm tides would have relatively little interest in the factors explaining Cretaceous sea levels; likewise, the geologists' glacioeustatic theories have little application to seasonal events. The problem of sorting out time scales and processes afflicts studies of climate change in general, and of the carbon cycle. The complexities multiply in any attempt to link the two together.

As a first step toward resolving these difficulties, it is useful to examine a very simplistic summary of mechanisms for sea-level and carbon cycle change over time scales ranging from decades to tens of thousands of years. This summary treats sea level and carbon cycling separately, and there is no guarantee that their mutual interaction does not entail distinct time characteristics of its own. But some intriguing conclusions emerge, particularly with regard to time scales of 1000 yr and longer.

TIME SCALES OF SEA-LEVEL CHANGE

Table 12.1 summarizes mechanisms of sea-level change by time scale and magnitude. Although these estimates are derived from several sources, they are somewhat subjective. For example, heat exchange is relatively rapid within the uppermost few hundred meters of the oceans. A one-dimensional treatment (e.g., Munk, 1966) implies that thermal expansion of these waters can occur on time scales of years to decades. Sea level will rise about 10 cm for every degree of temperature increase throughout the uppermost 500 m. (In this chapter, relationships between seawater volumes and sea-level changes are calculated assuming no isostatic adjustment.) Heat exchange with deep ocean waters is slower, occurring on the time scale of deep ocean mixing and probably longer (e.g., Hoffert *et al.*, 1980). If the deep ocean were to warm everywhere by about 10°C, as was perhaps the case during the early Tertiary (Brass *et al.*, 1982), sea level would rise by about 6 m.

The time scales and magnitudes of melting ice can be estimated from both historical data and mass balance considerations. The present Greenland and Antarctic ice sheets are remnants of the late Pleistocene ice sheets that increased sea levels about 100 m by melting over a period of several thousand years encompassing the end of the Pleistocene (see Matthews, Chapter 5, this volume). The mass balance estimates shown in Table 12.2 [taken largely from Lamb (1972) and Meier (1983)] suggest modern ice residence times on the order of 10^4 years (see also L'vovich, 1974). The Antarctic ice sheet contains enough water to raise sea level by about 60 m, and the Greenland ice sheet contains water equivalent to a 6-m sea-level rise. Moun-

TABLE 12.1 Mechanisms of Eustatic Sea-Level Change

	Time Scale (yr)	Order of Magnitude (cm)
Ocean Thermal Expansion		
Shallow (0 to 500 m)	10^0 to 10^2	10^0 to 10^2
Deep (500 to 4000 m)	10^2 to 10^4	10^2 to 10^4
Melting Ice		
Mountain glaciers	10^0 to 10^2	10^0 to 10^2
Greenland ice sheet	10^4+	$<10^3$
Antarctic ice sheet	10^4+	10^3 to 10^4
West Antarctic ice sheet	10^3?	$<10^3$
Crustal Deformation		
Glacial rebound	10^3 to 10^4	variable
Tectonism	10^6+	10^4+

TABLE 12.2 Polar Ice Mass Balance

Inputs (km³/yr)	Mass (km³)	Outputs (km³/yr)
Accumulation (1000 to 2000)	Antarctic ice sheet (30×10^6)	Surface ablation (0 to 100) Ablation under ice shelves (100 to 300) Icebergs (500 to 1500)
Accumulation (400 to 600)	Greenland ice sheet (2.6×10^6)	Surface ablation (100 to 300) Icebergs (200 to 300)

SOURCES: Lamb (1972) and Meier (1983).

tain glaciers are discussed by Meier (Chapter 10, this volume). The magnitude and time scales of their influence on sea level appear to be comparable to those for thermal expansion in the upper ocean.

The West Antarctic ice sheet has attracted much controversy as a potential source of relatively rapid and large sea-level rise. On the one hand, dynamical arguments suggest that it may be very sensitive to any climate change that might cause grounding-line retreat (Hughes, 1973; Weertman, 1974). On the other hand, its diminution during the Holocene appears to have been gradual, and perhaps nil for the last 1000 yr (Stuiver *et al.*, 1981; Bentley, 1983). On the basis of actual and anticipated ice stream flow rates, Bentley (1983, 1984) has estimated that 500 yr would be the minimum time required for disintegration of the West Antarctic ice sheet.

From worldwide correlations of coastal onlap and offlap sedimentary sequences, eustatic sea-level changes of hundreds of meters are inferred to have been caused by deformation of the Earth's crust (see, e.g., Chapter 7, this volume). These changes occurred over time scales of millions to hundreds of millions of years.

From the data in Table 12.1, it appears that sea-level changes on the order of meters or more require times on the order of hundreds of years or longer. In light of this relationship between larger sea-level changes and longer time scales, it is logical to subject the carbon cycle to a similar analysis, with a view toward estimating the time scale of the fossil-fuel CO_2 perturbation.

TIME SCALES OF CARBON CYCLE CHANGE

The carbon cycle can be broadly subdivided according to characteristic time scales. Figure 12.1 illustrates one such subdivision, emphasizing long-term effects. Using this scheme, the most rapid changes—on the order of hundreds of years—occur within the ocean–atmosphere–biosphere system. Over longer time scales, "reactive sediments" must be added to the system. These are sediments that can interact readily with the ocean–atmosphere–biosphere system on time scales of thousands to tens of thousands of years. Finally, over time scales approaching 100,000 yr or longer, the carbon cycle must be viewed as including interactions with the Earth's crust. (For a more complete and mathematically rigorous discussion of time scales of carbon cycle change, see Sundquist, 1985.)

From a consideration of both the historical record and the mechanisms of cycling carbon, it appears that there are limits to the magnitude of natural variations in atmospheric CO_2 within the system delineated in Figure 12.1 as the atmosphere, biosphere, oceans, and reactive sediments. It

has been hypothesized that atmospheric CO_2 could not have changed by greater than a factor of 2 within this system prior to man's activities (Sundquist, 1986). Therefore, if we are interested in geologic analogs of greater than twofold atmospheric CO_2 changes, we must study the record of processes having time scales longer than 100,000 yr.

Similarly, the relationships between magnitudes and time scales of sea-level change lead to questions about carbon cycle dynamics over time scales longer than the decades spanned by most CO_2 predictive studies. Once fossil-fuel CO_2 has been put in the atmosphere, how long will it stay there? Will high CO_2 levels persist long enough to approach the response times of the polar heat and water budgets? Answering these questions will require the development of a new generation of unified climate/carbon cycle models. In the meantime, some very general conclusions can be derived from a geochemical model of the ocean–atmosphere–sediment carbon cycle.

EXTENDING PREDICTIVE CO_2 MODELS TO LONGER TIME SCALES

Efforts to predict the effects of anthropogenic CO_2 have stimulated many advances in modeling the carbon cycle. These advances can be applied fruitfully to long-term aspects of the problem. However, predictive CO_2 models cannot be extended to long-term geochemical modeling by simply running them for longer times. A primary reason

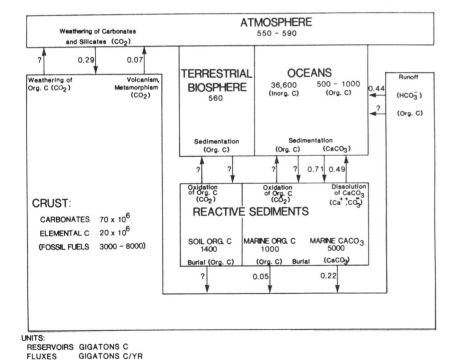

FIGURE 12.1 The carbon cycle, subdivided to emphasize the components and processes important to different time scales. Changes over time scales up to hundreds of years will occur within the ocean, atmosphere, and biosphere. Changes over time scales of thousands to tens of thousands of years will involve "reactive sediments." Changes over time scales longer than 100,000 yr will involve carbon in the Earth's crust. (From Sundquist, 1986)

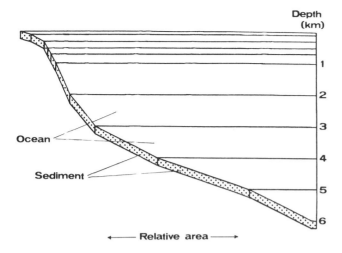

FIGURE 12.2 Basic geometry of an ocean-sediment box model, based on global seafloor hypsometry.

for this difficulty is that the predictive models usually do not incorporate specific interactions with marine sediments.

The most conspicuous feature in the worldwide distribution of marine pelagic sediments is the transition from carbonate-rich sediments at shallow depths to carbonate-depleted sediments in the deep ocean. This transition reflects the difference between shallow and deep seawaters in their state of saturation with respect to the carbonate minerals calcite and aragonite (Li *et al.*, 1969; Broecker and Takahashi, 1978; Plummer and Sundquist, 1982). As anthropogenic CO_2 is absorbed by the oceans, the undersaturated regions of the ocean will become more undersaturated, and some regions that are now supersaturated will become undersaturated. These changes, which can be predicted from fundamental chemical equilibrium calculations, imply that increasing atmospheric CO_2 will increase the extent of carbonate dissolution in marine sediments. Because adding dissolved carbonate to seawater increases its capacity to absorb CO_2 from the atmosphere, this mechanism is a principal sink for added CO_2 over time scales of thousands to tens of thousands of years. Other feedbacks—most notably biological—may also be important (see, e.g., Revelle and Munk, 1977), but they are not considered here because their mechanisms are poorly understood (particularly over such long time scales).

The importance of carbonate dissolution to long-term predictions can be illustrated by a simple calculation (Broecker, 1977; Sundquist, 1979). If all of the world's fossil-fuel resources were instantaneously added to the oceans as CO_2 and distributed in proportion to present dissolved inorganic carbon concentrations, the resultant atmospheric CO_2 concentration would be about three times its present value. However, if the same amount of CO_2 were allowed to react with an equal molar amount of

sedimentary calcium carbonate, the resultant atmospheric CO_2 concentration would be only about 30 percent above its present value. Thus, it is of considerable interest to develop a predictive model capable of examining the effectiveness of CO_2 buffering by dissolution of carbonate sediments.

To incorporate "reactive" marine sediments into carbon cycle models, it is necessary for the model oceans to have a specified bottom topography. Figure 12.2 shows how ocean hypsometry can be explicitly included in ocean box models. The areas of the surfaces between adjoining ocean boxes can be calculated from the global seafloor hypsometric curve. With linear interpolation, these areas imply a volume and a seafloor area for each box. The seafloor area is assumed to be the surface of contact between each ocean box and an associated sediment box.

This hypsometric ocean model permits a realistic approach to the relationships between the sedimentation and dissolution of carbonate particles. Calcite and aragonite, precipitated by organisms in the ocean surface, settle into the deep sea. Because nearly all of the aragonite dissolves before or soon after it reaches most of the seafloor, only calcite need be included in a model incorporating "reactive" sediments. Calcite appears to dissolve after it has reached the seafloor rather than while it is settling, which is relatively rapid (Vinogradov, 1961; Berger and Piper, 1972; Honjo, 1975). Thus, calcite sedimentation can be approximated by a flux from the ocean surface to the seafloor, with dissolution occurring only from those sediments that lie below the "saturation horizon" (Figure 12.3).

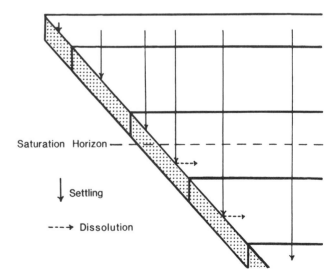

FIGURE 12.3 Carbonate sedimentation and dissolution in a hypsometric ocean-sediment box model. Carbonate particles are supplied to the sediments everywhere from the surface ocean box, but dissolution occurs only below the saturation horizon.

The depth of this horizon can be determined from thermo-dynamic calculations using the known distributions of alkalinity, dissolved inorganic carbon, and dissolved calcium as functions of depth. These distributions imply depth-dependent carbonate-ion concentrations and calcite solubilities. As illustrated in Figure 12.4, the saturation horizon corresponds to the depth at which the curve representing the carbonate-ion concentration intersects the curve representing carbonate-ion concentrations at equilibrium with calcite. Undersaturation occurs wherever the latter curve, termed the "critical carbonate-ion curve" by Broecker and Takahashi (1978), lies to the right of the carbonate-ion concentration profile. At any undersaturated depth, the distance between these curves represents the degree of undersaturation, a quantity needed to calculate dissolution rates. In ocean box models, both the depth of the saturation horizon and the degree of undersaturation can be approximated by interpolation between the carbonate-ion concentrations and critical carbonate-ion concentrations in adjacent boxes (Figure 12.4, inset).

Calcite dissolution adds alkalinity to seawater, and, in turn, variations in ocean alkalinities affect the carbonate-ion concentrations that control calcite dissolution. Whereas nearly all CO_2 predictive models treat alkalinity as a constant parameter, any ocean model that incorporates calcite dissolution must treat alkalinity as a time-dependent variable. Like dissolved inorganic carbon, alkalinity is conservative with respect to ocean mixing processes. Thus, model equations for alkalinity will include mixing terms very similar to those for dissolved inorganic carbon. Moreover, the stoichiometry of carbonate dissolution implies that dissolution fluxes of alkalinity (in equivalents) will be exactly twice as large as the corresponding dissolution fluxes of dissolved inorganic carbon (in moles).

Calcite dissolution rates depend not only on the degree of undersaturation but also on the amount of calcite available for dissolution. The hypsometric ocean model (Figure 12.2), together with interpolative determination of the saturation horizon (Figure 12.4), provides a straightforward representation of the relative seafloor areas exposed to dissolution. The most difficult aspect of modeling global calcite dissolution is approximating the time dependence of the calcite content relative to the noncarbonate components in sediments. This difficulty is illustrated by the scenario shown in Figure 12.5. At steady state (Figure 12.5a), sediments above the saturation horizon have a high calcite content, while dissolution causes those below the saturation horizon to have a low calcite content. If CO_2 is added to the oceans (Figure 12.5b), the consequent decrease in carbonate-ion concentrations causes the saturation horizon to rise. This process exposes calcite-rich sediments to undersaturated waters. The dissolution flux from these sediments may exceed the dissolution flux

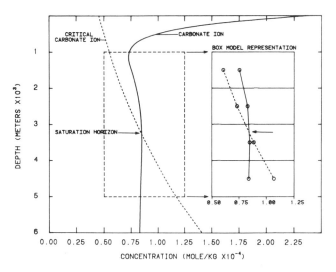

FIGURE 12.4 The carbonate saturation horizon in an ocean-sediment box model. The critical carbonate-ion curve agrees closely with the equilibrium carbonate-ion profile (Broecker and Takahashi, 1978). The inset shows how box model interpolation is a reasonable way of estimating the depth of the saturation horizon.

from calcite-depleted sediments that, although exposed to greater degrees of undersaturation, cannot supply calcite for dissolution at a rate greater than the flux of calcite settling to the seafloor. Dissolution, of course, decreases the calcite content of the sediments that were not previously exposed to undersaturated waters. If the amount of calcite dissolved is enough to cause carbonate-ion concentrations to increase (Figure 12.5c), the saturation horizon falls toward its original steady-state depth. As this occurs, some calcite-depleted sediments pass from undersaturated to oversaturated waters. The calcite content in these sediments increases as the continuous supply of settling calcite is no longer offset by loss due to dissolution.

This example demonstrates several fundamental problems in extending global atmosphere-ocean models to include calcite dissolution. The distribution of the dissolution flux is discontinuous; it exists only in undersaturated waters, which must be located by reference to a saturation horizon that can move through a wide range of depths. Care must be taken that the physical discontinuities implicit in the saturation horizon are neither ignored by spatial averaging nor allowed to generate exaggerated discontinuities in the numerical solutions to the model equations. The magnitude of the dissolution flux and the calcite content of the sediments must also be modeled to interact with each other. This interaction requires that the ocean model be coupled to a sediment model with its own additional time-dependent variables. Finally, the behavior of the model sediments must include a wide range of possible changes in the calcite content of sediments ex-

198

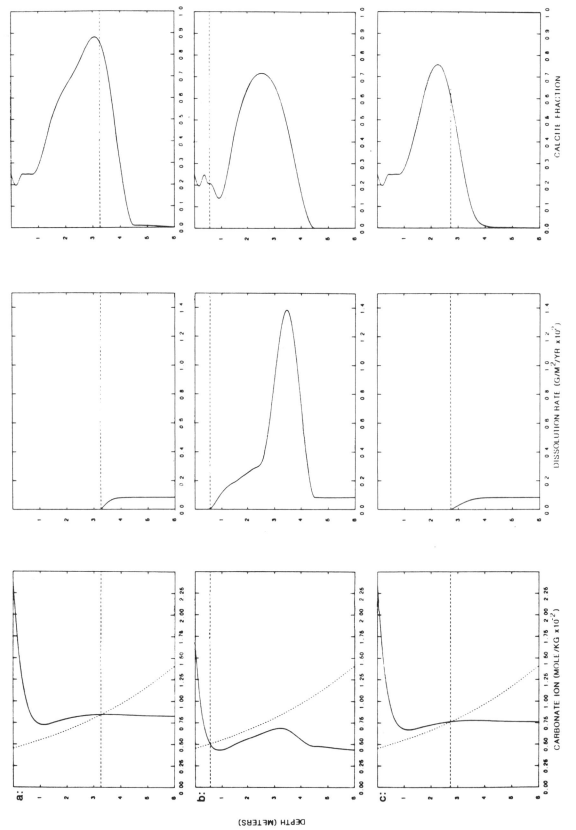

FIGURE 12.5 A scenario illustrating the response of the ocean-sediment model to added CO_2. The left-hand graphs show the critical carbonate-ion (solid) and carbonate-ion (dotted) concentrations. The middle graphs show the calcite dissolution rate. The right-hand graphs show the calcite fraction in the sediments. The saturation horizon is represented by the horizontal dashed line on each graph. In (a), the model is at steady state, with calcite sedimentation above the saturation horizon matching calcium-ion inputs. In (b), added CO_2 causes the carbonate-ion concentrations to decrease and the saturation horizon to shoal. In (c), continued carbonate dissolution causes carbonate-ion concentrations to increase and the saturation horizon to fall. The data for this figure came from years (a) 1700, (b) 2700, and (c) 11,700 of the model run for the "low fossil-fuel" case in Figures 12.9 to 12.17.

posed to both undersaturated and oversaturated waters. It is clear that the areal resolution of the coupled sediment model must be comparable to the spatial resolution of the ocean model, and that an ocean–sediment box model may require further resolution to distinguish between undersaturated and oversaturated behavior within a single ocean box.

A COUPLED ATMOSPHERE–OCEAN–SEDIMENT MODEL

The model shown in Figure 12.6 (from Sundquist, 1986) has been used to simulate the response of ocean and sediment chemistry to anthropogenic CO_2. The global ocean is divided into three regions representing waters south of latitude 50°S, north of latitude 50°N (55°N in the Pacific Ocean), and all waters between these polar regions. Ocean areas for these regions are taken from Baumgartner and Reichel (1975). Each polar region is divided into two vertical boxes by a boundary at 300 m. Temperate ocean boxes are separated by boundaries at 100, 300, 500, 700, 1000, 2000, 3000, 4000, and 5000 m. Each ocean box is coupled to a sediment box. The volume and area for each ocean box, and the area for each sediment box, are calculated using the hypsometric data of Menard and Smith (1966) (for the temperate and south polar regions) and Gorshkov (1980) (for the north polar region) to define the ocean bottom.

The model represents seawater chemistry, ocean mixing, and gas exchange using parameterizations similar to those employed in many CO_2 predictive models. For each ocean box, the speciation of dissolved inorganic carbon is calculated from total alkalinity and total dissolved inor-

ganic carbon using an iterative procedure and appropriate equilibrium constants from Lyman (1957), Culberson and Pytkowicz (1968), Mehrbach *et al.* (1973), Millero (1979), and Dickson and Riley (1979). Dissolved calcium and total borate are assumed to be proportional to salinity (Culkin, 1965).

In the ocean surface boxes, air–sea exchange of CO_2 is assumed to be proportional to the difference between the partial pressure of CO_2 in the atmosphere and the partial pressure of CO_2 at equilibrium with the surface mixed layer. CO_2 solubilities are taken from Weiss (1974), and the gas-exchange rate is assumed to be 7×10^{-4} moles $m^{-2}atm^{-1}yr^{-1}$ (Broecker *et al.*, 1980; Siegenthaler, 1983). Mixing across ocean boundaries is represented by both diffusive and advective terms. Diffusive terms, which appear only in the equations for the temperate ocean boxes, have the form

$$K_{vi} \frac{(X_{i+1} - X_i)}{(z_{i+1} - z_i)}, \qquad (12.1)$$

where the subscripts i and $i+1$ refer to vertically contiguous boxes, X represents the concentration of either total alkalinity or total dissolved inorganic carbon, z represents the box depth, and K_v represents the vertical diffusion coefficient. Values for K_v range from 1.7 cm²/s below the temperate ocean surface mixed layer (Li *et al.*, 1984) to 0.6 cm²/s in the deep ocean (Ku *et al.*, 1980). Advective fluxes are represented simply by $w_{ij} X_i$, where w_{ij} represents the flux of water from box i to box j. Values for w_{ij} are derived from the fluxes assumed for the polar production of cold deep and intermediate waters. Following Gordon and Taylor (1975), the formation of Antarctic bottom water is represented by advective flux terms total-

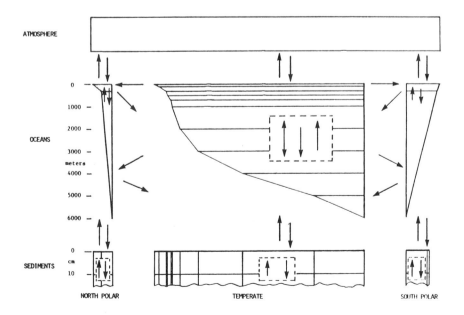

FIGURE 12.6 The atmosphere–ocean–sediment box model used to assess the carbonate dissolution response to fossil-fuel CO_2. The double arrow in the temperate ocean represents eddy diffusion; single arrows represent generalized exchange fluxes. See text for further details.

ing 40×10^6 m³/s from the deep south polar box to the temperate boxes deeper than 3000 m; the formation of North Atlantic deep water is represented by flux terms totaling 10×10^6 m³/s from the deep north polar box to the temperate boxes deeper than 1000 m; and the formation of Antarctic intermediate water is represented by flux terms totaling 20×10^6 m³/s from the south polar surface box to temperate boxes between 300 and 2000 m. These flux terms imply additional advective terms representing upwelling throughout the temperate water column.

Sediment coupling is based on the sediment model shown in Figure 12.7. The model assumes that the sediment box associated with each ocean box can be represented as a homogeneous bioturbated layer 10 cm thick (Berger and Heath, 1968; Peng et al., 1977; Sundquist et al., 1977; Peng and Broecker, 1978). Calcite and noncarbonate particles settle continuously to the surface of each sediment box, where they are incorporated as sedimentation fluxes into the homogeneous box. If the sediment surface is exposed to undersaturated waters, dissolution removes some of the calcite from the box. The sediment burial flux is equivalent to the difference between the sedimentation fluxes and the dissolution flux.

Model calcite and noncarbonate sedimentation fluxes are assigned constant values that are consistent with the known distribution of oceanic sedimentation rates. Following Broecker (1982), the calcite sedimentation flux is evaluated at 10 g/m²yr for sediments deeper than 300 m and shallower than 3000 m. For sediments deeper than 3000 m, the calcite sedimentation flux is assumed to be 8 g/m²yr. The calcite flux per area to sediments shallower than 300 m is assumed to be three to four times the flux per area to the deep sea. Noncarbonate sedimentation fluxes are assigned values that are consistent with the distribution of sediment carbonate fractions as a function of depth

(Milliman, 1974; Broecker and Takahashi, 1977) and with the global flux of suspended sediments delivered by rivers to the oceans (Milliman and Meade, 1983). Model noncarbonate sedimentation values are extremely depth-dependent, ranging from 120 g/m²yr for sediments shallower than 300 m to 1 g/m²yr for sediments deeper than 3000 m.

Calcite dissolution is modeled from laboratory rate measurements and a sediment pore water model. The exponential rate law determined experimentally by Keir (1980) is incorporated into a steady-state pore water model (Berner, 1980) to yield the dissolution flux term

$$K_{dis} \, [f(q/c)^n(d/q)^{n+1}]^{0.5}, \qquad (12.2)$$

where f is the sediment calcite fraction, q is the ratio of the carbonate-ion gradient to the total dissolved inorganic carbon gradient, c is the critical carbonate-ion concentration in the bottom water, d is the difference between the pore-water carbonate-ion concentration and the critical carbonate-ion concentration at the sea–sediment interface, and n is an experimental constant equal to 4.5 for calcite (Keir, 1980). The constant K_{dis} is derived from estimates for the sediment density and porosity, the pore-water diffusion coefficient for bicarbonate ion, the experimental dissolution rate constant, and the specific surface area of sedimentary calcite. Typical values for K_{dis} and q are 3300 moles/m²yr and 0.6, respectively. The above flux term, which represents the flux per unit area at the sea-sediment interface, must be integrated over the seafloor area exposed to undersaturated conditions. The value of the integral is estimated for variable c and d using the weighted mean value theorem (Apostol, 1967, p. 154).

As suggested in the discussion of Figure 12.5, it is possible that the dissolution flux may exceed the total sedimentation flux under certain conditions. This situation results in a "negative" burial flux; that is, previously buried sediments are exhumed and incorporated into the bioturbated layer. This possibility requires that the sediment model "remember" the properties of previously buried sediments.

Another modeling necessity suggested by Figure 12.5 is the distinction between sediments above and below a saturation horizon. In any box that is found to contain a saturation horizon, the model implements separate equations to distinguish sediments that are dissolving from those that are not. The calcite contents of these two classes of sediments are therefore treated independently. As the saturation horizon moves up or down, conservation of mass requires that some of the sediments of one class be transferred to those of the other class. As shown in Figure 12.8, the sediment model is modified to include this lateral flux whenever the associated ocean box contains a saturation horizon.

The calculation of steady-state solutions for ocean box

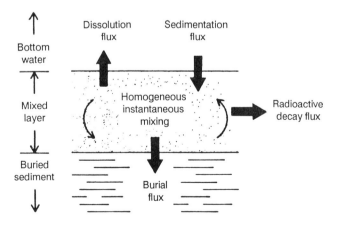

FIGURE 12.7 Sediment model used in the ocean–atmosphere–sediment model (after Sundquist et al., 1977). The homogeneous mixed layer is assumed to be 10 cm thick.

models is a significant modeling problem, requiring many compromises between model parameterizations and empirical observations (see, e.g., Wunsch and Minster, 1982; Bolin *et al.*, 1983). For this study, the flux terms for CO_2 gas exchange, ocean mixing, and calcite sedimentation and dissolution are assumed to conform to the parameterizations described above. The steady state for the entire model system is defined by the overall balance between inputs from rivers, volcanoes, and weathering of organic carbon and losses to sediment burial and CO_2 consumption during weathering. Additional terms in the steady-state equation account for the alkalinity sink and CO_2 source associated with deep-sea hydrothermal activity. A test of the coupled model's self-consistency is its representation of sedimentation and dissolution in a way that yields a reasonable value for the steady state global calcite burial flux. This flux is estimated to be 1.75×10^{13} moles/yr, equal to the flux of calcium ions delivered to the oceans by rivers (Holland, 1978), plus a small calcium-ion contribution from deep-sea hydrothermal reactions (Edmond *et al.*, 1979; Mottl, 1983). This burial rate is a net flux, representing the difference between calcite particle fluxes to the global ocean bottom and calcite dissolution fluxes from those areas of the seafloor where calcite dissolves. The model can be tuned to yield this global burial flux exactly, with only minor adjustment of the shallow-water calcite sedimentation fluxes. Likewise, ocean-surface dissolved inorganic carbon concentrations can be tuned slightly to yield air–sea exchange fluxes that conform exactly to the steady-state equation for atmospheric CO_2.

Thus it is assumed that the model parameterizations are a reasonable approximation of steady-state behavior. In practice, this assumption requires the introduction of residual terms to satisfy the steady-state equations for alkalinity and dissolved inorganic carbon in each ocean box. These residuals, which are held constant throughout each modeling experiment, represent processes (such as organic carbon cycling) that are not represented by the parameterizations described above. The model equations therefore represent perturbations relative to an assumed network of steady-state processes, many of which are poorly understood. Steady-state concentrations are calculated using an iterative procedure that assures that the model concentrations for the year 1973 agree with the volume-weighted GEOSECS measurements (Takahashi *et al.*, 1981) and the seasonally adjusted atmospheric measurement from Mauna Loa (Keeling and Bacastow, 1977).

THE LONG-TERM PERSISTENCE OF FOSSIL-FUEL CO₂

This model is grossly oversimplified, but nevertheless it suggests an answer to our question about the persistence

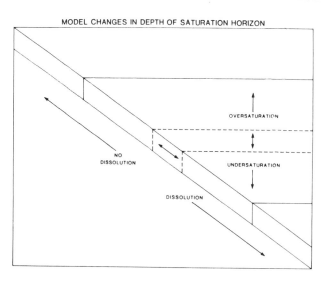

FIGURE 12.8 The relationship between sediment dissolution and the position of the saturation horizon in the atmosphere–ocean–sediment box model. Sediments above and below the saturation horizon must be categorized and modeled separately even though they may be associated with a single ocean box. Changes in the depth of the saturation horizon require a corresponding transfer of sediments from the dissolving to the nondissolving category, or vice versa.

of fossil-fuel CO_2. This answer emerges from the results of two modeling experiments representing the addition of different amounts of fossil-fuel CO_2. These amounts are selected to be well within the range of fossil-fuel resource estimates shown in Table 12.3, which shows both identified and ultimately recoverable resources as an index of the uncertainty in the estimates. The two modeling experiments simulate the addition of 2500 and 5000 billion tons of carbon as CO_2. The lower number is somewhat less than the world's total identified resources, while the high number is near the median between the identified and ultimately recoverable resource estimates.

The time dependence of these scenarios for CO_2 release is shown in Figure 12.9. CO_2 production through the year 1980 is based on the yearly production estimates of Rotty (as compiled by Watts, 1982). Extrapolation beyond 1980 is based on the logistic resource-depletion function suggested by Perry and Landsberg (1977). The peakedness of these curves is perhaps steeper than recent growth of fossil-fuel consumption would indicate, but the primary factor in long-term effects is the total integral under the curves rather than their short-term derivatives.

The model equations are solved numerically using fifth- and sixth-order Runge–Kutta formulas implemented in the subroutine DVERK (IMSL, 1982). Repetitive runs, using different error control options, established that the global numerical error never exceeds 0.5 percent for any model variable.

TABLE 12.3 World Remaining Fossil-Fuel Resources
(gigatons equivalent carbon content)

	Identified	Ultimately Recoverable
Coal	3226[a]	6743[a]
Crude oil	97[b]	253[a]
Natural gas	41[c]	133[a]
Oil shale	23[d]	288[a]
Oil sands/heavy crude	97[a]	97[a]

[a]World Energy Congress (1980).
[b]Halbouty and Moody (1980).
[c]Nehring (1981).
[d]Ovcharenko (1981).

Calcite dissolution is represented in Figure 12.10 by the model output for sediments at about 3000 m depth. The lower curve, from the high fossil-fuel case, shows a total depletion of calcite and a response time of tens of thousands of years. The upper curve shows less intense dissolution and a more rapid return to initial conditions for the low fossil-fuel case.

In both cases, model sedimentation rates (Figure 12.11) for the same water depth fall to negative values for about 1000 yr. That is, the rate of dissolution exceeds the rate of incoming sediments, and chemical erosion occurs. The lower curve is the high fossil-fuel case, again showing much more intense dissolution. Once calcite is depleted in the burrowed sediment layer, the sedimentation rate re-

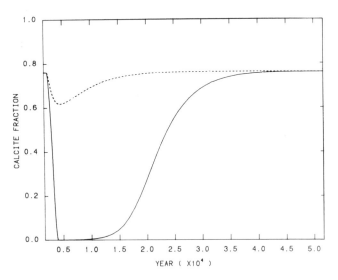

FIGURE 12.10 Model sediment calcite fractions in sediments at 3000 m depth. The solid and dashed curves correspond to the fossil-fuel consumption scenarios shown in Figure 12.9.

turns to a positive value representing the noncarbonate sedimentation rate. The sedimentation rate in the high fossil-fuel case eventually rises to its initial value as calcite is replenished in the burrowed layer, over about 25,000 years.

The importance of different processes to different time scales is perhaps best illustrated in this model by the behavior of its calcite saturation horizon. On a time scale of 1000 yr (Figure 12.12), in both the high and low fossil-fuel cases, water at depths corresponding to the dissolved

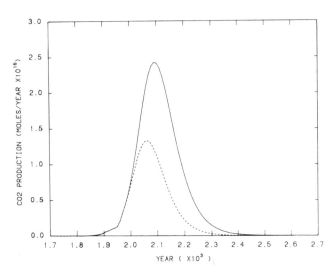

FIGURE 12.9 Fossil-fuel consumption scenarios used in modeling experiments. The solid curve represents the "high fossil-fuel" case; the dashed curve represents the "low fossil-fuel" case. See text for details.

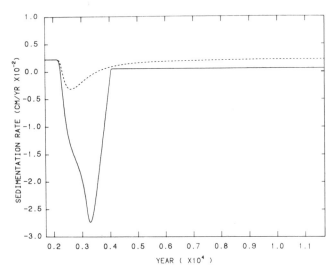

FIGURE 12.11 Model sedimentation rates at 3000 m depth. The rates shown are for total sediments, including noncarbonate. The solid and dashed curves correspond to the fossil-fuel scenarios shown in Figure 12.9.

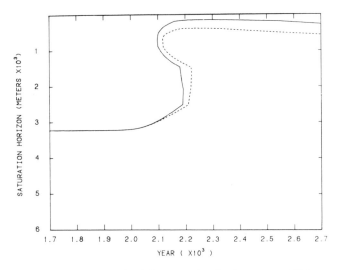

FIGURE 12.12 Model calcite saturation horizons over 1000 yr. The solid and dashed curves correspond to the fossil-fuel consumption scenarios shown in Figure 12.9.

oxygen minimum become undersaturated shortly after the time of peak CO_2 additions. This occurs because CO_2 is already abundant at these depths, where much of the organic matter settling from the ocean surface is oxidized. For a short time, the waters immediately below the oxygen minimum zone remain supersaturated, causing the model ocean to have three saturation horizons instead of one. (A problem with the model is also apparent in this figure. The sudden jump in the deepest saturation horizon occurs because the model's interpolating equations cannot "see" the saturation reversal between two of its average box depths.)

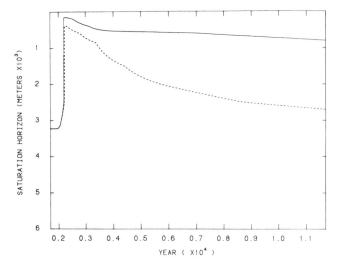

FIGURE 12.13 Model calcite saturation horizons over 10,000 yr. The solid and dashed curves correspond to the fossil-fuel consumption scenarios shown in Figure 12.9.

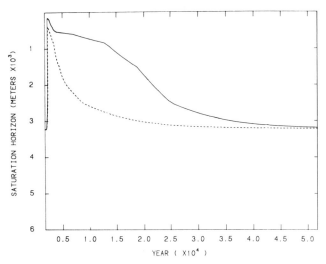

FIGURE 12.14. Model calcite saturation horizons over 50,000 yr. The solid and dashed curves correspond to the fossil-fuel consumption scenarios shown in Figure 12.9.

Over a time scale of 10,000 yr (Figure 12.13), the high and low fossil-fuel cases show drastically different influences on the calcite saturation horizon. The high fossil-fuel case maintains undersaturation up to depths shallower than 1000 m, while the low fossil-fuel case returns to a near-normal saturation state. There is simply not enough calcite available in this model to buffer the high fossil-fuel CO_2 additions. Instead, the buffering in this case is paced by the much slower return to a balance between the input of dissolved bicarbonate in river water and the sedimentation of calcite. This balance assures that, after 50,000 yr, the model saturation horizon for both cases has returned to its initial depth (Figure 12.14).

Given the exhausted buffering capacity in the high fossil-fuel case, it is not surprising that its atmospheric CO_2 level stays higher for a longer period. After 1000 yr, its atmosphere contains about four times as much CO_2 as it did initially, while the low fossil-fuel case has buffered its atmospheric CO_2 increase to less than twofold (Figure 12.15). After 10,000 yr (Figure 12.16), the high fossil-fuel atmosphere is still at about twice the initial CO_2 concentration, while the low fossil-fuel atmosphere contains about 400 ppm CO_2, a level that we will probably approach during the next few decades. Over a time scale of 50,000 yr (Figure 12.17), the model atmospheres approach their new steady-state values of about 380 ppm for the low fossil-fuel case and 450 ppm for the high fossil-fuel case.

CONCLUSIONS

Before discussing the implications of the results described above, it is important to reemphasize the shortcomings of the model. Several important feedbacks are

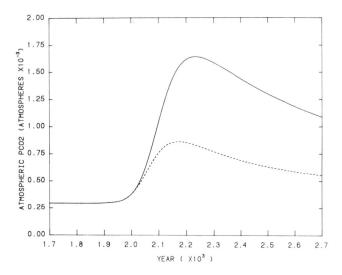

FIGURE 12.15. Model atmospheric CO_2 concentrations over 1000 yr. The solid and dashed curves correspond to the fossil-fuel consumption scenarios shown in Figure 12.9.

ignored. Massive production of fossil-fuel CO_2 will almost certainly alter the global cycling of organic carbon, both on land and in the sea. Long-term interactions between the climate system and the carbon cycle are so pervasive that any model that separates them is inherently inadequate. For example, ocean heating will probably be a positive feedback. The solubility of CO_2 in seawater decreases with increasing temperature. Although this effect is relatively minor when considered for the ocean surface only, it will be amplified to the extent that warm deep water replaces cold water. This will also profoundly

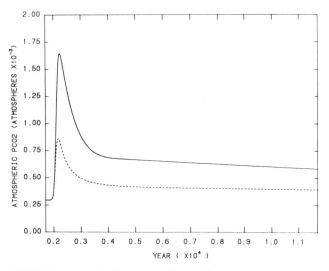

FIGURE 12.16. Model atmospheric CO_2 concentrations over 10,000 yr. The solid and dashed curves correspond to the fossil-fuel consumption scenarios shown in Figure 12.9.

affect ocean density stratification and therefore circulation. Warming will also affect calcite solubility, decreasing its effectiveness as a buffer. Another important feedback is the effect of high atmospheric CO_2 on chemical weathering rates. This feedback is effected in soils, where CO_2 is delivered to rocks by organisms that respond to environmental changes in notoriously complex ways. These and other processes must be better understood before the long-term persistence of fossil-fuel CO_2 can be reliably predicted.

However, principal features of the model results described above are maintained throughout a number of important sensitivity tests. Ocean mixing and air–sea exchange are relatively rapid, so wide variations in their parameters have little effect on model results beyond a

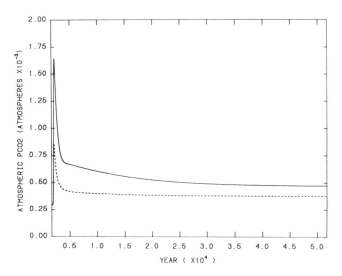

FIGURE 12.17. Model atmospheric CO_2 concentrations over 50,000 yr. The solid and dashed curves correspond to the fossil-fuel consumption scenarios shown in Figure 12.9.

few thousand years. The model is likewise insensitive to large variations in the shape of the CO_2 production curve for a given total integral under the curve. Calcite dissolution rate parameters also have little influence on the model results beyond a few thousand years. In short, the model results appear to be relatively insensitive to errors in the processes included in the model; the model's principal shortcomings derive from the processes it does not include.

The model results suggest that significantly elevated atmospheric CO_2 concentrations may persist for a time long enough to approach the response times of the polar heat and water budgets. Moreover, the magnitude of the persistent long-term CO_2 increase will depend on the total amount of fossil fuels consumed during coming decades and centuries. More specifically, the model indicates that ocean-sediment interactions may not be an effective buffer for massive amounts of fossil-fuel CO_2. Other feedbacks, ignored in this study, may be very important, and the time scales projected here certainly extend beyond our ability to anticipate technological influences on the ultimate consumption of fossil fuels. But it is clear that our questions about fossil-fuel CO_2—both scientific and societal—must be extended to a very broad continuum of time scales and effects.

REFERENCES

Apostol, T. M. (1967). *Calculus: One-Variable Calculus, with an Introduction to Linear Algebra*, Blaisdell Publishing Co., Waltham, Mass., 666 pp.

Baumgartner, A., and E. Reichel (1975). *The World Water Balance*, Elsevier, New York, 179 pp.

Bentley, C. R. (1983). The West Antarctic ice sheet: Diagnosis and prognosis, in *Proceedings of the Carbon Dioxide Research Conference, Carbon Dioxide, Science and Consensus*, U.S. Dept. of Energy, pp. IV.3–IV.50.

Bentley, C. R. (1984). Some aspects of the cryosphere and its role in climatic change, in *Climate Processes and Climate Sensitivity*, J. E. Hansen and T. Takahashi, ed., Geophysical Monograph Series 29, American Geophysical Union, Washington, D.C., pp. 207–220.

Berger, W. H., and G. R. Heath (1968). Vertical mixing in pelagic sediments, *J. Mar. Res. 26*, 134–143.

Berger, W. H., and D. J. Piper (1972). Planktonic foraminifera: Differential settling, dissolution and redeposition, *Limnol. Oceanogr. 17*, 275–287.

Berner, R. A. (1980). *Early Diagenesis*, Princeton University Press, Princeton, N.J., 241 pp.

Berner, W., H. Oeschger, and B. Stauffer (1980). Information on the CO_2 cycle from ice core studies, *Radiocarbon 22*, 227–235.

Berner, R. A., A. C. Lasaga, and R. M. Garrels (1983). The carbonate-silicate geochemical cycle and its effect on atmospheric carbon dioxide over the past 100 million years, *Am. J. Sci. 283*, 641–683.

Bolin, B., A. Bjorkstrom, and K. Holmen (1983). The simultaneous use of tracers for ocean circulation studies, *Tellus 35B*, 206–236.

Brass, G. W., J. R. Southam, and W. H. Peterson (1982). Warm saline bottom water in the ancient oceans, *Nature 296*, 620–623.

Broecker, W. S. (1977). Recommendations of the working group on carbonate dissolution, in *The Fate of Fossil Fuel CO₂ in the Oceans*, N. R. Andersen and A. Malahoff, eds., Plenum, New York, pp. 207–212.

Broecker, W. S. (1982). Ocean chemistry during glacial time, *Geochim. Cosmochim. Acta 46*, 1689–1705.

Broecker, W. S., and T. Takahashi (1977). Neutralization of fossil fuel CO_2 by marine calcium carbonate, in *The Fate of Fossil Fuel CO₂ in the Oceans*, N. R. Andersen and A. Malahoff, eds., Plenum, New York, pp. 213–241.

Broecker, W. S., and T. Takahashi (1978). The relationship

between lysocline depth and in situ carbonate ion concentration, *Deep-Sea Res. 25*, 65–95.

Broecker, W. S., T.-H. Peng, G. Mathieu, R. Hesslein, and T. Torgersen (1980). Gas exchange rate measurements in natural systems, *Radiocarbon 22*, 676–683.

Chamberlin, T. C. (1898). The influence of great epochs of limestone formation upon the constitution of the atmosphere, *J. Geology 6*, 609–621.

Culberson, C., and R. M. Pytkowicz (1968). Effect of pressure on carbonic acid, boric acid, and the pH in seawater, *Limnol. Oceanogr. 13*, 403–417.

Culkin, F. (1965). The major constituents of sea water, in *Chemical Oceanography 1*, J. P. Riley and G. Skirrow, eds., Academic Press, London, pp. 121–161.

Delmas, R. J., J. M. Ascenico, and M. Legrand (1980). Polar ice evidence that atmospheric CO_2 20,000 yr BP was 50% of present, *Nature 284*, 155–157.

Dickson, A. G., and J. P. Riley (1979). The estimation of acid dissociation constants in seawater media from potentiometric titrations with strong base. II. The dissociation of phosphoric acid, *Mar. Chem. 7*, 101–109.

Edmond, J. M., C. Measures, R. E. McDuff, L. H. Chan, R. Collier, B. Grant, L. I. Gordon, and J. B. Corliss (1979). Ridge crest hydrothermal activity and the balances of the major and minor elements in the ocean: The Galapagos Data, *Earth Planet. Sci. Lett. 46*, 1–18.

Gordon, A. L., and H. W. Taylor (1975). Heat and salt balance within the cold waters of the world ocean, in *Proceedings of the Symposium on Numerical Models of Ocean Circulation*, National Academy of Sciences, Washington, D.C., pp. 54–56.

Gorshkov, S. G. (1980). *Ocean Atlas Reference Tables*, (in Russian), Department of Navigational Oceanography, Ministry of Defense, USSR, 156 pp.

Halbouty, M. T., and J. D. Moody (1980). World ultimate reserves of crude oil, in *Proceedings of the World Petroleum Congress, 10th*, Heyden and Son, Philadelphia, Penna., pp. 291–301.

Hoffert, M. I., A. J. Callegari, and C.-T. Hsieh (1980). The role of deep sea heat storage in the secular response to climatic forcing, *J. Geophys. Res. 85*, 6667–6679.

Holland, H. D. (1978). *The Chemistry of the Atmosphere and Oceans*, John Wiley, New York, 351 pp.

Honjo, S. (1975). Dissolution of suspended coccoliths in the deep-sea water column and sedimentation of coccolith ooze, in *Dissolution of Deep-Sea Carbonates*, W. V. Sliter, A. W. H. Bé, and W. H. Berger, eds., Cushman Foundation Foraminiferal Research, Spec. Publ. 13, U.S. National Museum, Washington, D.C., pp. 114–128.

Hughes, T. (1973). Is the West Antarctic ice sheet disintegrating? *J. Geophys. Res. 78*, 7884–7910.

IMSL (1982). Chapters A to D, in *The IMSL Library: Reference Manual*, Vol. 1, 9th ed.

Keeling, C. D., and R. B. Bacastow (1977). Impact of industrial gases on climate, in *Energy and Climate*, Geophysics Study Committee, National Research Council, National Academy Press, Washington, D.C., pp. 72–95.

Keir, R. S. (1980). The dissolution kinetics of biogenic calcium carbonates in seawater, *Geochim. Cosmochim. Acta 44*, 241–252.

Ku, T. L., C. A. Huh, and P. S. Chen (1980). Meridional distribution of ^{226}Ra in the eastern Pacific along GEOSECS cruise tracks, *Earth Planet. Sci. Lett. 49*, 293–308.

L'vovich, M. I. (1974). The Earth's water balance, in *1974 World Water Resources and Their Future*, English translation, American Geophysical Union, Washington, D.C., pp. 51–59.

Lamb, H. H. (1972). *Climate Present, Past and Future*, Methuen & Co., Ltd., London, 613 pp.

Lasaga, A. C., R. A. Berner, and R. M. Garrels (1985). An improved geochemical model of atmospheric CO_2 fluctuations over the past 100 million years, in *The Carbon Cycle and Atmospheric CO_2: Natural Variations Archean to Present*, E. T. Sundquist and W. S. Broecker, eds., Geophysical Monograph Series 32, American Geophysical Union, Washington, D.C., pp. 397–411.

Li, Y.-H., T. Takahashi, and W. S. Broecker (1969). Degree of saturation of $CaCO_3$ in the oceans, *J. Geophys. Res. 74*, 5507–5525.

Li, Y.-H., T.-H. Peng, W. S. Broecker, and H. G. Ostlund (1984). The average vertical eddy diffusion coefficient of the ocean, *Tellus 36B*, 212–217.

Lyman, J. (1957). Buffer Mechanism of Seawater, Ph.D. thesis, Univ. of California, Los Angeles, 196 pp.

Mehrbach, C., C. H. Culberson, J. E. Hawley, and R. M. Pytkowicz (1973). Measurement of the apparent dissociation constants of carbonic acid in seawater at atmospheric pressure, *Limnol. Oceanogr. 18*, 897–907.

Meier, M. F. (1983). Snow and ice in a changing hydrological world, *Hydrol. Sci. J. 28*, 3–22.

Menard, H. W., and S. M. Smith (1966). Hypsometry of ocean provinces, *J. Geophys. Res. 71*, 4305–4325.

Millero, F. J. (1979). The thermodynamics of the carbonate system in seawater, *Geochim. Cosmochim. Acta 43*, 1651–1661.

Milliman, J. D. (1974). *Marine Carbonates*, Springer-Verlag, New York, 375 pp.

Milliman, J. D., and R. H. Meade (1983). World-wide delivery of river sediment to the oceans, *J. Geology 91*, 1–21.

Mottl, M. J. (1983). Metabasalts, axial hot springs, and the structure of hydrothermal systems at mid-ocean ridge, *Geol. Soc. Am. Bull. 94*, 161–180.

Munk, W. H. (1966). Abyssal recipes, *Deep-Sea Res. 13*, 707–730.

National Research Council (1983). *Changing Climate*, Carbon Dioxide Assessment Committee, National Academy Press, Washington, D.C., 496 pp.

Neftel, A., H. Oeschger, J. Schwander, B. Stauffer, and R. Zumbrunn (1982). Ice core sample measurements give atmospheric CO_2 content during the past 40,000 years, *Nature 220*, 223.

Nehring, R. (1981). The outlook for conventional petroleum resources, in *Long Term Energy Resources*, UNITAR, Pitman, Marshfield, Mass., pp. 315–327.

Ovcharenko, V. A. (1981). Reassessment of oil shale prospects, in *Long Term Energy Resources*, UNITAR, Pitman, Marshfield, Mass., pp. 451–484.

Peng, T.-H., and W. S. Broecker (1978). Effect of sediment

mixing on the rate of calcite dissolution by fossil fuel CO_2, *Geophys. Res. Lett. 5*, 349–352.

Peng, T.-H., W. S. Broecker, G. Kipphut, and N. Shackleton (1977). Benthic mixing in deep sea cores as determined by ^{14}C dating and its implications regarding climate stratigraphy and the fate of fossil fuel CO_2, in *The Fate of Fossil Fuel CO_2 in the Oceans*, N. R. Andersen and A. Malahoff, eds., Plenum, New York, pp. 355–373.

Perry, H., and H. H. Landsberg (1977). Projected world energy consumption, in *Energy and Climate*, Geophysics Study Committee, National Research Council, National Academy Press, Washington, D.C., pp. 35–50.

Plummer, L. N., and E. T. Sundquist (1982). Total individual ion activity coefficients of calcium and carbonate in seawater at 25°C and 35 o/oo salinity, and implications to the agreement between apparent and thermodynamic constants of calcite and aragonite, *Geochim. Cosmochim. Acta 46*, 247–258.

Revelle, R., and W. Munk (1977). The carbon dioxide cycle and biosphere, in *Energy and Climate*, Geophysics Study Committee, National Research Council, National Academy Press, Washington, D.C., pp. 140–158.

Shackleton, N. J., and N. G. Pisias (1985). Atmospheric carbon dioxide, orbital forcing, and climate, in *The Carbon Cycle and Atmospheric CO_2: Natural Variations Archean to Present*, E. T. Sundquist and W. S. Broecker, eds., Geophysical Monograph Series 32, American Geophysical Union, Washington, D.C., pp. 303–317.

Siegenthaler, U. (1983). Uptake of excess CO_2 by an outcrop-diffusion model of the ocean, *J. Geophys. Res. 88*, 3599–3608.

Stuiver, M., G. H. Denton, T. J. Hughes, and J. L. Fastook (1981). History of the marine ice sheet in West Antarctica during the last glaciation: A working hypothesis, in *The Last Great Ice Sheets*, G. H. Denton and T. J. Hughes, eds., John Wiley and Sons, New York, pp. 319–436.

Sundquist, E. T. (1979). Carbon Dioxide in the Oceans: Some Effects on Sea Water and Carbonate Sediments, Ph.D. thesis, Harvard University, Cambridge, Mass., 215 pp.

Sundquist, E. T. (1985). Geological perspectives on carbon dioxide and the carbon cycle, in *The Carbon Cycle and Atmospheric CO_2: Natural Variations Archean to Present*, E. T. Sundquist and W. S. Broecker, eds., Geophysical Monograph Series 32, American Geophysical Union, Washington, D.C., pp. 397–411.

Sundquist, E. T. (1986). Geologic analogs: Their value and limitations in carbon dioxide research, in *The Changing Carbon Cycle: A Global Analysis*, Trabalka and Reichle, eds., Springer-Verlag, New York.

Sundquist, E. T., and W. S. Broecker, eds. (1985). *The Carbon Cycle and Atmospheric CO_2: Natural Variations Archean to Present*, E. T. Sundquist and W. S. Broecker, eds., Geophysical Monograph Series 32, American Geophysical Union, Washington, D.C., 627 pp.

Sundquist, E. T., D. K. Richardson, W. S. Broecker, and T.-H. Peng (1977). Sediment mixing and carbonate dissolution in the southeast Pacific Ocean, in *The Fate of Fossil Fuel CO_2 in the Oceans*, N. R. Andersen and A. Malahoff, eds., Plenum, New York, pp. 429–541.

Takahashi, T., W. S. Broecker, and A. E. Bainbridge (1981). The alkalinity and total carbon dioxide concentration in the world oceans, in *Carbon Cycle Modeling*, SCOPE 16, B. Bolin, ed., John Wiley and Sons, New York, pp. 271–286.

Vinogradov, M. Y. (1961). Food sources of the deep-water fauna: Speed of decomposition of dead Pteropoda, *Soviet Oceanogr. 136/141*, 39–42, Trans. from Russian, *Dokl. Akad. Nauk SSSR Oceanol.*

Walker, J. C. G., and P. B. Hays (1981). A negative feedback mechanism for the long-term stabilization of Earth's surface temperature, *J. Geophys. Res. 86*, 9776–9782.

Watts, J. A. (1982). The carbon dioxide question: A data sampler, in *Carbon Dioxide Review: 1982*, W. C. Clark, ed., Oxford University Press, New York, pp. 429–469.

Weertman, J. (1974). Stability of the junction of an ice sheet and an ice shelf, *J. Glaciol. 13*, 3–11.

Weiss, R. F. (1974). Carbon dioxide in water and seawater: The solubility of a non-ideal gas, *Mar. Chem. 2*, 203–215.

World Energy Congress (1980). *Survey of Energy Resources*, Federal Institute for Geosciences and Natural Resources, Hanover, Federal Republic of Germany.

Wunsch, C., and J.-F. Minster (1982). Methods for box models and ocean circulation tracers: Mathematical programing and nonlinear inverse theory, *J. Geophys. Res. 87*, 5647–5662.

Sea Level and the Thermal Variability of the Ocean

13

DEAN ROEMMICH
Scripps Institution of Oceanography

ABSTRACT

The time variability of deep-ocean temperature and the relationship of temperature changes to changes in steric height and sea level are addressed using data from the subtropical North Atlantic and the eastern North Pacific spanning several decades. This is the longest period for which high-quality hydrographic time-series measurements are currently available. In the Atlantic, data consisted of 27 yr of repeated deep hydrographic stations near Bermuda, sea level at Bermuda, and two large spatial-scale hydrographic surveys of the subtropical North Atlantic Ocean that were separated in time by 23 yr. In the Pacific, a number of coastal sea-level stations were considered together with a grid of hydrographic stations sampled to 500 m depth in the period 1950 to 1978. By concentrating on these data-rich regions, the intention was to obtain a non-aliased view of the scales of time variability that should be applicable elsewhere in the oceans. The following conclusions were reached.

1. Although seasonal and interannual temperature fluctuations are concentrated in the upper ocean, decadal time-scale variations are not, at least in the Atlantic. Rather, they extend deep into the ocean. In the Atlantic study area, the upper 1000 m cooled by up to 0.5°C, and the interval from 1000 to 3000 m warmed by as much as 0.2°C.

2. The long time-scale temperature changes resulted in steric height change that was closely related to sea-level variability. The estimated power spectra of steric height near Bermuda and of sea level at Bermuda are indistinguishable and equally red at long periods. Coherence between steric height and sea level is very high at longer periods. In the eastern North Pacific, the areally averaged interannual sea-level and steric height changes were also indistinguishable within the limits of sampling errors.

3. The trends in sea-level records are typically 1 to 2 cm/decade, whereas the coherent oscillations of sea level and steric height at 10- to 20-yr periods have amplitudes of the order of 5 cm. A time series of about 50-yr length of sea level and steric height is required to determine whether the sea-level trend is accompanied by a similar trend in steric height.

4. Long time-scale temperature changes are spatially variable on the scale of the oceanic gyres, particularly for depths in and above the main thermocline. In terms of sea level, this means that fluctuations in mid-ocean sea level may be quite independent of those at coastal stations, even over a period of 20 yr.

INTRODUCTION

This study focuses on changes in the steric height of the ocean surface on time scales of 10 yr or longer and the relationship of steric height changes to changes in sea level. (The steric height of the sea surface is defined as the integral of the specific volume from a specified pressure level to the ocean surface.) Global sea-level fluctuations may be due to steric height change (e.g., temperature or salinity variation), to changes in the total mass of water in the ocean because of melting or freezing of ice caps, or to changes in the volume or shape of the ocean basin. It has been shown in regional studies that steric height changes account very well for the sea-level variations (on the order of 10 cm) observed on seasonal and interannual time scales (e.g., Shaw and Donn, 1964; Schroeder and Stommel, 1969; Reid and Mantayla, 1976). On the other hand, sea-level fluctuations over periods of thousands of years appear to be so large—for example, the 150-m increase of the past 15,000 yr (Moore, 1982)—that steric expansion can be ruled out as a major contributor. A warming of the entire ocean from 0°C to the currently observed temperatures would involve a thermal expansion of only a few meters. What then of the intermediate range of time scales from tens to hundreds of years? Can the secular trends currently observed in many sea-level records be attributed to steric expansion and contraction of the water column? Is there some residual sea-level rise that cannot be due to steric expansion?

For the reader's convenience, some values of the thermal expansion coefficient of seawater at a level of 35 practical salinity units are listed in Table 13.1. These are based on the 1980 equation of state (UNESCO, 1981). A point of interest is that the expansion coefficient increases significantly both with increasing temperature and with increasing pressure. The steep increase with increasing temperature has sometimes been used as an argument that deep cold water may be ignored in steric height calculations relative to warm surface water. This is not so; a parcel of water at 4°C at 2000 dbar has a thermal expan-

sion coefficient that is 60 percent as large as that of a parcel at the ocean surface at 20°C.

Barnett (1983) found that the global hydrographic data base is inadequate for determining long-term trends in steric height. He considered pairs of stations from nearly the same location and the same month of the year, but separated in time by more than 30 yr. In each of 11 regions of the world ocean, between 1 and 68 pairs were found. But the standard deviation of the difference in steric height of the sea surface relative to 1000 dbar was very large (20 dynamic centimeters in several regions) and none of the areas had a statistically significant trend. This is because the "noise" consisting of mesoscale eddies and other energetic low-frequency phenomena is very large compared to a hypothetical signal of the order of 1 dynamic centimeter per 10 yr. Thus, the sampling problem is formidable.

The approach used here is to concentrate on two relatively data-rich regions, the subtropical North Atlantic and the eastern North Pacific, rather than to seek a global solution. The objective is to discover how much variability in steric height is contained in time scales of decades or longer and to determine what vertical and horizontal scales of variability correspond to these long-term changes. These essential questions must be answered before we can ask how long the ocean must be sampled at a single location, how many locations must be sampled, and to what depths the sampling should extend to determine the relationship of steric height to the long-term trend in sea level.

BERMUDA SEA LEVEL AND THE PANULIRUS DATA

The longest regular time-series of deep (>1000 m) hydrographic stations is the Panulirus series (32°10'N, 64°30'W) near Bermuda (Figure 13.1). This program of measurements was initiated by Henry Stommel and was carried out by the Bermuda Biological Station beginning in 1954. Weather permitting, and barring occasional other problems, two casts per month were made at a location where the water depth is about 3000 m. Nearly all casts extend to 2000 m and many to 2500 m or deeper. From 1954 to 1981, 392 casts were made to depths greater than 2000 m. The time series is not very long for our purposes, and its limited depth precludes a study of the abyssal water, but it is all we have.

A number of investigators have studied the Panulirus data, sometimes in combination with Bermuda sea level. Shaw and Donn (1964) compared monthly means of sea level data (corrected for atmospheric pressure variations) with monthly means of the steric height of the sea surface relative to 2000 dbar and found them to be indistinguishable. Schroeder and Stommel (1969) noted that this sea-

TABLE 13.1 Thermal Expansion Coefficient (10^{-7}/K) of Sea Water at Salinity of 35 Practical Salinity Units for Different Temperatures and Pressures

| Pressure (dbar) | Temperature | | |
	20°C	10°C	4°C
0	2572	1669	1022
2000	2784	2012	1470
4000	2978	2323	1872

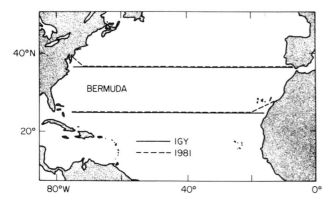

FIGURE 13.1 Map of the North Atlantic, showing Bermuda and ship tracks of IGY and 1981 transatlantic hydrographic sections.

sonal signal was mainly confined to the upper 200 dbar and was due to local buoyancy fluxes across the sea surface. In addition, they found that year-to-year variations in steric height were mirrored in the sea-level record; they attributed these variations to vertical motion of the main thermocline. Wunsch (1972) considered 8 yr of Bermuda sea-level data and 13 yr of Panulirus station data. He found that the steric height spectrum tapered off at periods longer than a year, an inference that we will contradict (Figure 13.8) after considering the longer records now available. Pocklington (1972) constructed time series of temperature at a number of standard depths using data from 1955 to 1969. He observed a trend of decreasing temperature down to 400 m with a suggestion that it continued to decrease as deep as 800 m. Below this there was no apparent trend. Frankignoul (1981) compared temperature at 10 standard levels in the first 12 yr of Panulirus data with temperature in the next 12 yr. He too found a temperature decrease down to 1000 m with a slight increase at 1600 and 2000 m. [The text and Table 1 of Frankignoul (1981) appear to reverse the sign of the changes, but they are given correctly in his Figure 2].

Here, 27 yr of Panulirus data will be used for two purposes. First, it will be shown that in contrast to seasonal and interannual variations, which are confined to the upper 1000 m, a significant 27-yr trend in the data extends down to the level of the deepest observations. Then this trend in time will be compared to a large space-scale temperature difference between two deep hydrographic surveys, each of which covered the subtropical North Atlantic at the same two latitudes (Figure 13.1).

All of the Panulirus casts were sorted by month and year and interpolated to 24 standard depths that were 50 m apart down to 300 m and then 100 m apart down to 2000 m. Casts in a given month were averaged over all years to estimate a mean annual cycle. Then, for each cast the mean for that month (over all years) was subtracted to

remove the annual cycle and then all casts in a year were averaged to estimate a mean anomaly for the year. We are interested in seeing how, on different time scales, the variability extends to different depths. Figure 13.2 shows steric depth for each of the 12 mean monthly profiles. By steric depth, we mean that the specific volume anomaly is integrated downward from the ocean surface to the pressure level indicated by the ordinate. Since the slope of each curve is the specific volume anomaly, it can be seen that the month-to-month changes in specific volume are large in the upper 200 m and small below this level. Figure 13.3 is of steric depth for the 26 yearly anomalies. The year-to-year changes in steric depth are slightly larger in magnitude than the annual cycle and clearly are distributed over a much greater depth range. This depth difference is summarized by examining empirical orthogonal functions (EOFs) of the specific volume profiles. The first EOF is, by definition, a best fitting shape, in a least squares sense, to all of the profiles in a group. Figure 13.4 shows the first EOF based on the 12 mean monthly profiles and the first EOF based on the 26 profiles of yearly anomalies. The much greater depth range of the yearly anomalies is apparent. The secondary maximum in the yearly anomalies at 800 m is in the main thermocline. These EOFs describe 90 percent and 62 percent of the total variance in the profiles of the annual cycle and the yearly anomalies in specific volume, respectively.

But our main interest is in longer time scales. The yearly temperature anomalies are smoothed with a 5-yr running mean filter (Figure 13.5). Successive standard levels are offset from one another by 0.25°C. In and above

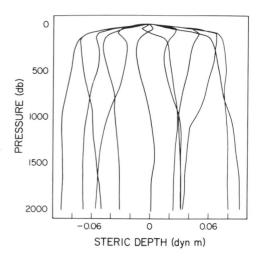

FIGURE 13.2 Mean monthly values of steric depth at the Panulirus station, averaged over 1954 through 1981, with the average over all months removed. Steric depth is the integral of specific volume from the ocean surface down to the pressure level indicated on the ordinate.

the thermocline, down to 1000 m, one can see decreasing temperature between 1957 and 1970, as described by Pocklington (1972). However, the temperature then began to rise again, reaching a maximum in 1975 to 1977 that was still generally cooler than the initial values. Below 1000 m, the pattern is qualitatively different. In the deep levels, the 20-yr oscillation seen above is not apparent. Rather, there is a more regular rise amounting to between 0.1 and 0.2°C. The long time-scale variations below 1500 m appear not to be coupled to those in the thermocline.

In order to extract the longest time scale in the Panulirus data, a linear trend in specific volume over 26 yr was computed at each standard depth. For comparison purposes, the trends were normalized in the same way as the EOFs and are plotted in Figure 13.4. The pattern is similar to the EOF of yearly anomalies in that it has a maximum in the main thermocline. An important difference though is that the trend does not become negligibly small below 1000 m. Rather it extends to the deepest levels sampled and to some undetermined depth beyond, with a change in sign at about 1000 m. In viewing the trends in relation to the running mean values in Figure 13.5, it is clear that the oscillations are as large or larger than the apparent trend and that the apparent trend in another 25 yr of data could be quite different.

So far we have mentioned only temperature change although the density of seawater depends on salinity as well. On the time scales of interest here, tens to hundreds of years, the effect of salinity changes on global sea-level changes are slight. The total salt content of the oceans remains approximately constant and sea-level changes due to vertical redistribution of salt in the water column are likely to be small. Heat, unlike salt, is freely exchanged with the atmosphere and even the combined heat content

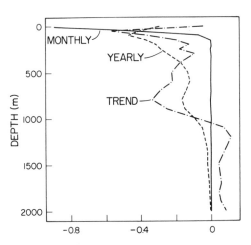

FIGURE 13.4 The first empirical orthogonal function of the mean monthly profiles of steric depth and the first EOF of the yearly anomaly profiles, normalized to have unit length. The third profile is of the linear trend in time of steric depth over the period 1955 to 1981, also normalized to unit length.

of the ocean-atmosphere system may change significantly over a few decades or centuries. Thus we would argue that global expansion or contraction of ocean waters must be dominated by temperature change rather than by salinity change. Of course, this may not be true in any particular region, so a word about salinity changes in the Panulirus data is in order. In general, temperature changes are accomplished by salinity changes such that a fairly tight temperature-salinity correlation is maintained. A temperature increase of 1°C in the thermocline at the Panulirus station is accompanied by a salinity change of approximately 0.1 practical salinity units. This temperature change produces a decrease in specific volume whose magnitude is more than three times the corresponding increase in specific volume caused by the salinity change. Thus the effects are of opposite sign, and although salinity is not negligible, the temperature effects dominate. Talley and Raymer (1982) studied temporal changes in the temperature-salinity relation at the Panulirus station and found that the slope of the relation changed very little between 1954 and 1971. There was a slope change between 1972 and 1975 in waters at the top of the thermocline, but the effect on steric height of the observed temperature-salinity shift is much less than that due to the observed temperature changes.

The trends observed at the Panulirus station are presumably not strictly local, and one would like to know the spatial extent of the signal. For this purpose, we compare the Panulirus trends to two large space-scale surveys separated in time by 22 yr. As part of the International Geophysical Year (IGY) hydrographic survey, transatlantic sections were occupied along latitudes 36°15'N from the

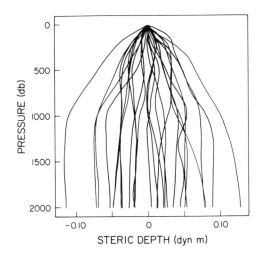

FIGURE 13.3 Yearly anomalies in steric depth at the Panulirus station for the years from 1955 to 1981.

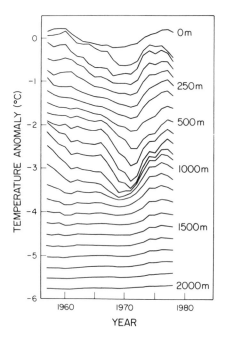

FIGURE 13.5 The 5-yr running mean of temperature anomaly at the Panulirus station at 24 standard depths. Standard depths are offset from one another by 0.25°C.

United States to Spain in 1959 and 24°30'N from Morocco to the Bahamas in 1957. With similar ship tracks, these sections were repeated by the R/V Atlantis II in 1981, as shown in Figure 13.1. The two IGY sections have a total of 99 hydrographic stations while the 1981 sections, made with a continuously profiling conductivity–temperature–depth (CTD) recorder, have 191 stations to the ocean bottom. Large-scale differences in water mass volumes were discussed by Roemmich and Wunsch (1984). Contoured profiles of temperature and salinity are displayed for the IGY data by Fuglister (1960) and for the 1981 data by Roemmich and Wunsch (1985).

Here, we examine the average temperature difference as a function of depth along that part of the 36°N section that is east of the Gulf Stream and west of the mid-Atlantic ridge (this is about half of the total section and is roughly 3000 km in length). This difference is shown in Figure 13.6 together with the linear trend in temperature in the Panulirus data scaled to the same time frame. The standard error of the linear trend is shown by the error bars. Although there are differences in detail, these two profiles of temperature difference are remarkably similar considering that one is the difference between two large space-scale surveys and the other is a trend in time at a single point. They show a cooling trend, or uplifting of isotherms, down to the base of the thermocline at about 1000 m and a warming or downward displacement below. In the large space-scale surveys, the warming extended down

to a depth of about 3000 m. Moreover, a similar pattern existed along 36°N, east of the mid-Atlantic ridge, and along 24°N for the width of the ocean. At 24°N, the cooling was in a more restricted layer down to only 500 m with warming from 500 to 3000 m (Roemmich and Wunsch, 1984). Just as the thermocline showed more structure than the deep water in the time domain, there also appears to be greater spatial variability in the 22-yr difference above 1000 m than in the deep water (see Roemmich and Wunsch, 1984, for temperature difference sections). The existence of both the Panulirus time series and the two large space-scale surveys allows us to confirm that the secular trends in the Panulirus data are indicative of change over much of the subtropical gyre.

Of central importance here is the relationship between the apparent trends in specific volume and sea level. In Figure 13.7, the 5-yr running mean of Bermuda sea level is shown together with that of the Panulirus steric height of the sea surface relative to 2000 dbar. For reference, we also show sea level at Charleston, South Carolina. It is at nearly the same latitude as Bermuda, and the record there is typical of the east coast of the United States. The correlation of Bermuda sea level and Panulirus steric height is very high—the correlation coefficient is about 0.9 for the unsmoothed yearly values. Charleston sea level, on the other hand, does not show the fall and rise that is associated with the thermocline variations at Bermuda (Figure 13.5) but rather has a more steady rise. If the Charleston record is detrended, then the residual has substantially less variance than the Bermuda record. A plau-

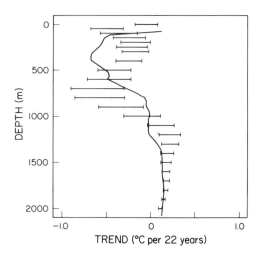

FIGURE 13.6 The smooth curve expresses the temperature difference between the IGY (1959) and 1981 surveys along 36°N west of the mid-Atlantic ridge and east of the Gulf Stream. A positive value indicates that the ocean was warmer in 1981. The horizontal bars show the linear trend of temperature in time, with the standard error of the estimate, at the Panulirus station.

sible explanation is as follows. The base of the thermocline slopes sharply upward across the Gulf Stream shoreward from the Sargasso Sea. The 7°C isotherm, found at about 1000 m in the Sargasso Sea, is at around 200 m shoreward of the Gulf Stream. Thus, if all water above 7°C was cooled by a uniform amount, the effect on sea level would be 5 times as great on the offshore side of the Gulf Stream. Thus, because of the greater thickness of the warm layer on the offshore side, the lower sea-level variance at the coast could be explained by lower sensitivity to temperature fluctuations in and above the thermocline. An interesting question, which requires a long time series of deep hydrographic data near the coast, is whether all or part of the sea-level rise observed along the eastern United States can be attributed to the subthermocline warming seen throughout the subtropical North Atlantic.

A power-density spectrum of Bermuda sea level was computed using monthly mean values from 1955 to 1978. The high-frequency end is undoubtedly affected by aliasing, but here we are concerned only with the longest periods. A comparable series of monthly steric height of the sea surface relative to 2000 dbar was obtained from the monthly means of the Panulirus data, also from 1955 to 1978. Missing months were filled in by linear interpolation. The two spectra are displayed together in Figure 13.8. They are indistinguishable at periods of a year and longer. Indeed the only appearance of a difference is in a band from about 3 to 5 months, where the sea-level spectrum appears elevated. A prominent annual signal is seen in both records. In the lowest frequency bands, the spectra slope more steeply than (frequency)$^{-1}$, indicating that the

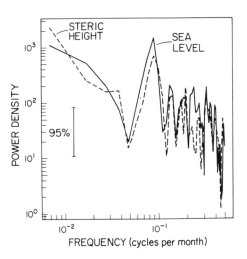

FIGURE 13.8 Power-density spectra, band averaged over three frequency bands, of Bermuda sea level and Panulirus steric height (0 to 2000 dbar).

contribution per unit frequency to the total record variance increases with increasing period.

Estimates of coherence amplitude and phase from the cross spectra of the sea-level and steric height series are displayed in Figure 13.9. The estimates have been band-averaged over 5 adjacent frequency bands. The amplitudes are high at low frequencies, well above the 95 percent confidence limit, with corresponding phases near zero. At higher frequencies, the coherence drops. This is likely to be due to the sampling noise in the Panulirus steric height series, that is, in forming monthly averages from pairs of hydrographic stations.

Another way of illustrating the strong similarity of the sea-level and steric height series at low frequency is showing the difference of steric height (0 to 2000 dbar) subtracted from sea level. That is shown, again as a 5-yr running mean, in the bottom line of Figure 13.7. Time variations in this difference amount to a few centimeters. For reference, a line sloping upward at 1 cm/decade is also shown. There may be a trend in the difference, a rise in sea level relative to steric height, but the ≈10-yr variations are larger than the apparent trend. Clearly a longer time series is required to resolve the question of whether the difference between sea level and steric height is growing. Further, data from the 1981 and IGY surveys of the subtropical Atlantic suggest that if the warming from 2000 to 3000 dbar were considered, the apparent trend in the sea-level residual would be reduced.

Another point to note in Figure 13.7 is that whereas there is an upward trend in the overall sea-level record at Bermuda (1933 to 1980), there is no trend in the period overlapping the Panulirus observations (1954 to 1980). Thus, several decades more of sea-level data together with

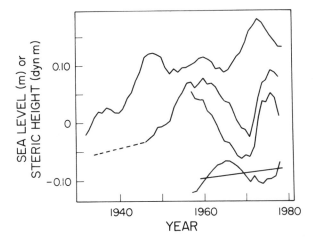

FIGURE 13.7 From top to bottom, 5-yr running means of Charleston sea level, Bermuda sea level, Panulirus steric height (0 to 2000 dbar), and the residual of steric height subtracted from sea level. The height of each curve is arbitrary. The straight line drawn through the bottom curve has an upward slope of 1 dynamic centimeter per decade.

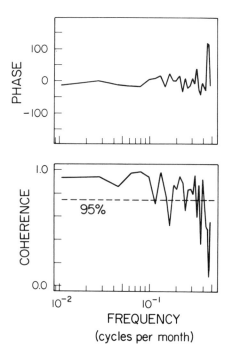

FIGURE 13.9 Coherence amplitude and phase from cross spectra of monthly mean sea level at Bermuda and monthly interpolated Panulirus steric height (0 to 2000 dbar) from 1955 to 1978, band averaged over five frequency bands.

steric height are required in order to see whether the upward trend in sea level is reestablished, and if so, whether it is accompanied by a similar rise in steric height. An additional valuable piece of information would be direct measurement of the time rate of change in the geopotential height of the Bermuda benchmark used for reference for the sea-level station. Only then could we separately estimate the three possible contributions to change in relative sea level, i.e., steric height change, change in the mass of seawater per unit area, and change in the height of the land in relation to the geoid.

THE EASTERN NORTH PACIFIC

On a global basis, the great majority of sea-level stations are located along continental coastlines, as opposed to the mid-ocean island station discussed in the previous section. It is anticipated that direct comparisons of steric height with sea level near continental boundaries are subject to additional difficulties not encountered in mid-ocean. These difficulties include the presence of time-dependent boundary currents that have a substantial effect on sea-surface height and the presence of continental shelves that dictate that deep water for steric height calculations is located some significant distance away from shore and the sea-level station. We elected to look at the California

Cooperative Fisheries (CALCOFI) hydrographic data, which has some unique advantages and some unfortunate disadvantages for our purposes. The principle drawback is that the maximum depth of routine sampling is only 500 m. We can say nothing about changes below this level. The great advantage is that the data consist of a grid of stations extending about 2000 km along the west coast of the United States and Baja California and from very near the coast to several hundred kilometers offshore. The same basic grid has been occupied at irregular intervals since 1950. For statistical reasons, we decided to use only those stations occupied more than 50 times. The resulting subset of stations is shown in Figure 13.10.

The first problem is in deciding what sea-level record or records should be compared with the CALCOFI steric height. Sea-level stations in the area of interest and their trends and standard errors (in parentheses) are at San Diego (1.6 ± 0.4), La Jolla (1.5 ± 0.4), Los Angeles (-0.1 ± 0.4), San Francisco (1.5 ± 0.4), and Alameda (0.1 ± 0.5)(Hicks *et al.*, 1983). Thus, even nearby stations such as San Francisco and Alameda, and also Los Angeles and Santa Monica, have very different trends. The average trend for the region is not well estimated. We will make an arbitrary choice of San Diego, Los Angeles, and San Francisco

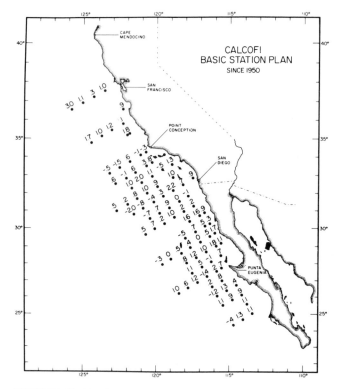

FIGURE 13.10 Map of the eastern North Pacific showing CALCOFI stations occupied on more than 50 occasions between 1950 and 1978. The trend in steric height (0 to 500 dbar) is indicated for each station (see text).

as representative, bearing in mind that the average of these is a poor estimate of the true average sea-level change. Yearly mean sea level for each of these stations, and the yearly mean of the three-station average, are shown in Figure 13.11. Year-to-year changes in sea level are clearly well correlated along the coast, and the trend of the three-station average is 1.0 cm/decade.

The CALCOFI grid yielded 90 stations where hydrographic casts were made to a depth of 500 m on 50 or more occasions between 1950 and 1978. For each cast, we computed the steric height of the sea surface relative to the 500-dbar surface. Casts were sorted by month and year. At each station, an annual cycle was estimated by averaging all casts in each month, regardless of year. This annual cycle in steric height was then removed by subtraction and yearly anomalies were computed by averaging all casts at a given station in a given year. We then computed a trend for each station based on the yearly anomalies. The trends, in centimeters per decade, are shown on the station map, Figure 13.10.

There is considerable variability in the trends from station to station and from line to line of the grid. This small-scale variability is interpreted as noise due to undersampling in time. There are no apparent large-scale variations in the trend. That is, if the California Current had strengthened in the period 1950 to 1978, we would see steric height at the offshore stations increasing relative to the nearshore stations. This is not the case. Similarly, there is no apparent difference in the steric height trends in the northern half of the grid as compared to the southern half, indicating that the alongshore pressure gradient did not change greatly. The station-to-station changes in trend

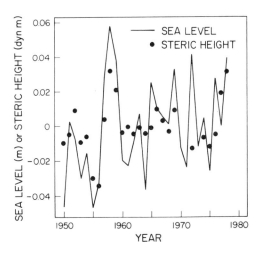

FIGURE 13.12 Averaged annual anomalies in sea level and steric height (0 to 500 dbar) using hydrographic stations of Figure 13.10 and the sea-level stations of Figure 13.11.

are as random as can be expected given the fact that the grid is occupied on a line-by-line basis, that is, adjacent stations in a given line are usually very close together in time.

If there are no strong spatial patterns, what about the overall trend? We averaged the yearly anomalies for all 90 stations, and these yearly averages are shown in Figure 13.12 together with the average sea level. The zero level for each set of data is arbitrary and has been adjusted so that they approximately coincide. In general, the interannual variations in sea level are well reproduced by the steric height series. The large sea-level oscillation of the 1950s and the rise in the mid-1970s are examples. There are also some notable disagreements, for example, 1972. Examination of the records reveals that whereas the 1972 CALCOFI data were collected early in the year, the large sea-level rise associated with the 1972 El Niño episode occurred late in the year. Thus, the undersampling in time can generate large errors in estimating annual mean steric height.

The linear trend in steric height (0 to 500 dbar) for the averaged data of Figure 13.12 is 0.5 cm/decade, compared to 1.0 cm/decade for the averaged sea-level data. Within the error bounds established by the sampling noise, these are indistinguishable, but also, the steric height rise is not statistically different from zero. Once again, it appears from Figure 13.12 that a time series of about twice the length of the present CALCOFI data is required to overcome the sampling noise. Unfortunately, the ambiguity of the result is severalfold:

1. Because of the large station-to-station changes in the coastal sea-level trend, the areally averaged coastal sea-level trend is not well determined.

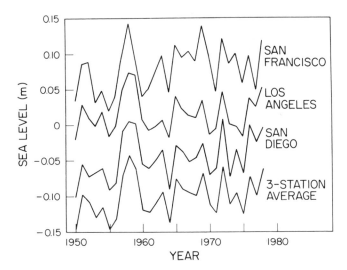

FIGURE 13.11 Annual mean sea level at San Francisco, Los Angeles, and San Diego and the average of the three stations. The vertical offset is arbitrary, for display purposes.

2. For the 29-yr period over which data were available, the sampling noise in the steric height record was of the same order as the signal.

3. Even if the sea-level and steric height records were long and error free, the lack of data below 500 m would still be troubling. Unlike the Atlantic Ocean, there is no deep water formed at high latitudes in the North Pacific. Therefore, one might suspect that the deep North Pacific is less sensitive to climatic fluctuations than the Atlantic. But without supporting data, this is pure conjecture.

DISCUSSION

A principal lesson of the extensive North Atlantic and eastern North Pacific data sets is that steric height variations occur over such a variety of space and time scales that possible trends cannot be identified in the approximately 30-yr time series, no matter how dense the sampling. In the 27 yr of overlapping steric height and sea-level data at Bermuda, there was no evidence of a "spectral gap" that would allow separation of the time scale spanned by the data from a much longer time scale that would appear as a trend. Rather, the sea-level and steric height spectra were equally red in the lowest frequency bands. But the news is not all bad. Many longer sea-level records, such as the Charleston record in Figure 13.7 (for which there is no accompanying steric height time series), do show evidence of such a gap. That is, the 15-cm trend over 50 yr is greater than the 5-cm fluctuations over 10 to 20 yr at that location. The expectation then is that about 50 yr of sea-level data together with steric height data at Bermuda and off the California coast would be sufficient to resolve the question of whether the sea-level trend is accompanied by a trend in steric height (at those locations).

Steric changes in the water column on time scales of a decade and longer are not confined to or concentrated in the upper ocean. In the subtropical North Atlantic such changes extend to depths of at least 3000 m. Maxima in temperature change appeared in the thermocline at depths of from 300 to 700 m and below the thermocline at about 1800 m. For observational programs the large vertical extent of the signal means that the entire depth of the ocean should be sampled. The Panulirus data are not sufficiently deep, though this deficiency can be made up by a few ancillary deep measurements elsewhere. For climate modeling studies, it is clear that the commonly used one-dimensional vertical advection-diffusion models not only miss the essential physics of the problem but also do not reproduce the observed patterns of change. Critical processes are air-sea interactions at middle and high latitudes, which determine volume and composition of the

mid-depth and deep water masses, and the spreading of those waters to lower latitudes.

Temperature variations had a somewhat different character in and above the thermocline than in the deep water in the subtropical North Atlantic. The thermocline fluctuations had greater variance in the time domain than the deeper changes and also more variability around the subtropical gyre. Over 27 yr, the thermocline fluctuations made a greater contribution to steric height change than to the deep changes, but on a longer time scale, it is not clear whether mid-depth or deep variations would dominate. The thermocline comes to the surface along the northern rim of the subtropical gyre, and these waters are therefore subject to atmospheric forcing relatively near to the area of our study. The deep water is farther removed from contact with the atmosphere, which can account for the more gradual changes and larger spatial scale of the deep signal.

In combination, the Atlantic and Pacific data sets highlight several problems. If we are to unambiguously determine the relationship between sea-level rise, steric height change, and changes in the mass of water in the ocean the following are necessary.

1. Density measurements in the ocean should extend to the ocean bottom at least occasionally.

2. Sampling in the time domain should be sufficient, at least in selected locations, so that variations with periods of months to years are not aliased into much longer periods.

3. At key sea-level stations, the rate of change of the height of reference benchmarks relative to the geoid needs to be determined.

ACKNOWLEDGMENTS

I thank Walter Munk and S. Tabata for making helpful comments. The work was supported by the National Science Foundation through grants OCE-8121262 and OCE-8317389 to UCSD.

REFERENCES

Barnett, T. (1983). Long-term changes in dynamic height, *J. Geophys. Res. 88*, 9547–9552.

Frankignoul, C. (1981). Low-frequency temperature fluctuations off Bermuda, *J. Geophys. Res. 86*, 6522–6528.

Fuglister, F. (1960). *Atlantic Ocean Atlas of Temperature and Salinity Profiles and Data from the International Geophysical Year of 1957–1958*, Woods Hole Oceanographic Institution Atlas Series 1, Woods Hole, Mass., 209 pp.

Hicks, S., H. Debaugh, Jr., and L. Hickman, Jr. (1983). *Sea Level Variations for the United States 1855–1980*, U.S. Department of Commerce, National Ocean Survey, 170 pp.

Moore, W. (1982). Late Pleistocene sea level history, in *Uranium Series Disequilibrium: Applications to Environmental Problems*, M. Ivanovich and R. Harmon, eds., Clarendon, Oxford.

Pocklington, R. (1972). Secular changes in the ocean off Bermuda, *J. Geophys. Res. 77*, 6604–6607.

Reid, J., and A. Mantayla (1976). The effect of geostrophic flow upon coastal sea elevations in the northern North Pacific Ocean, *J. Geophys. Res. 81*, 3100–3110.

Roemmich, D., and C. Wunsch (1984). Apparent changes in the climatic state of the deep North Atlantic Ocean, *Nature 307*, 447–450.

Roemmich, D., and C. Wunsch (1985). Two transatlantic sections: Meridional circulation and heat flux in the subtropical North Atlantic Ocean, *Deep-Sea Res. 32*, 619–664.

Schroeder, E., and H. Stommel (1969). How representative is the series of monthly mean conditions off Bermuda? *Prog. Oceanogr. 5*, 31–40.

Shaw, D., and W. Donn (1964). Sea level variations at Iceland and Bermuda, *J. Marine Res. 22*, 111–122.

Talley, L., and M. Raymer (1982). Eighteen degree water variability, *J. Marine Res. 40*, 757–775.

UNESCO (1981). Tenth report of the joint panel on oceanographic tables and standards, *UNESCO Technical Papers in Marine Science 36*, UNESCO, Paris.

Wunsch, C. (1972). Bermuda sea level in relation to tides, weather, and baroclinic fluctuations, *Rev. Geophys. Space Phys. 10*, 1–49.

FUTURE   MEASUREMENTS

Strategy for Future Measurements of Very-Low-Frequency Sea-Level Change

14

WALTER MUNK, ROGER REVELLE, PETER WORCESTER, and
MARK ZUMBERGE
Scripps Institution of Oceanography

INTRODUCTION

Tide-gauge records of long duration (on the order of 1 century) appear to be dominated by two processes: (1) noisy fluctuations with time scales on the order of 1 decade and root-mean-square (rms) amplitudes of 50 mm, and (2) a long-term trend on the order of 1 mm/yr (Figure 14.1). The decade fluctuations have horizontal coherence scales commensurate with the size of ocean basins (Figure 14.2), and they appear to be associated with what is loosely called El Niño. The long-term trend appears to be global and has been thought to be associated with climate changes of glacial periods. Table 14.1 summarizes some possible numbers, and Figure 14.3 sketches an associated spectrum.

We have taken $a = 30$ m for the rms amplitude of the ice-age-like fluctuations and typical period scales of 10,000 yr. This corresponds to a frequency of $f = 10^{-4}$ cycles per year (cpy) and an rms rate of $2\,fa = 20$ mm/yr, which is 20 times the present measured rate. It is hard to account for this 20:1 discrepancy. The geologic record of sea level and of glaciation prevents us from taking a much smaller amplitude than 30 m. Perhaps the current rate of rise is unusually small because we are approaching a climatic optimum; in fact, there is some evidence that the rate of sea-level rise was an order of magnitude larger around 10,000 yr ago.

DETECTABILITY OF LONG-TERM TREND

There have been many discussions about the long-term 1-mm/yr sea-level rise. The question is asked whether this is now accelerating or decelerating. The following very simple calculation shows the difficulty in making such a prognosis. The difficulty has to do with the large amplitude of the decadal fluctuations. Suppose we form two averages, one over each half of the past century, and compare the two numbers for an indication of the long-term trend. For any single measurement of sea level, the high-frequency "noise" associated with the decadal fluctuations is 50 mm rms. The band width is such that we can form independent samples every 5 yr, or 10 samples in 50 yr. The rms error in a 50-yr average is then 50 mm/$(10)^{0.5}$ = 15 mm. The rms error in the difference between the two 50-yr averages is $15 \times (2)^{0.5} = 21$ mm. The expected difference in sea level associated with the long-term rise is 50 mm, about twice the noise level. This rough calculation is in general accord with what is shown in Figure 14.1.

There are a number of ways for improving the estimate of the long-term rise. The usual way is to form a global average. From Figure 14.2 we estimate that one can get 10 spatially independent samples, and in this way the decadal noise in the difference between the two 50-yr averages is reduced from 21 mm to 21/$(10)^{0.5}$ = 7 mm, as compared to

221

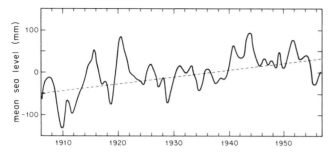

FIGURE 14.1 Honolulu mean sea level after removal of all tidal components and all noise above 2 cpy (after Munk and Cartwright, 1966).

CRUSTAL NOISE

The trouble with the above procedure is that the "solid" crust to which the tide gauges are anchored moves up and down at rates that are comparable to the rate of sea-level rise. Such being the case, the "crustal noise" rather than El

the long-term trend of 50 mm. This gives a quite respectable signal to noise ratio of 17 db, but the existence of pronounced local biases makes it difficult to realize a $n^{-0.5}$ gain from spacial averaging.

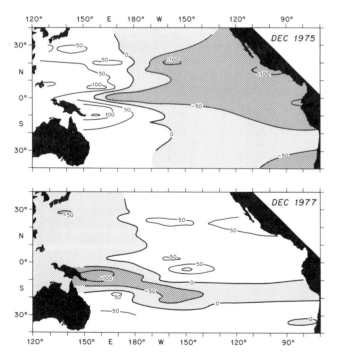

FIGURE 14.2 Maps of sea-level anomaly (from Wyrtki and Nakahoro, 1984) for December 1975 and December 1977. Contours show sea-level anomalies in millimeters after removal of seasonal cycle. The two cases were selected for their great contrast.

TABLE 14.1 Periods, Amplitudes, and Rates of Sea-Level Fluctuations

	Climatological Fluctuations	El Niño-like Fluctuations
Period	10,000 yr	10 yr
rms amplitude	30 m	50 mm
rms rate	20 mm/yr	20 mm/yr
Current rate	1 mm/yr	—

Niño-like noise becomes the limiting factor in a global estimate. If the crustal movement could be independently measured, then the global estimates of sea-level rise could be vastly improved.

THERMAL EXPANSION VERSUS CHANGES OF OCEAN MASS

In considering fluctuations in sea level $h(t)$, we distinguish between external processes involving variations in the total mass per unit area and those associated with internal changes in the density distribution (without much altering the total mass). The former processes are associated with hydrostatic pressure fluctuations ($\rho g h$) on the seafloor; the latter, steric processes, give bottom pressure fluctuations that are smaller by 2 or 3 orders of magnitude or are absent altogether.

A warming by 1°C of the entire water column would raise sea level by about 0.2 m. A warming by 10°C would

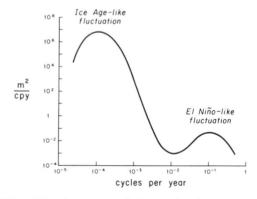

FIGURE 14.3 A cartoon of the sea-level spectrum. The climatological (ice-age-like) oscillations are taken to have an rms amplitude of 30 m and a time scale of 10,000 yr. The El Niño-like events have an rms amplitude of 50 mm and a time scale of 10 yr. The inferred rms rate of sea-level change is 20 mm/yr for the climatological oscillations (20 times the present rate) and also 20 mm/yr for the El Niño-like fluctuations. There is no firm evidence for the spectral gap at 10^{-2} cpy, or for the drop-off below 10^{-4} cpy; the ice-age-like and El Niño-like fluctuations may be plateaus on a monotonic red spectrum.

raise sea level by about 8 m (much more than 10×0.2 m because the coefficient of thermal expansion increases with temperature). Implication is that a 30-m variation in sea level cannot be attributed to thermal expansion. The conclusion is that the rise over the last few thousands of years must have been largely due to external processes, mainly owing to the decay of continental ice sheets.

The situation is not clear with regard to the rise by approximately 1 mm/yr during this century. The glaciological evidence (Meier, 1984) seems to rule out a contribution by much more than 0.3 mm/yr from mountain and alpine glaciers; an 8 percent decrease in the current glacial volume would correspond to a 50-mm rise in global sea level. A 50-mm rise could also be brought about by a 0.3 percent decrease in the volume of the Antarctic ice sheet (corresponding to a lowering of the ice-sheet surface by 6 m). Most Antarctic glaciologists, however (Robin, 1985; National Research Council, 1985), do not believe that such a change is taking place.

At the same time, there is evidence of a warming by 0.4°C of global surface temperature during the twentieth century (Gornitz et al., 1982), and this yields a steric rise of perhaps 0.6 mm/yr when interpreted in terms of a model of vertical ocean diffusivity. But the evidence is not convincing, and the model is demonstrably poor. In summary, the ice-age-like fluctuations are probably largely external, but the associated long-term trend over the last 50 yr may include an important steric component.

Now with regard to the decadal fluctuations, Roemmich and Wunsch (1984) demonstrated that they are largely steric. Independent measurements could be used to subtract the steric (El Niño-like) noise from the sea-level measurements and so obtain a better estimate of the external component in the long-term trend.

MEASUREMENT OF g

The position of the sea surface relative to a benchmark on land is affected both by sea-level change and by vertical motions in the crust beneath the benchmark. Tide-gauge measurements alone may be insufficient to monitor true changes in sea level—the heights of the tide gauges themselves must be monitored with respect to a worldwide reference frame.

One technique that can be brought to bear on the problem is the absolute measurement of g, the acceleration due to the Earth's gravity. As an observer moves away from the center of the Earth, g decreases at the fractional rate of 3 parts in 10^{10}/mm. The local value of g can currently be determined by portable instruments (Zumberge et al., 1982) to within an uncertainty of about 1 part in 10^8, providing a sensitivity to height changes on the order of 30 mm. Thus, periodically making accurate determinations of g near tide-

gauge installations can provide a cost-effective means of separating sea-level change from tectonic motion.

The nature of the method used to make absolute gravity measurements makes them well suited to a search for very-low-frequency signals such as those related to sea-level change. In a modern absolute gravity meter, a mass is made to fall freely in a vacuum while its position as a function of time is determined by a laser interferometer. By calculating the acceleration of the falling mass, g is determined in terms of the wavelength of a stabilized laser and the frequency of an atomic time standard. Both are referenced to absolute standards having stabilities surpassing 1 part in 10^9. Other factors limit the accuracy obtained to about 1 part in 10^8, but we believe that this can be improved by a factor of 2 or 3 in the next decade (Faller et al., 1983). The important point is that g is determined absolutely, and thus the data are not likely to become contaminated by instrumental drift.

Incorporating gravity data into an analysis of a tide-gauge record can be a complicated problem. A rise in sea level will affect gravity because of the additional water mass and through deformation of the crust by the added load. Conversely, a local expansion of the underlying crust will move the observation point and displace sea water, both of which affect g.

An order-of-magnitude assessment of the contribution to change in g from the two mechanisms—sea-level change as opposed to tectonic motion—relies on two simple numbers. The first (mentioned above) is the free air gradient of gravity. The fractional change in g for a height displacement z is

$$\frac{\Delta g/g}{\Delta z} = 3 \times 10^{-10} /\text{mm}. \qquad (14.1)$$

This will be roughly the size of the gravity signal accompanying a vertical displacement caused by crustal deformation. The other value we need is the gravitational attraction of a laterally infinite layer of water having thickness t. Relative to the nominal value of g (about 9.8 m/s²), the attraction is

$$\frac{\Delta g/g}{t} = 4 \times 10^{-11} /\text{mm}. \qquad (14.2)$$

If we ignore the crustal loading effect (its sign and magnitude are comparable to the infinite layer effect) (Warburton and Goodkind, 1977; Goad, 1980), we see that the change in gravity accompanying an apparent change in sea level due to crustal deformation is an order of magnitude larger than the gravity change caused by a real variation in sea level. Thus repeated gravity measurements can be useful in distinguishing one mechanism from the other.

Other geodetic methods can be used to monitor the vertical positions of tide gauges, and any strategy designed to address the sea-level problem should take advantage of them. For example, electromagnetic distance measurements to satellites using radio waves or lasers can determine relative geodetic heights with an uncertainty ranging from 20 to 50 mm between stations separated by 100 km or more (Christodoulidis and Smith, 1983; Strange, 1984). Over intercontinental distances, very-long-baseline interferometry can determine heights to 100 mm or better (Shapiro, 1978; Herring, 1984). As these methods improve, they are bound to exert a profound influence on the strategy for assessing global sea level.

PRESSURE ON THE SEA FLOOR

The best long-term measurements of seafloor pressure made to date have relied on a careful characterization of the drift characteristics of the pressure sensor from laboratory measurements so that the pressure record could be corrected for sensor drift after completion of the experiment. Wearn (Wearn and Baker, 1980; Wearn and Larson, 1982), for example, believed that he was able to correct data obtained using Paroscientific quartz-crystal pressure sensors with sufficient precision to reduce the residual drift to between 1 and 3 mbar (approximately 10 to 30 mm) per year at an ambient pressure of about 500 dbar (approximately 500 m). This corresponds to a relative accuracy of 0.002 to 0.006 percent, or 2×10^{-5} to 6×10^{-5}. To determine long-term trends in sea level, one would like to be able to make measurements with substantially greater precision; an increase of 1 mm in 4000 m of water, for example, represents a change of 2.5×10^{-7}. One does not want to make measurements at much shallower depths because the pressure change then has both barotropic and baroclinic components. In order to measure such trends, one therefore needs to either (1) devise a pressure sensor with lower drift or (2) calibrate the sensor against a primary standard in situ. A pressure sensor with significantly lower drift than the Paroscientific quartz sensor used by Wearn does not exist; a transducer with lower drift would therefore have to employ an entirely new technology that is unknown at this time. In the remainder of this section we will therefore discuss the precision that is achievable with in situ calibrations.

One could imagine configuring the system in two distinct ways.

1. A valve could be used to switch the input port of the pressure sensor between ambient ocean pressure and a reference pressure used to calibrate the sensor. This technique requires a sensor with adequate short-term stability, which may be a problem. The reference pressure would need to be close to ambient pressure in order to minimize hysteresis effects in the sensor.

2. A reference pressure could be used with a differential pressure sensor of much lower full-scale pressure than required to directly measure ambient ocean pressure. This configuration takes advantage of the fact that drift is typically proportional to the full-scale pressure of the sensor, whereas the absolute size of the fluctuating signal is the same for either absolute or differential sensors.

In either case, the best candidate for providing the reference pressure is a high-quality, oil-operated, piston-gauge deadweight tester. In such a unit a calculable pressure is generated in a fluid by placing a known mass on a piston of known diameter. The absolute accuracy achieved with a high-quality piston gauge is 0.01 percent, or 1×10^{-4}. But it is important to remember that what is required is relative, not absolute pressure changes. The important parameter is the reproducibility or stability of the pressure generated by a piston gauge. The National Bureau of Standards (now National Institute for Standards and Technology—NIST) routinely finds short-term repeatabilities at the 1 ppm (1×10^{-6}) level when intercomparing high-quality piston gauges, but long-term (years) differences in some cases have significantly exceeded this level (D. R. Johnson and C. Tilford, NIST, personal communication). NIST believes that 10 ppm is a more realistic expectation for long-term stability; they are not able to verify the absolute stability at any better level since 10 ppm is the irreproducibility of the dimensional measurements of the pistons and cylinders.

This is a rather disappointing state of affairs. A relative stability of 1×10^{-5} corresponds to 40 mm out of 4000 m. This is a factor of 2 to 6 better than Wearn claims to have achieved by measuring and removing the drift of his pressure sensors, but is substantially less stable than desired. It is conceivable that one might achieve better than 10 ppm stability in the deep-sea environment; the temperature is relatively stable (fluctuations are measured in millidegrees centigrade in the deep Pacific), the instrument would not be handled, and the piston weights would never be changed. Testing this speculation would require a substantial research project in which a number of piston gauges would be maintained in a simulated deep-sea environment for periods in excess of a year. Such a test would reveal noise that is uncorrelated between units. Systematic errors can be determined only by comparison to a better standard; no such standard exists. The best that one would be able to do is compare piston gauges from several manufacturers.

An alternative approach would be to deploy the sensor at significantly shallower depths, decreasing the fractional accuracy required. At shallower depths one might also expect the drift problem to be more tractable, since one would be dealing with lower pressures. The disadvantage would be that the pressure signal would then have a significant baroclinic component. However, even at 100 m, corresponding to 1-mm precision for a stability of 1×10^{-5}, one would expect the baroclinic component to be substantially attenuated relative to that measured by a surface tide gauge (since the largest ocean temperature changes occur in the upper ocean). If lower drift could be achieved at pressures corresponding to a few hundred meters depth, one might be able to move the sensor somewhat deeper. But this is clearly not the most desirable approach.

If one were to pursue the use of piston gauges on the seafloor for long time periods, a number of practical problems would need to be addressed. Some of the less obvious ones include the following:

1. Although NIST is of the opinion that a high-quality, oil-operated piston gauge would function continuously for a year, there is no evidence of anyone that has done it for longer than two weeks (D. R. Johnson and C. Tilford, NIST, personal communication).

2. The axis of the gauge must be aligned to the vertical to better than 1 milliradian.

3. The gauge would have to be installed on the seafloor in such a way that it would settle by less than 1 mm/yr.

4. The external pressure applied to the piston by the gas in the pressure vessel used to house it must either be measured or eliminated by housing it in a vacuum.

5. The temperature of the piston and cylinder must be monitored to allow compensation for the approximately 9 ppm/°C temperature coefficient of the effective area.

Many other detailed considerations would be required for making high-precision measurements; the ones given above are only examples. Success with this approach would prompt the desire to construct instruments to be deployed for multiyear periods, with real time data readouts on shore. Fiber-optic technology currently under active development should make it economically feasible to connect the instruments via cables to shore-based data recorders.

INVERTED FATHOMETER

Consider a fathometer looking upwards from the seafloor. The one-way acoustic travel time is Sh, where $S = 1/C$ is the sound slowness and h is the total water thickness. In a homogenous ocean, the departure in travel time is $\delta t = S(z_{surface} - z_c)$, as a result of departures $z_{surface}$ and z_c in the surface and crustal elevations, respectively. We define $z_{IF} - t/S$ so that $z_{IF} = z_{surface} - z_c$.

For an external contribution to the water budget $z_{IF} = z_b - z_c$. The effect of steric disturbances is twofold: it leads to an additional component, z_s, in the surface elevation, and it changes the sound speed in the water column.

In considering the effect of a long baroclinic wave in a

two-layer ocean, the upper ocean is characterized by a thickness h_1, density ρ_1, temperature θ_1, and salinity s_1, with a corresponding h_2, ρ_2, θ_2, and s_2 for the lower ocean.

With a change in layer thickness from h_j to $(h_j + \delta h_j)$, $j = 1,2$, but leaving θ_j and s_j unchanged, the condition of constant mass per unit area requires that $\rho_1 \delta h_1 + \rho_2 \delta h_2 = 0$, or

$$\delta h_2 = -(\rho_1/\rho_2)\delta h_1. \qquad (14.3)$$

The steric change in sea levels equals

$$z_s = \delta h_1 + \delta h_2 = \frac{\rho_2 - \rho_1}{\rho_2} \delta h_1. \qquad (14.4)$$

The change in acoustic travel time (one-way) is

$$\delta t = S_1 \delta h_1 + S_2 \delta h_2 - S_0 z_{IF}. \qquad (14.5)$$

This can be written

$$\delta t = S_2 \delta h_1 \left(\frac{\rho_2 - \rho_1}{\rho_2} \frac{S_2 - S_1}{S_2} \right). \qquad (14.6)$$

The first term is $S_2 z_s$, which is very nearly the increase in travel time associated with the steric rise in sea level. Typical values are $(\rho_2 - \rho_1)/\rho_2 = 0.001$ and $(S_2 - S_1)/S_2 = 0.01$ so that the second term dominates. In terms of the temperature differential $\delta\theta$, we have $\delta\rho/\rho = -a\delta\rho$ and $\delta S/S = -\alpha\delta\rho$, with $a \cong 0.13 \times 10^{-3}$/°C and $\alpha = 3 \times 10^{-3}$/°C. Thus the ratio of the second to the first term in brackets is

$$\rho = \alpha/a \cong 23. \qquad (14.7)$$

The result is that the second (negative) term dominates. A thickening of the upper layer (positive δh_1) raises the sea level by $z_s = [(\rho_2 - \rho_1)/\rho_2]\delta h_1$ and thus increases the one-way travel time by Sz_s; but the relative thickening of the warm (high speed) upper layer decreases travel time, and the latter effect dominates. We can write the result of a purely steric disturbance in the form

$$z_{IF} = z_s(1 - 2), \quad \rho \gg 1. \qquad (14.8)$$

TOMOGRAPHY

The inverted fathometer method is associated with a steep acoustic path and by its nature depends on a combination of surface elevation and the variable interior field of sound speed. Tomography depends on near-horizontal refracted paths that do not intersect the surface and so can be used to estimate z_s directly. An advantage of the method is that it provides a spatial average (order 1000 km). This reduces the mesoscale noise (100 km, month scale) in the estimate of the decadal fluctuations (10^4 km, decade scale). Munk and Wunsch (1985) estimated an rms noise level of 5 mm in the determination of the mean annual steric sea level, using tomographic transmissions.

INTERPRETATION

What could be learned when some of these measurements are combined with the tidal recording?

Let z_c designate the crustal height relative to some suitable reference; changes in z_c can be estimated from the gravity measurements.

Let z_s be the steric sea level; an increase in z_s can arise from thermal expansion and haline contraction. The former effect is dominant. (The annual tide, with a typical amplitude of 10 cm, is largely the result of such temperature changes.) An important consideration is that the coefficient of thermal expansion increases appreciably with increasing temperature. Accordingly, most of the changes in z_s are associated with the warm ocean water above the thermocline, both because the shallow temperature changes are relatively large and because the coefficient of thermal expansion of the upper ocean waters is relatively large. A distinction can be made between the effects of a net change in heat content (e.g., due to radiation, evaporation) and the local change in temperature associated with flow divergence, as in baroclinic waves.

Let z_b designate the barotropic sea level. Here a distinction can be made between the effects of a global change in the ocean's mass arising from glacial melting, and the local changes associated with flow divergence. An example of the latter is the buildup of the gyre center at the expense of the flanks, with the net gyre mass remaining unchanged.

We consider the records obtained from a tide gauge, a bottom-pressure recorder, an inverted fathometer, and a gravimeter:

$$z_{TG} = z_b + z_s - z_c, \qquad (14.9)$$
$$z_{BP} = z_b - z_c,$$
$$z_{IF} \cong z_b + z_s(1 - \rho) - z_c,$$
$$z_{GR} = z_c.$$

Without bottom pressures, which are difficult to measure, there are three equations in the three unknowns z_b, z_s, and z_c. With bottom pressure, the problem is overdetermined.

The system of equations is oversimplified. As an example, a rise in global sea level due to global melting of ice will lead to an increase in (due to the added water mass beneath the gravimeter) and a further increase due to the compression of the crust from loading, in addition to any effect from tectonic crustal uplift.

The system of equations is intended as an illustration of how a particular set of observations can be used. There are many other ways too. Repeated hydrographic surveys yield a direct measure of the steric level (the classical

dynamic height). Changes in the crustal elevation can also be surmised by the perturbation of satellite orbits.

A PROPOSED STRATEGY

We propose that two existing tide stations be instrumented to include a gravimeter, a bottom-pressure gauge, and an inverted fathometer.

There is much to be said for combining this new instrumentation with a program proposed by NOAA (1985) as a contribution to the World Ocean Circulation Experiment (WOCE). This program consists of three principal components: (1) new tide gauges that record digitally and have a high degree of linearity, (2) very-long-baseline interferometry (VLBI), and (3) the Global Positioning System (GPS). VLBI and GPS provide a highly accurate (±1 cm) global reference frame with a role similar to that of the gravimeter.

We propose Hawaii and Bermuda as the two sites for developing the advanced concepts of sea-level measurements. Both stations are relatively free of meteorological noise (Brown *et al.*, 1975), and have long-time series of tidal measurements. Hawaii is a site for the proposed NOAA sea-level program. Bermuda has been the site of the Panulirus measurements of steric sea level since 1954 (see Roemmich, Chapter 13, this volume).

As a second phase of the proposed program of enhanced sea-level measurements, one should establish a limited global network using the experience gained at the prototype stations (Hawaii and Bermuda). If local adjustment for crustal movement and steric sea-level fluctuations is successful, then presumably the steric component of the long-term trend can be estimated from measurements at a relatively small number of stations. The steric component (if any) of the long-term trend will be difficult to extract from the 50-mm rms background noise of the El Niño-like fluctuations. For 10 spatially independent stations, each forming 5 independent samples (this takes 25 yr for a 5-yr decorrelation time), we thus have 50 degrees of freedom in the global mean, or ±7 mm standard deviation. During this 25-yr period we can expect a change of 10 to 25 mm in steric level resulting from the long-term trend. The problem is made even more difficult by the expectation that the long-term trend itself will change during this 25-yr interval.

REFERENCES

Brown, W., W. Munk, F. Snodgrass, H. Mofjeld, and B. Zetler (1975). MODE bottom experiment, *J. Phys. Oceanogr.* 5(1), 75–85.

Christodoulidis, D. C., and D. E. Smith (1983). *The Role of Satellite Laser Ranging through the 1990's*, NASA/Goddard

Space Flight Center Technical Memorandum TM-85104, Greenbelt, Maryland.

Faller, J. E., Y. Guo, J. Gschwind, T. Niebauer, R. Rinker, and J. Xue (1983). The JILA portable absolute gravity apparatus, paper presented at the 18th Assembly of the IUGG, Hamburg, Germany.

Goad, C. C. (1980). Gravimetric tidal loading computed from integrated Green's functions, *J. Geophys. Res. 85*, 2679–2683.

Gornitz, V., L. Lebedeff, and J. Hansen (1982). Global sea level trend in the past century, *Science 215*, 1611–1614.

Herring, T. A. (1984). Precision of vertical position estimates from VLBI, in *Proceedings from the Chapman Conference on Vertical Crustal Motion*, Harpers Ferry, W. Va.

Meier, M. F. (1984). Contributions of small glaciers to global sea level, *Science 226*, 1418–1421.

Munk, W., and D. Cartwright (1966). Tidal spectroscopy and prediction, *Phil. Trans. Roy. Soc. A 259*, 533–581.

Munk, W., and C. Wunsch (1985). Biases and caustics in long-range acoustic tomography, *Deep-Sea Res. 32*, 1317–1346.

National Research Council (1985). *Glaciers, Ice Sheets, and Sea Level: Effects of a CO_2-Induced Climatic Change*, Committee on Glaciology, National Academy Press, Washington, D.C., 330 pp.

NOAA (1985). Global Sea Level Program, draft program development plans, National Oceanic and Atmospheric Administration, U.S. Department of Commerce, Washington, D.C.

Robin,. de Q. (1985). Changing the sea level, in *International Assessment of the Impact of an Increased Atmospheric Concentration of Carbon Dioxide on the Environment*, WMO/ICSU/UNEP.

Roemmich, D., and C. Wunsch (1984). Apparent changes in the climatic state of the deep North Atlantic Ocean, *Nature 307*, 446–450.

Shapiro, I. I. (1978). Principles of very long baseline interferometry, in *Proceedings of the Ninth GEOP Research Conference*, I. I. Mueller, ed., The Ohio State University, Columbus, pp. 29–33.

Strange, W. E. (1984). The accuracy of Global Positioning System for strain monitoring, *EOS 65*, 854.

Warburton, R. J., and J. Goodkind (1977). The influence of barometric-pressure variations on gravity, *Geophys. J. Roy. Astron. Soc. 48*, 281–292.

Wearn, R. B., Jr., and D. J. Baker, Jr. (1980). Bottom pressure measurements across the Antarctic Circumpolar Current and their relation to the wind, *Deep-Sea Res. 27A*, 875–888.

Wearn, R. B., Jr., and N.. Larson (1982). Measurements of the sensitivities and drift of Digiquartz pressure sensors, *Deep-Sea Res. 29*, 111–134.

Wyrtki, K., and S. Nakahoro (1984). Monthly Maps of Sea Level Anomalies in the Pacific 1975–1981. Hawaii Institute of Geophysics Report HIG-84-3.

Zumberge, M. A., R. Rinker, and J. Faller (1982). A portable apparatus for absolute measurements of the Earth's gravity, *Metrologia 18*, 145–152.

Index